T0296141

THE EXPERIMENTAL
BASIS OF CHEMISTRY

THE EXPERIMENTAL BASIS OF CHEMISTRY

SUGGESTIONS FOR A SERIES OF EXPERIMENTS ILLUSTRATIVE OF THE FUNDAMENTAL PRINCIPLES OF CHEMISTRY

BY

IDA FREUND

SOMETIME STAFF LECTURER AND ASSOCIATE OF
NEWNHAM COLLEGE, CAMBRIDGE

EDITED BY

A. HUTCHINSON, M.A.

FELLOW OF PEMBROKE COLLEGE,
UNIVERSITY DEMONSTRATOR OF MINERALOGY

AND

M. BEATRICE THOMAS

LECTURER IN CHEMISTRY, GIRTON COLLEGE, CAMBRIDGE

CAMBRIDGE
AT THE UNIVERSITY PRESS
1920

CAMBRIDGE
UNIVERSITY PRESS

University Printing House, Cambridge CB2 8BS, United Kingdom

Cambridge University Press is part of the University of Cambridge.

It furthers the University's mission by disseminating knowledge in the pursuit of education, learning and research at the highest international levels of excellence.

www.cambridge.org
Information on this title: www.cambridge.org/9781107511552

© Cambridge University Press 1920

First published 1920
First paperback edition 2015

A catalogue record for this publication is available from the British Library

ISBN 978-1-107-51155-2 Paperback

PREFACE

IDA FREUND had been for many years before her death a naturalised British subject, but was Austrian by birth. Left an orphan while still quite young, she was brought up by her grandparents in Vienna, and received her early education at a Bürgerschule in that city. Afterwards she took the diploma of a State Training College for teachers, and the experience there gained in the study of continental methods broadened her outlook and was possibly the origin of the interest in the profession of teaching and sympathy with teachers which were to become marked characteristics of her later career. She then came to England to make her home with her uncle, the violinist Ludwig Straus, well known to music lovers as a member of the Joachim quartet. Her uncle sent her to Cambridge, where as a student of Girton College she took the complete honours course in Natural Sciences, and in 1886 was placed in the first class of the second part of the Natural Sciences Tripos for her knowledge of chemistry.

In the following year she began her life's work as a teacher at Newnham College where she laboured till her retirement in 1912. At that time women students were not admitted to the University Chemical Laboratory until they had passed Part I of the Tripos, and thus Miss Freund was entirely responsible for the laboratory training of the majority of her students, many of whom came up to College with little or no knowledge of chemistry.

Triumphing over disabilities due to physical infirmities and indifferent health such as would have daunted a less intrepid spirit, she devised and elaborated for her first year students a course of practical work supplemented by short lectures, demonstration experiments and discussions, and these form the basis on which this book rests. In 1904 she published a considerable work entitled *The Study of Chemical Composition*, which carried her influence as a teacher far beyond the limits of her own laboratory. The orderly arrangement of the book, the fulness of its historical references, and the quotations often of considerable length from the original papers in which the fundamental laws of chemistry were enunciated and established by their discoverers give it permanent value as a students' "source book" of chemical theory, and secured for it a favourable reception. This encouraged Miss Freund to attempt to bring to the notice of other teachers her views as to the manner in which students might be helped to realise that chemistry is a science based on experiment and that the logical interpretation of experiment leads directly to the generalisations known as the laws of chemistry. Miss Freund had a dread of thoughtless experimenting and slipshod thinking. She felt strongly that much that passes for training in science has little relation to scientific method and is of small educational value. The scheme of practical work which she arranged for her students was designed to include not only the performance of many of the experiments usually found in an elementary course, but also the repetition in a simple form of experiments historically interesting and of fundamental importance to the theory of chemistry, and such that the manipulative difficulties involved were not too great to allow of the attainment of a reasonable degree of accuracy in the hands of beginners. By directing special attention to the sources of error inherent in the methods employed, by distinguishing carefully between what was taken for granted and what was really proved, and by getting her students to compare the accuracy attained in their illustrative experiments with that of the most trustworthy work on the subject, she was able to arouse the critical faculty and to give some insight into the methods and aims of the science. To quote her own words, "I aimed at giving by means of class teaching not only a common ground of knowledge, but also a common standard concerning the nature of scientific proof and the meaning of real accuracy to a number of students differing greatly in knowledge of chemical facts and manipulative experience. As was to be expected, I find that in order to make a connected and fairly proportioned book, gaps have to be filled in between the experiments, and a connecting story has to be supplied so as to make it clear where and how the experiments fit in to the fabric of the science and to establish a sequence ; as a matter of fact I have to write what I used to give in every demonstration as a half hour's introductory lecture. This connecting story has turned out very long in the chapters already written ...but I have got to feel convinced that this is a necessity, and even to think it possible that it may prove an advantage, raising the book above the scope of a mere laboratory manual."

After her retirement from active teaching she began to arrange the material collected in her laboratory note books and students' records with a view to describing a series of illustrative experiments such as she had found specially suited to her needs. The Syndics of the Cambridge University Press, the publishers of her first book, having expressed their willingness to bring out a second work from her pen, an agreement was signed in November 1913, and almost up to the time of her death, which followed an operation in May 1914, Miss Freund was busily engaged in preparing the manuscript for the Press. The book was planned to consist of twenty chapters; the first ten are those which appear here, the rest were to have dealt with the detailed study of water, oxygen and hydrogen, and with the consideration of acids, bases and the classification of oxides, and were to have included a discussion of the law of mass action, of oxidation and reduction and of the conditions which modify chemical change.

The first ten chapters were left by Miss Freund almost ready for the Press. They would have formed the larger part of the book and the part in which she was most interested, and are not only complete in themselves but also give a clear idea of her views and aims. Further they exhibit many and characteristic differences from the ordinary text book of experimental chemistry. For these reasons it was decided to proceed with the printing of this section, and to us as personal friends of Miss Freund of long standing, and well acquainted with her ideas, was entrusted the duty of seeing the book through the Press. As editors our task has been a light one; we have corrected a few obvious slips, made a few verbal changes, and here and there slightly altered the construction of a sentence, where we deemed that by so doing Miss Freund's meaning would be more clearly expressed. We have scrupulously refrained from making any omission, addition or alteration which should in any way conflict with or obscure her intentions. To Mr Peace, the University Printer, and to his staff, we desire to express our best thanks for the patient kindness with which they have endeavoured to carry out our views as to the way in which justice should be done to the somewhat complicated system of headings and subheadings indicated in the manuscript, and for the care which they have devoted to the reproduction and arrangement of the diagrams.

All teachers worthy of the name strike out lines of their own and devise their own schemes, and it is unlikely that many will feel inclined to conduct their students through the whole of the work here detailed. But feeling as we do that Miss Freund's criticisms of methods still current are just and that many valuable suggestions are to be found in the following pages, we commend them to students and teachers alike, in the belief that much may be learnt by examining methods which have stood the test of practical experience in the laboratory of a teacher richly endowed with the critical faculty, keenly sensitive to fallacious reasoning, and quick to detect an unwarrantable assumption.

A. HUTCHINSON.
M. B. THOMAS.

August 1920.

TABLE OF CONTENTS

INTRODUCTORY

The figures in black type refer to pages

CHAPTER I

THE NATURE AND RECOGNITION OF CHEMICAL CHANGE

CHAPTER II

THE CLASSIFICATION OF SUBSTANCES INTO COMPLEX AND SIMPLE (ELEMENTS)

CHAPTER III

CLASSIFICATION OF COMPLEX SUBSTANCES INTO MIXTURES AND COMPOUNDS

CHAPTER IV

THE PART WHICH AIR PLAYS IN COMBUSTION

CHAPTER V

THE CONSERVATION OF MASS

CHAPTER VI

THE LAW OF FIXED RATIOS

CHAPTER VII

THE LAW OF MULTIPLE RATIOS

CHAPTER VIII

THE LAW OF PERMANENT RATIOS

CHAPTER IX

COMBINING OR EQUIVALENT WEIGHTS

CHAPTER X

THE LAW OF COMBINING VOLUMES

LIST OF EXPERIMENTS

INTRODUCTORY

I. Method of treatment indicated by title.

In the title "Suggestions for a series of experiments illustrative of the fundamental principles of chemistry," the introduction of the terms 'suggestions' and 'illustrative' has been intentional and deliberate.

1. Suggestions.

In the first place the word *suggestions* is intended to convey that in the great majority of experiments dealt with, the method adopted and the experimental procedure to be followed will be indicated in outline only, minor details and elaborate descriptions of apparatus being avoided. It is hoped that the figures, which generally are at least approximately to scale, together with the tabulated records of the experimental work involved, will convey *per se* a considerable amount of information as to the nature, size and arrangement of the apparatus employed, the special kinds and quantities of materials used, and the degree of accuracy aimed at in the various measurements involved, thereby supplementing the intentionally restricted amount of verbal direction and description. Moreover, it is assumed that the students have access to standard manuals of practical work (such as Clowes' *Practical Chemistry*, Clowes and Coleman's *Quantitative Analysis*), and that they will be encouraged to consult these whenever the instructions given in this book do not suffice to make clear, *before* the experiment is begun, the principle and the technique of what it is proposed to do : how to obtain an answer, as satisfactory and conclusive as is possible under the circumstances, to the definite question about to be put to Nature. The additional information required will vary in amount according to the average attainments of the class and the individual needs of specially backward or specially advanced students; how much of this must be given by the teacher and how much can be left to the pupil's individual effort depends, of course, on the special conditions. Obviously the greater the call on the learner's independence, originality and ingenuity, the better.

In each experiment, with the object of accomplishing a desired effect, directions are given for the carrying out of certain processes ; the word *suggestions* is used in order to convey the fact that the apparatus depicted, the procedures described, do not represent the only or even the best possible means of accomplishing this effect, but merely that these schemes of work lend themselves to the obtaining of results such as those quoted, and that they have been selected because of their simplicity. It will tend to increase the efficiency of the work all round if variety is encouraged, some members of the class, guided by books and their own ingenuity, carrying out the experiment with modifications in detail or even in principle. Thus, for showing the increase in weight of iron on being heated in air (*post*, pp. 55 and 86), one student could use finely divided iron held by a horseshoe magnet suspended from a hook of a balance, while another could pursue the more common plan of placing the iron in a dish, crucible or open tube ; or again, the means of collecting for purposes of measurement the hydrogen evolved by the action of magnesium on dilute acid can be made the subject of a competition, the pupils being set to discover how many different simple methods they might employ.

2. Illustrative Experiments.

In the second place, the substitution of 'illustrative experiments' for the current term 'experimental proofs' is intended to define from the outset the attitude of strong disapproval taken by the author towards the use of the words *research, discovery, proof* in connection with the work done and the results obtained by the average student working in a school or college laboratory.

(1) It is considered of the utmost importance to impress on the students that the result of one experiment, or even of three or four fairly concordant ones, does not constitute a *proof* unless the amount of preliminary work, the nature of the relation investigated, the precautions taken, the prestige due to the personal skill and experience of the experimenter, have, separately or conjointly, created a special and exceptional case. 'Special cases' of that kind actually occurred in Lavoisier's experiment with mercury calx and in Pasteur's experiment by which laevo-tartaric acid was produced, but even in those realms of scientific research where the giants reign, such instances are rare. These two examples so well emphasise the point that they are worthy of fuller description.

(*a*) Lavoisier, by *one* experiment, involving the quantitative synthesis and complete analysis of mercury calx[1], supplied irrefutable deductive proof of the true nature of combustion, including the elucidation of the part played by the air and the composition of air.

(*b*) When Pasteur first announced that, starting from racemic acid, he had been able to prepare a substance in every way identical with ordinary tartaric acid, except that it rotated the plane of polarisation to the left instead of to the right, he was required to produce *proof* of this unexpected and startling phenomenon by making a specimen of the new laevo-tartaric acid under the personal supervision of Biot, at that time the *doyen* of French scientists, who himself supplied all the necessary materials. This episode in the history of Pasteur's great discoveries deserves to be told in detail, with such explanations as are necessary for the proper understanding of the points at issue.

Tartaric acid and racemic acid are both obtained from grape juice, and are substances which even before Pasteur's classical research had already played an important part in the development of chemical theory. As far back as 1829 Berzelius had shown that these two substances, though differing in important physical and chemical properties, had the same

---

[1] Freund, *Study of Chemical Composition*, pp. 51 *et seq.*; Muir, *Heroes of Science, Chemists*, pp. 87, 88; Lewis, *Inorganic Chemistry*, 1907, pp. 243, 244.

composition, *i.e.* contained the same constituent elements—carbon, hydrogen and oxygen—united in the same ratio. This was one of the fundamental observations from which was built up the doctrine of isomerism. Later on it was shown that the differences between tartaric acid and racemic acid extend to their optical properties, in that whereas a solution of tartaric acid rotates the plane of polarisation to the right, racemic acid is optically inactive. Pasteur, who in 1848 took up the study of these acids and their salts, found the property of optical activity definitely related to the crystallographic property termed 'hemihedrism,' which consists in the exhibition by the crystal of a number of faces all turned in the same direction, either all to the left or all to the right (fig. 1).

Tartaric acid and all its optically active salts exhibited hemihedrism, the crystals showing a number of faces all turned to the right, whilst optically inactive racemic acid and its salts showed no signs of such unsymmetrical crystallographic development. But the generalisation thus arrived at from the study of a large number of salts seemed at first to break down when applied to the double salts sodium-ammonium tartrate and sodium-ammonium racemate. Experiment showed that the optically active tartrate was hemihedral, but that, contrary to expectation, the optically inactive racemate was hemihedral also[1],

...Only, the hemihedral faces which in the tartrate were all turned the same way, were in the racemate inclined sometimes to the right and sometimes to the left....I carefully separated the crystals which were hemihedral to the right from those hemihedral to the left, and examined their solutions separately in the polarising apparatus....The crystals hemihedral to the right deviated the plane of polarisation to the right, and those hemihedral to the left deviated it to the left; and when I took an equal weight of each of the two kinds of crystals, the mixed solution was indifferent towards the light in consequence of the neutralisation of the two equal and opposite deviations.

The announcement of the above facts naturally placed me in communication with Biot, who was not without doubts regarding their accuracy.... He made me come to him and repeat before his eyes the decisive experiment. He handed over to me some racemic acid which he had himself previously studied with particular care, and which he had found to be

[1] Sodium-ammonium racemate is inactive and the crystals are symmetrical (fig. 1 (*f*)), but when crystallised from water at ordinary temperatures it splits up into dextro and laevo tartrates, the former identical with ordinary sodium-ammonium tartrate, the other the new substance discovered by Pasteur. From these salts the two corresponding tartaric acids (fig. 1 (*c*) and (*d*)) can be prepared.

perfectly indifferent to polarised light. I prepared the double salt in his presence, with soda and ammonia which he had likewise desired to provide.

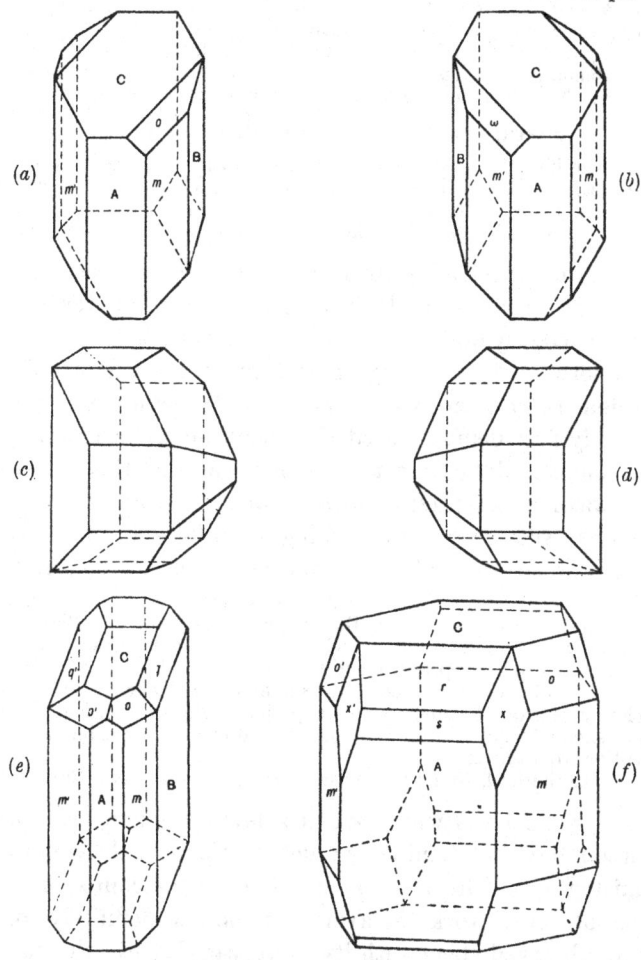

Fig. 1.

(a) and (b). Dextro and laevo sodium ammonium tartrate tetrahydrate (Pasteur). Note in fig. (a) the presence of a face o between C and m and the absence of such a face between C and m'. (b) is the mirror image of (a).

(c) and (d). Dextro and laevo tartaric acid (Pasteur). The crystals are mirror images of one another.

(e) and (f). Potassium racemate dihydrate (Pasteur) and sodium ammonium racemate monohydrate (after Scacchi). In these two crystals the faces m, o, q; m, x, o are symmetrically arranged on either side of a plane at right angles to the faces A and C.

The liquid was set aside for slow evaporation in one of his rooms. When it had furnished about 30 to 40 grams of crystals, he asked me to call at the *Collège de France* in order to collect them and isolate before him, by recognition of their crystallographic character, the right and the left crystals, requesting me to state once more whether I really affirmed that the crystals which I should place at his right would deviate to the right, and the others to the left. This done, he told me that he would undertake the rest.

The dramatic *dénouement* is thus described :

Then the illustrious old man, who was visibly affected, took me by the arm and said : " Mon cher enfant, j'ai tant aimé les sciences dans ma vie, que cela me fait battre le cœur."

(*Researches on Molecular Asymmetry*, Alembic Club Reprints, No. 14.)

(*c*) Another striking illustration of the meaning carried by the words *proof* and *discovery*, when these terms are applied correctly and legitimately, is afforded by a chapter in the history of that youngest and most marvellous of all the sciences, radio-activity. When Sir William Ramsay and Mr Soddy first demonstrated the spontaneous production of the element helium from the element radium, this transmutation was not accepted as *proved* until the same result had been observed by others working in different places, with different apparatus and with different samples of material.

The production of helium from radium was soon confirmed by a number of investigators. P. Curie and Dewar placed about 400 milligrams of radium bromide in a quartz tube. The salt was heated to fusion, and the tube exhausted to a low pressure and then sealed off. The spectrum of the gases in the tube was examined 20 days later by Deslandres, and gave the complete spectrum of helium. This experiment showed conclusively that the presence of helium in old radium preparations could not be due to its absorption from the air.

(Rutherford, *Radio-active Substances and their Radiations.* 1913.)

(2) A tendency which set in a few years ago, and which unfortunately still dominates much of the school teaching of chemistry, would have us believe that in the course of some couple of hours' work the average pupil can definitely correlate an observed effect with its cause, can *discover* the nature of a chemical relationship, or can prove a law. We are told that " the composition of water is *discovered* by burning the hydrogen obtained by the action of acids on metals "..." the girls should be *discovering* the composition and properties of soap." Compare and contrast with this the true evaluation of the work of pupil and teacher in the acquisition of scientific knowledge as given by Ostwald. In the most delightful of

elementary text books[1], the main facts of inorganic and physical chemistry are dealt with in the form of dialogues between a teacher and a pupil. The method is heuristic in the truest and best sense, but there is no make-believe, no pretence about what the pupil really accomplishes himself and what is done for him. Thus in the investigation of the effect of varying pressure on the volume of a definite quantity of air confined over mercury, a number of corresponding measurements are made in the usual way[2] and the results recorded in tabular form.

Pupil. What is the use of that?

Teacher. I want to show you *how to discover a law of nature.*

And when, after a number of explanations, directions and trials, the relation $pv =$ constant has been formulated:

Teacher. Right. Now you have found the law which connects the pressure and the volume of air with each other, or makes them dependent upon each other.

Pupil. I should never have found that out without your help.

Teacher. I quite agree.

Pupil. I say, did *you* find it out by yourself?

Teacher. No. An English physicist named Boyle discovered it more than 250 years ago, and it goes by the name of Boyle's Law.

Moreover, explicit and emphatic protests come from teachers of high standing and experience, who in the strongest possible terms condemn an attitude which, regarding the learning of elementary science as the making of a series of discoveries, is nothing better than a make-believe fraught with grave intellectual danger.

The discovery of physical laws by the average pupil seems to me an achievement quite outside the range of the possible; no heuristic process can do away with the prerogative of genius. Of course we can arrange matters so as to produce in the pupils the delusion that they have newly discovered an old truth; but by such a practice we are deliberately deceiving them about matters so fundamentally important as the growth of knowledge, the nature of research and their own abilities. (P. Johannesson, *Die physikalischen Uebungen am Sophienrealgymnasium zu Berlin.*)

It is of the utmost importance that in the school teaching of chemistry, especially in its elementary stages, the interpretation of complicated processes should be avoided; otherwise the result is almost bound to be the encouragement of loose thinking and megalomania on the part of the pupils. This is a very real danger, because again and again we are confronted by a pedagogical requirement which indicates a very low standard

---

[1] Ostwald, *Schule der Chemie.* Authorised translation by E. C. Ramsay, entitled "*Conversations on Chemistry,*" pub. by Wiley and Son, New York.

[2] Glazebrook, *Hydrostatics*, 1904, pp. 159 *et seq.*

of psychological insight. " The pupils must themselves discover the causes
of chemical phenomena, they must devise experiments suitable for attaining
this object, and must themselves elaborate the details of these experiments."
Well, if the pupils were capable of doing all this, they would need no
teaching, nay, they would in mental ability excel a Faraday, a Bunsen. What
humanity through the foremost of its representatives has laboured at
painstakingly for centuries, young people are now to play at discovering in
the course of a few lessons....[All that] can be accomplished in the school
teaching of science is practice in logical thinking, the pupils being *made* to
follow lines of thought similar to those traversed by the great discoverers.
Note, *similar* lines, not by any means *identical* lines. We guide our pupils
along roads which might have led to great discoveries, and at all stages we
must tell them as facts a great deal of what has been found by mere fortuitous trial and not as the result of any thinking process, simply because
these special facts are outside the scope of logical inference. What logical
considerations could suggest the heating of mercury calx[1], when the calces
themselves are produced by heating in air ? Where does logic come in
when in testing for acids and bases we use litmus, a substance unknown to
the majority of people ? (R. Winderlich, *Logik in der Chemie.*)

Surely, therefore, the more honest, intellectually more
bracing and eventually more fruitful course is to sweep away
all delusions as to what the pupils can discover for themselves, and further to impress on them at as early a stage as
possible the fundamental difference between the ' illustrative
experiments ' they perform and real research work.   In the
practical working of the so-called " research method" of
teaching, what really happens is : under the direction of
somebody who knows (practically always through other
people's work) how a certain desired result can be brought
about, a certain favourable combination of conditions is
created, the effect thereby produced on some specific kind
of matter is observed in its qualitative aspect and measured
in its quantitative aspect, and the *assumption* is then made
that, except for those intentionally modified, all the conditions
under which the action proceeds exert no influence on the
final result.   And finally, whatever may be pretended about
' discovering ' and ' proving,' the result so obtained is compared with that *expected* according to the standard work on
the subject, accepted when it agrees with it, rejected when
it differs.   And yet it is only by the judicious use of apparent
failure, only when some serious attempt is made to trace the
cause of deviation from the standard result, that the element
of discovery comes within the scope of the work.   But as

1 See *ante*, p. 3.

things are, the attitude of many teachers of elementary chemistry who are considered most progressive and most truly scientific has much in common with that of the Alchemists of an earlier age. They cry now as then, "Follow Nature"; but then, as now, it was a following in which "the vision was seen before the following began."

II. Fundamental difference between students' illustrative experiments and research work.

What is it, then, which characterises work intended to lead to and culminating in real discovery, real proof, real standard measurements? For one thing, as many as possible of the conditions under which the experiment is carried out are tested as to their influence on the final result : the experiment is performed a considerable number of times, varying those conditions which are thought uninfluential, and making agreement of the various results obtained a criterion of the legitimacy of this assumption. Even so, in research work many deliberate assumptions are made, but always with the limitation that what is assumed has itself been the subject of previous investigations at least as trustworthy and accurate as those of the special research just being done. The best way for the student to realise in its full extent the difference between an 'illustrative experiment' and a 'research' would be to read some good original paper and to compare it with the corresponding experiment adapted for students' use as described in any good text book of practical chemistry. A fundamental difference exists between the two, and though what begins as an 'illustrative experiment' may of course at any time develop towards a 'proof' or really become one, this necessitates a fundamental change of method and procedure. Good opportunities for learning to appreciate this fundamental difference are supplied by the following cases :

(1) The demonstration by synthesis and by analysis of the qualitative composition of water.

Compare and contrast the directions for students' experiments contained in the ordinary text books with Cavendish's account[1] of his own work which culminated in the statement :

> When a mixture of inflammable and dephlogisticated air [our hydrogen and oxygen] is exploded...the condensed liquor...seems pure water, without any addition whatever ;

and the complementary achievement of Sir Humphry Davy[2] summarised by him in the words :

> Water chemically pure is decomposed by electricity into gaseous matter alone, into oxygen and hydrogen.

But in certain syllabuses we are told : "The composition of water is *discovered* by"—etc., leaving the impression that the achievement of such a discovery is well within the reach of the average pupil of school age.

(2)   The determination of the gravimetric composition of water.

The short abstract contained in *Nature*[3] of Morley's work on this subject should be read side by side with the directions usually given for the use of Dumas' method in a students' experiment[4].

(3)   The determination of the normal density of hydrogen chloride, *i.e.* the weight in grams of 1 litre of the gas at 0° C. and 760 mm.

This being an experiment which is quite frequently done by students towards the end of their school or the beginning of their college course, it may be useful to deal with it in somewhat greater detail under each of the two aspects which it is intended to contrast.

(*a*)   Students' illustrative experiment.

The necessary instructions for the experimental procedure are given on p. 11, and a suitable apparatus is shown in fig. 2[5].

---

[1] *Experiments on Air* (Alembic Club Reprints, No. 3) ; Thorpe, *Essays in Historical Chemistry*, 1911, pp. 92 *et seq.*

[2] *Study of Chemical Composition*, p. 10.

[3] *Nature*, **53**, 1896, p. 428.

[4] Ramsay, *Experimental Proofs of Chemical Theory*, § 38 ; Lewis, *Inorganic Chemistry*, p. 48.

[5] The appendix at the end of this chapter gives a description of an arrangement which in the writer's practice has been found to work extremely well, and to repay fully the somewhat greater trouble involved in the setting up.

The table on p. 13 indicates the number and kind of measurements to be made, together with the calculation involved, and gives the actual results obtained in a series of students' experiments.

Apparatus for determining the Density of Hydrogen Chloride.

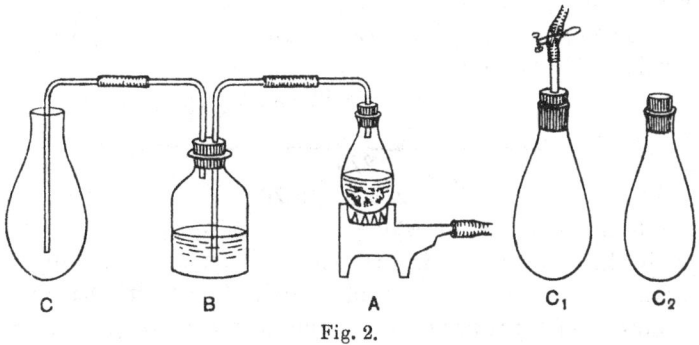

C   B   A C_1 C_2

Fig. 2.

The above apparatus is recommended by Ramsay in *Experimental Proofs of Chemical Theory*. The substitution of C_2 for C_1 represents a further simplification of the arrangements there proposed.

A. Generating flask containing sodium chloride, preferably in the form of lumps of rock salt, which is covered with strong sulphuric acid.

B. Drying flask containing strong sulphuric acid.

C. Carefully dried density flask, round bottomed and of about 300 c.c. capacity, which is filled with hydrogen chloride by downward displacement, the appearance of copious fumes at the open neck indicating complete expulsion of air.

C_1. The same flask fitted for weighing. When the flask, placed as in C, is supposed to be full, the delivery tube is *slowly* withdrawn, the well-fitting good rubber cork quickly and firmly inserted, and the clip opened for a second to allow the excess of gas compressed by insertion of the cork to escape. With this arrangement the volume entering into the calculation is that of the water filling the flask and glass tube when the cork is inserted and carefully pushed in to the amount indicated by a file-mark made on the neck of the flask.

C_2. The same flask with unbored cork (arrangement alternative to C_1). After slowly withdrawing the delivery tube, the cork is quickly put on flush with the opening of the flask, and pushed in firmly to any desired but not fixed amount. With this arrangement the volume entering into the calculation is that of the *whole* flask, assuming that the gas dealt with is that which at atmospheric pressure completely filled the flask, and was compressed by the insertion of the cork.

Let us now turn to the main object for which this experiment has been introduced here, namely to show in their proper relation the experimental work done and the assumptions made. As a matter of fact these latter are so numerous that it may conduce to clearness to deal with them classified according as they are related to: (α) the antecedent data, (β) the correctness and the accuracy of the measuring instruments used, (γ) the chemical purity of the substance investigated.

(α) The expression giving the weight of 1 litre of hydrogen chloride at 0° and 760 mm.

$$\frac{\left(w' - w + \dfrac{d_{\text{air}} \times a \times 273 \times p}{(273 + t) \times 760}\right) \times 1000}{a \times \dfrac{273 \times p'}{(273 + t') \times 760}}$$

is made up of terms of which some, viz. a, w, w', t, t', p, p', have in the experiment above described been made the subject of actual measurement (see table), whilst others are assumed as known from previous work and are accordingly called *antecedent data*. In the present instance d, the weight of 1 litre of air $= 1\cdot29$ grams, is *assumed* as known from previous independent work, and it is usual, even if the student has himself found and measured this value, to employ the standard number, which represents the result of a large amount of very accurate research. It is further *assumed* that hydrogen chloride and air behave as perfect gases, *i.e.* that the relations between volume, temperature and pressure are represented by the equation $v_0 = \dfrac{v \cdot p \cdot 273}{760 \cdot (273 + t)}$ (laws of Boyle and Charles), which, since $d = \dfrac{w}{v}$, makes it possible to reduce the density found for temperature and pressure t and p to what it would be at 0° and 760 mm., that is, under the conditions which for convenience of comparison are called normal. It is obvious, therefore, that the antecedent data used involve the making of a large number of *assumptions*.

(β) The next set of *assumptions* relates to the instruments used in the measurements of weight, volume, temperature and

The weight of 1 litre of Hydrogen Chloride at 0° C. and 760 mm. pressure.

The principle of the method consists in finding the weight of a vessel of known volume filled first with air and then with the gas (or vapour) examined.

The density of air at 0° C. and 760 mm. $= d = {\cdot}00129$ gram per 1 c.c. is assumed as known.

	I	II	III	IV
Total volume of empty flask **C** (measured by filling completely with water and delivering contents into graduated cylinder)...... $= a$	347 c.c.	346 c.c.	347 c.c.	346 c.c.
Wt. of carefully dried flask **C₂** filled with air $= w$	84·865 gr.	84·690 gr.	84·883 gr.	84·813 gr.
Temp. at place of inserting cork .... $= t$	21° C.	21° C.	21° C.	21° C.
Barometer .... $= p$	766 mm.	774 mm.	766 mm.	768 mm.
Wt. of air filling flask **C₂** at temp. t and pressure $p = d_{\text{air}} \times a \times \dfrac{273 \times p}{(273 + t)\,760}$ .... $= W$	·420 gr.	·422 gr.	·420 gr.	·419 gr.
$\therefore$ Wt. of vacuous flask .... $= w - W$	84·445 gr.	84·268 gr.	84·463 gr.	84·394 gr.
Wt. of flask supposed to be completely filled with dry hydrogen chloride gas .... $= w'$	84·991 gr.	84·814 gr.	85·008 gr.	84·938 gr.
Temp. at place of inserting cork .... $= t'$	24° C.	22° C.	24° C.	22° C.
Barometer .... $= p'$	766 mm.	774 mm.	766 mm.	768 mm.
Wt. of hydrogen chloride filling flask of capacity a c.c. at $t'°$ and p' mm. .... $= w' - w + W$	·546 gr.	·546 gr.	·545 gr.	·544 gr.
Wt. of 1000 c.c. of hydrogen chloride measured at 0° and 760 mm.................. $= \dfrac{(w' - w + W)\,1000}{a \times \dfrac{273\,p'}{(273 + t')\,760}}$	1·699	1·675	1·696	1·682

Mean of Exps. I—IV............................. **1·688**

Standard Values: Gray and Burt 1909...... 1·63915

Scheuer   1909...... 1·6394

pressure actually made. The arms of the balance may not be of equal length[1], the weights may not be correct[2]; the measuring cylinder used in the determination of the volume of water filling the flask C may not indicate the true volume[3]; the fixed points of the thermometer may have shifted owing to slow volume changes in the glass[4]; the scale of the barometer may require a correction[5]. Most likely the magnitude of these constant errors is in no case so great as to tell in comparison with the experimental error (*post*, p. 24) involved in the readings made, but the fact remains that some ideal conditions are *postulated* from which the real conditions differ by a greater or less, but always appreciable amount. Moreover in the students' experiment the difference between w and w', the weights of the flask C when filled with air and with hydrogen chloride respectively, is taken as due *entirely* to change in the contents of the flask, which involves yet another *assumption*, viz. that the buoyancy of the air—the weight of air displaced which has to be added to the observed weight in order to get the true weight—was exactly the same on the separate occasions of the two weighings. But this is very unlikely to be the case, and with a flask of about 300 c.c. capacity a change of as little as 1°C. in the temperature of the air will produce a change of 1 mg. in the weight observed (*post*, p. 31). The simple device of counterpoising by a vessel of equal or at least very nearly equal volume, which should be resorted to whenever the object is to ascertain small changes in weight of apparatus of large volume, is not sufficiently made use of in students' experiments.

(γ) And finally, there is a whole host of tacitly made *assumptions* concerning the purity of the substance investigated. The hydrogen chloride is made in A by the action of strong sulphuric acid on sodium chloride, and it is *assumed* that what passes over into the flask C is hydrogen chloride

[1] Glazebrook, *Statics*, 1904, p. 153.
[2] Findlay, *Practical Physical Chemistry*, 1906, pp. 24 *et seq.*
[3] Clowes and Coleman, *Quantitative Analysis*, 1911, pp. 142—144; Newth, *Manual of Chemical Analysis*, 1903, p. 306.
[4] Glazebrook, *Heat*, 1908, p. 18.
[5] Glazebrook, *Hydrostatics*, 1904, p. 148.

only, all the impurities, chief among which is aqueous vapour, having been retained by passage through the strong sulphuric acid in *B* ; no attempt is made to settle by actual investigation whether the provision made for purification has been adequate or not. Again, the directions given in the book quoted are that the current of hydrogen chloride should be allowed to pass until copious fumes issue from the mouth of *C*, it being *assumed* that when this happens, all the air in the flask has been replaced by the specifically heavier hydrogen chloride. Such an *assumption* is probably not justified, and it is not an invariable rule in students' experiments to test how much air has been left by opening the flask after the last weighing under water (or dilute soda), when the inrushing water absorbs all the easily soluble hydrogen chloride, leaving behind the air, the amount of which can then be measured and allowed for in the final calculation[1]. Again, the formation of fumes at the mouth of the flask tells its own tale ; these fumes are the liquid produced by the combination of the excess of the escaping hydrogen chloride with the moisture in the air, and owing to their comparatively high specific gravity they are likely to sink into the flask and to contribute to the weight *w'*, supposed to be that of a flask of volume *a* filled with *gaseous* hydrogen chloride *only*. Yet another *assumption* made is that the glass vessel plays a passive part in the whole transaction, that neither air nor hydrogen chloride condenses on its surface; but considering the very great solubility of hydrogen chloride in water and the difficulty of completely freeing glass from the film of moisture which clings to it most pertinaciously, it is quite certain that in the second case, at any rate, this *assumption* is not in accordance with fact. These various deviations from the *postulated* ideal state of purity of the hydrogen chloride gas would of course produce different errors in the value required: contamination with air (which is less dense than hydrogen chloride) would lower the density ; contamination with aqueous vapour, condensation of hydrogen chloride on the walls of the glass vessel, would

[1] Fenton, *Outlines of Chemistry*, Part I, 1909. Determination of density of carbon dioxide, p. 12 ; of ammonia, p. 17.

make the density too high. What the absolute and relative values of these various errors would be, and how far they would tell when compared with the experimental error attaching to the measurements made, depends altogether on the special conditions and cannot be evaluated *a priori*, but must be determined experimentally. What however is certain and should be impressed on the student is that in the case of his determination of the density of hydrogen chloride, fair agreement with the standard value cannot be taken as a proof of the absence of large errors in the measurements involved, since these may have happened to compensate one another. All the same, it is not intended to depreciate unduly the value of an illustrative experiment such as the above, but rather to increase this value by transference from the realm of the semi-fictitious to that of the real: to make the student realise how great is the number and how fundamental the nature of the *assumptions* he has made, sometimes consciously, more often still unconsciously; to describe the special features of standard work on the same subject, and thereby to supply data for evaluating the probable effect on the students' results of the discrepancy between the actual conditions of their work and the ideal ones *assumed* by them. How the first of the above two desiderata may be realised has already been shown in the classification and critical discussion of the large number of *assumptions* actually made in the students' experiment on the density of hydrogen chloride just described; a short account of the pertinent portions of a recent standard determination of the same physical constant will now be used for the purpose of the comparison recommended above, between students' work and research work.

(*b*) Standard determination. Gray and Burt's recent (1908) determination of the density, *i.e.* the weight of a normal litre, of hydrogen chloride.

The principle of the method used was the same as that of the students' experiment, viz. the determination of the weight W of gas filling a known volume V at the measured temperature t and pressure p. Regnault, who was the first to measure gaseous densities with considerable accuracy, worked on this principle, and devised a comparatively simple method involving great reduction in the experimental errors. It was Regnault who first taught us to eliminate corrections for changes in the buoyancy of the air (see p. 30) through the ingenious device of counterpoising by another vessel of as nearly as possible the same volume, so that any change in the weight of the density globe due to an external cause, viz. change in the density of the air displaced, is compensated for by a similar change in the weight of the counterpoise used; filling the globe whilst it was surrounded by ice did away with the necessity for any temperature correction; the retention of any air was avoided by several successive evacuations of the density globe followed by admission of the gas examined. The actual measurements consisted in finding w_1, the weight of the globe of internal volume V holding some of the gas investigated at the low pressure p_1, and then w_2, the weight of the globe when holding the gas at a pressure p_2 very nearly equal to that of the atmosphere; the required value is then given by the equation

$$D = \frac{(w_2 - w_1) \cdot 760}{V \cdot (p_2 - p_1)}$$

Regnault's method for the determination of gaseous densities long reigned supreme, and still continues to be used in work of the highest accuracy, though of course there have been modifications in technique, the result of new discoveries and inventions. First in order of time came the perfecting of apparatus for producing vacua[1] so high that for purposes of calculation p_1 was eliminated from the above equation; this was followed by the invention of comparatively simple machinery for the rapid production of large quantities of liquid air[2], which, together with the use of cooled charcoal as an additional agent for producing very high vacua[3], has

[1] For description of Töpler's pump see Glazebrook, *Hydrostatics*, 1904, pp. 196, 197.

[2] For description of Hampson's apparatus for liquefying air, see Newth, *Inorganic Chemistry*, 1907, p. 77.

[3] Charcoal, especially the porous kind obtained by the carbonisation (strong heating out of contact with air) of cocoanut, is a very effective agent for the absorption of gases, and to Sir James Dewar we owe the discovery that this power increases rapidly as the temperature is lowered. Hence by surrounding

revolutionised the technique of preparing, storing, transferring and weighing gases in a state of great purity. Gray and Burt's determination of the density of hydrogen chloride is a case in point.

Fig. 3 shows the essential features of the apparatus used by these investigators[1], the procedure for obtaining the data required being as follows: the carefully purified hydrogen chloride stored in a suitable vessel was allowed to flow into the completely evacuated measuring globe A, kept at the temperature of $0°C.$; V, the volume of A, had been previously ascertained by weighing A at first full of air and then again with the distilled water which had completely filled it at $0°C.$, a suitable arrangement having been made for the accommodation of the overflow due to the expansion of the water when the temperature rose from $0°$ to that of the balance case. The pressure of the gas filling A could be brought with great precision to that of the atmosphere, p, the value of which could be measured with great accuracy by a standard barometer. W, the weight of the hydrogen chloride filling the globe A (*i.e.* occupying the volume V measured at $0°$ and p mm.), was then ascertained.

This is the mere outline; with regard to the details of the work which in the case of the students' experiment led to the numerous assumptions already referred to, it will conduce to clearness if we deal with the various points arising in this connection according to the classification made use of above (p. 12).

(α) Antecedent data.

The density of air does not enter into the calculation, but on the other hand the method of indirectly measuring the volume of A by finding the weight of a substance of known density occupying that volume leads to the substitution of $\dfrac{W_{\text{water}}}{D_{\text{water } 0°}}$

a so-called 'absorption tube,' *i.e.* a vessel filled with cocoanut charcoal, with liquid air, the action can be made so complete as to produce in the vessels communicating with the absorption bulb a vacuum as perfect as any that it had been found possible to produce hitherto by evacuation with a suitable pump.

[1] *J.C.S.* 1909, **95**, p. 1633.

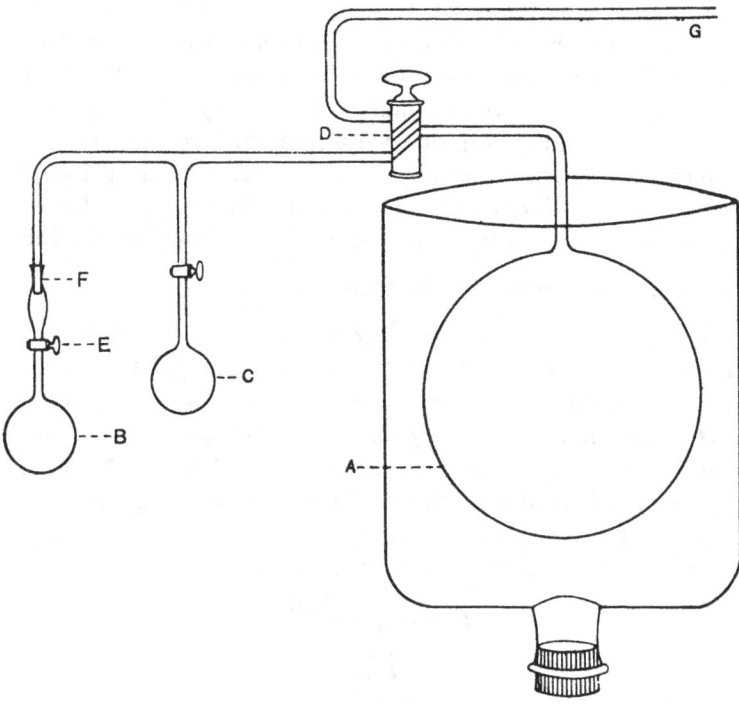

Fig. 3.

A. Measuring globe of accurately determined capacity (465·857 c.c.), surrounded by vessel holding ice and distilled water.

B. Absorption bulb (capacity about 20 c.c.) filled with cocoanut charcoal. Slow cooling of **B** by liquid air, followed by opening of the taps **D** and **E**, produces *complete* transference to **B** of the gas filling **A** at the temperature of melting ice and atmospheric pressure. When, after closing **E** and detaching at **F**, the temperature of **B** is allowed to rise to that of the room, the pressure of the absorbed hydrogen chloride does not exceed a few centimetres, a consequence of the high value of the coefficient of absorption of this gas.

C. Auxiliary charcoal bulb, used for the initial complete evacuation of **A** and of the capillary connection tubes.

D. Three-way tap for connecting **A** either with the manometer and the vessels used for purifying and storing the gas, or with either of the charcoal absorption bulbs **B** and **C**.

F. Joint at which **B** can be detached for purposes of weighing.

G. Tube leading to rest of apparatus, consisting of:

    (i) manometer,

    (ii) vessels for generating, purifying and storing the hydrogen chloride.

for V, and hence to the introduction of the antecedent datum $D_{\text{water }0°}$; but this is a value known to a degree of accuracy so high[1] as to exceed that likely to be reached in the other measurements involved in finding the density of the gas investigated. Filling the globe A at $0°$, the standard temperature, which it is possible to produce and maintain with the greatest accuracy, introduces another great simplification, in that it does away alike with the *assumption* represented in the students' experiment by the factor $\dfrac{273}{273+t}$ (Charles' Law) and the alternative of a special determination of the coefficient of expansion of hydrogen chloride between $0°$ and $t°$. As regards the coefficient of compressibility, assumed in the students' experiment to be adequately represented by $760/p$, in Gray and Burt's work the deviation from Boyle's Law between 760 mm. and p mm. formed the subject of a special investigation, the results of which made it possible to apply the necessary correction to the factor $760/p$ in the equation

$$D = \frac{(w_2 - w_1) \cdot 760}{\dfrac{W_{\text{water}}}{D_{\text{water }0°}}} \cdot p$$

(β) The correctness of all the measuring instruments used was tested, and the degree of accuracy attainable with them actually measured; moreover the number of measuring instruments used was only half that required in the students' experiment. The elimination of a temperature correction produced not only the simplification already dealt with, but also did away with the use and the consequent testing of a thermometer; besides which, the substitution of $\dfrac{W_{\text{water}}}{D_{\text{water }0°}}$ obviated the use of a graduated vessel. There remained therefore the measurement of weight, involving the use of a balance and of weights, and the determination of the atmospheric pressure, involving the measurement of a length on the scale of a barometer. The *relative* value of the weights

---

[1] 0·999878 (Volkmann).
0·999871 (Rossetti).
0·999868 (Thiesen, Scheele and Diesselhorst).

used was ascertained by careful calibration according to an accepted method[1], which also took into account the air displacement of the weights. No correction was needed for the absolute values of the weights, nor for errors due to a possible inequality in the length of the arms of the balance, because these would all be represented by a factor affecting alike $(w_2 - w_1)$, the weight of the hydrogen chloride and W_{water}, which appear in the numerator and denominator of a fraction and hence cancel one another. Variations in the buoyancy of air were eliminated in the usual way by means of counterpoises of volume equal to that of the vessel weighed. The probable error of the weighings was ascertained to be not greater than ·03 mg. The scale of the barometer could be read to ·03 mm. and was tested at the Standards' Office[2].

(γ) Very great precautions were taken in the preparation of the hydrogen chloride used; all conceivable[3] sources of contamination were critically examined, and definite tests—

---

[1] Clowes and Coleman, *Quantitative Analysis*, 10th ed. 1914, p. 6; Newth, *Manual of Chemical Analysis*, 1907, p. 192; Findlay, *Practical Physical Chemistry*, 1906, pp. 24 *et seq.*

[2] The standard of length is defined as the distance at 62° F. between the centres of the transverse lines on the two gold plugs in a bronze bar deposited in the office of the Exchequer.

In accordance with the Weights and Measures Act of 1878, the British Standards are now kept at the Standards' Office of the Board of Trade at Westminster.

[3] 'Conceivable' is of course not equivalent to 'possible'; the term is deliberately used to emphasise the fact that the sum total of errors existing probably includes far more than those of which we have cognisance, the number of the latter being representative of the state of knowledge reached, and being added to as that knowledge grows. Thus Regnault made no allowance for the effect of a change in pressure of less than 1 atmosphere on the volume of solids. It was reserved for Lord Rayleigh to point out that this was the source of a constant error in the determination of gaseous densities by Regnault's method, in which a glass globe is weighed, first vacuous, and then full of gas, being both times counterpoised by another glass globe of exactly the same external volume as the full globe. But in the first case the external pressure of the atmosphere produces a shrinkage of the vacuous globe which can be found by direct experiment, and which amounts to between ·04 and ·016 per cent. of the volume of the globe. The effect of this shrinkage is to make the vacuous globe displace a smaller volume of air than when filled with a gas under normal pressure. Consequently the weight of the gas, found from the difference between the weight of the globe when full and when vacuous, is smaller than the true value.

qualitative and quantitative—applied in order to ascertain how far the various precautions taken had served their end. The gas, prepared from material as pure as possible, after being dried by passing over phosphoric oxide, was liquefied (B.P. $-82\cdot9^\circ$ C.) and freed from impurities by fractional distillation. The isolation and retention of the portion distilling over first, served to remove impurities of higher boiling point, such as water; a repetition of the process, in which the first portion of the distillate was rejected, served to remove the impurities of greater volatility, such as oxygen and nitrogen. As in all quantitative measurements aiming at a high degree of accuracy, the purity of the substance investigated became a factor of the utmost importance ; a series of determinations of some physical property was therefore carried out with material obtained in different ways, agreement of the results being made the test and the measure of the purity of the substance dealt with. The reasoning followed in the establishment of this correlation runs thus : If the results obtained, say the densities of two specimens of hydrogen chloride prepared from different sources, were the *same* (by which is meant that they do not differ from one another by more than do the results of two separate determinations with material from the same source), either the two specimens are identical because both are pure, or they both contain impurities of an amount and kind such as will affect the true value of the density to exactly the same degree. Of these alternatives the first is much simpler, representing merely realisation of the conditions aimed at, and is therefore more probable than the second, which would involve some not very likely chance coincidences ; and if the variation in the mode of obtaining the hydrogen chloride is carried further still, and a third and a fourth specimen, each differently prepared, continue to give the same results, the assumption of identity of material, due to the absence of all appreciable amount of impurity, becomes increasingly justified. In the special investigation now under consideration the hydrogen chloride was prepared in three different ways and the density of the products determined ; the results obtained were as follows :

Mode of Preparation	No. of determinations	Extreme values of density. Wt. in grams of 1 litre	Mean value. Wt. in grams of 1 litre
(1) Sulphuric acid and sodium chloride	4	1·64053 and 1·63986	1·64016
(2) Sulphuric acid and ammonium chloride	10	1·64069 ,, 1·63950	1·64016
(3) Silicon tetrachloride and water	7	1·64083 ,, 1·63976	1·64023

Moreover it was not assumed that the walls of the containing vessel played an absolutely passive part, but the problem was specially investigated, with the following results :

...the density of the gas in a new bulb decreases with each filling, and finally reaches a minimum value which varies fairly regularly about a mean. In all probability this decrease is due to the presence on the surface of the bulb of a film of moisture which absorbs the gas in relatively large quantities; this film being partly absorbed by the charcoal in the weighing bulb, together with the gas itself, the apparent density is too great. With each successive filling the film should diminish and reach a constant value. As a matter of fact...a limit of desiccation was certainly reached after the first few fillings ...[but] in the absence of all avoidable moisture there was [still] a small but appreciable amount of gas adsorbed by the glass....That even this small amount of gas is taken up by cooled cocoanut charcoal is certain from the fact that on letting a little silver nitrate solution into the exhausted volume bulb at the end of each series of determinations, no trace of opalescence was ever observed. Blank experiments carried out with the same volume of silver nitrate solution showed that as little as 2 cub.mm.[1] of the gas could be readily detected.

A specially devised method for accurately measuring the amount of hydrogen chloride adsorbed by glass gave 0·00013 gram for the amount to be subtracted from the weight of one litre of the gas[2].

Another possible source of error was the solution of traces of hydrogen chloride in the grease used for the stop-cocks. An attempt was made to measure this. It was found that the amount of gas dissolved by the grease[3] on the bulb stop-cock did not exceed 0·01 ccm. [that is, about 1/46,585 of the quantity of gas involved], and since this amount of gas represents an error smaller than the mean probable error in any of the series, it was decided to neglect it.

The concluding statement about neglecting the error thus measured provides a suitable transition to another point which requires to be dealt with at the outset of a course of practical work, much of which is of a quantitative nature.

[1] The volume of the measuring bulb having been 465·856 c.c., 2 cub.mm. (= ·002 c.c.) is about 1/230,000 of the quantity of gas involved.

[2] The mean value for the weight of 1 litre being 1·64006, this represents a correction of about 1/10,000.

[3] Pure paraffins, for which it had been shown that they were not chemically acted upon by hydrogen chloride.

III.   Error of experimental measurements.

Under this heading it is proposed to discuss : the nature and the evaluation of what in quantitative work is termed the *experimental error* or the *probable error* ; the application of knowledge concerning the magnitude of the experimental error to the theoretical interpretation of experimental results; and the practical bearing of this subject on the degree of accuracy which it is rational to aim at in each of a number of connected measurements required in one and the same experiment.

1.   The nature and the evaluation of the experimental error.

To start with, it is necessary to emphasise the fact that the term 'experimental error' or 'probable error' applies only to what under the special conditions of the experiment are inevitable errors. These must be differentiated from errors which can be avoided and which are *mistakes* due to wrong registration, such as when in adding up the weights used one is accidentally omitted, or a ·02 gram weight is mistaken for ·01 gram ; or when in a scale reading, owing to want of care in ascertaining whether the zero is at top or at bottom, 159 c.c. is written down as 161 c.c., etc., etc.   Of course *mistakes* frequently occur in students' experiments ; and whilst repetition of the experiment or comparison of the results with those obtained by other members of the class are means for detecting the presence of mistakes (without however revealing their origin), practice, experience and care are the means for reducing their number.   Whenever it becomes clear that a mistake has been made, it is essential that confidence should be regained at once by repeating the experiment, not merely once but two or three times, so as to have sufficient data for contrasting the degree of coincidence between the later values obtained with the divergence[1] exhibited by the

---

[1] See numbers enclosed in brackets in the various tables giving the results obtained by a whole class for a special measurement, *e.g.* pp. 47 and 71.

result supposed to be tainted by a mistake; thus, and thus only, can we succeed in justifying the rejection of one or more special results. But when mistakes have in this manner been eliminated, there still remains the fact that all our measurements, however much we increase the perfection of our measuring instruments, however great is the care we take in the use of these instruments, give only an approximation to the true value, and what that true value is we can never know. It is therefore of the utmost importance that any student who attempts serious quantitative work in science, work controlled and dominated by common sense and intelligence, should always try to get, before starting the experiment, or at least at the earliest possible stage, a clear conception of what he can and what he cannot expect to achieve ; what is due to imperfections in the measuring instruments used, and what to the limited sensitiveness of our sense-organs ; what would be the effect of changes in physical conditions so small as to be imperceptible to our senses ; what disturbances might arise from the presence of small amounts of impurities in the substances dealt with.

Errors due to imperfections in measuring instruments may be illustrated by means of the pipette, the balance and the barometer.

(1) Pipettes (fig. 4, p. 39) are measuring vessels supposed to deliver definite specified volumes of liquid. Of course, owing to the difference in the coefficient of expansion for glass and for water[1] (or any other liquid), the graduation of any such measuring vessel can be correct for one particular temperature only; but if for the present we leave out of consideration errors due to changes in temperature, how far are we justified in considering the volumes delivered on different occasions as identical? Since at the same temperature weight and volume are strictly proportional, measurement of the weights delivered can be made a measure of the corresponding volumes. In an experiment made with an ordinary 50 c.c.

---

[1] Coefficient of expansion of glass between 0° and 100° C. $= 0.000024$;
,, ,, water, at about 18° C. $= 0.000185$.

pipette the weights of pure distilled water delivered successively were :

(i) 49·890 grs.

(ii) 49·860 „

(iii) 49·870 „

the extreme difference being ·030 or 0·06°/₀[1], which represents the experimental error of this special volume measurement, an error due chiefly to the fact that varying quantities of liquid are retained clinging to the walls of the vessel, but also to the uncertainty of exact adjustment of the level of the liquid to the mark on the stem. The diameter of the stem, the size of its open end, the shape given to the point of the jet, all these will influence the magnitude of the experimental error, which is the average of the maximum difference between successive measurements made at the same temperature. This experimental error must not be confounded with the error due to faulty calibration, which latter is given by the difference between 50 c.c., the volume marked on the pipette, and the *average* volume actually delivered by it.

(2) The use of the balance is attended by two errors of the same kind as those discussed above. There is the constant error produced by the balance not being true[2] (the arms being of unequal length), which corresponds to the error in the calibration of the pipette, and then there is the experimental error represented by the difference in successive weighings of the same body by the same sensitive balance. These differences are due partly to the instrument (want of rigidity of the beam, insufficient sharpness of the knife edges, which produce variations in the distance between the fulcrum and the points at which the weights act), and partly

---

[1] This is the value for the probable error in the case of an ordinary pipette used without taking special precautions for securing uniformity, such as careful cleaning of the glass, choice of the best method of uniform delivery. The error need not necessarily be anything like as high. Thus in Ostwald's *Physicochemical Measurements* it is said that by giving to 1 and 2 c.c. pipettes an improved form, it is possible to measure quantities of water exact to 1—2 mg. With a 1 c.c. pipette the following numbers were obtained for consecutive measurements :—0·9995, 0·9990, 0·9998, 0·9997.

[2] Glazebrook, *Statics*, p. 153.

to defective manipulation (jerky release of the beam, causing variation in the exact position of suspension of the scale pans, local heating and air currents, etc., etc.). In a special research of Landolt's, to which more detailed reference will be made later, a test was made of the efficiency of the balance used by accurately measuring on different days the small difference in weight—about 4 mgs.—between two brass cylinders each weighing about 400 grams. The results were :

Date of weighing	Difference in weight between the two cylinders
May 13th	4·273 mgs.
17th	4·260
23rd	4·250
24th	4·242
27th	4·260
	Mean = 4·257

Greatest difference between two weighings =　0·031 mg.
Average deviation from mean　　　　= ± 0·009

(3) Similarly readings of the height of the mercury column in a standard barometer taken in quick succession are not the same, as is shown by the following numbers :

Successive barometer readings over period of 20 minutes

29·540 in.
29·534
29·534
29·530
29·530
29·526

As regards the causes of these discrepancies, there is the difficulty of exact setting, not only of the zero point[1] but also of the vernier scale, the lower edge of which is supposed to indicate the height of the mercurial column. Differences in the results of measurements such as those described in the above examples are termed *experimental errors*.

2. The application of knowledge concerning the magnitude of the experimental error to the theoretical interpretation of experimental results.

Examples from within the scope of the student's own work will best serve our purpose.

(1) Take the experiment often dealt with in an elementary course in which the liquid obtained by burning hydrogen in

---

[1] Glazebrook, *Hydrostatics*, p. 147.

air is compared with distilled water, and the identity of the two substances shown to be very probable by a comparative determination of certain of their physical constants, viz. the density, the freezing point, the boiling point. When, as is generally the case, rapidity of execution is a great consideration, the comparison of the densities is best made by weighing in previously accurately counterpoised vessels volumes of the two liquids which are supposed to be equal because they have been obtained by delivery from the *same* pipette ; obviously the absolute value of the volume held and delivered by the pipette does not enter into consideration, and any error of calibration, however great, would not affect this special experiment. In one such experiment the results were :

| | Weight of liquid delivered from same 10 c.c. pipette | |
	I. Distilled water	II. Liquid obtained by burning hydrogen in air
(1)	10·185	10·170
(2)	10·160	10·192
(3)	10·187	10·178
Mean	**10·177**	**10·180**
Greatest difference between two determinations	0·027	0·022
Difference between the two means 0·003		

Inspection of the figures in the last line shows that it would be wrong to make a difference between I and II of the order of magnitude 3/10,000 a reason for declaring the densities of the two liquids to be different, since this value is less than the experimental error of the volume measurement. On the other hand identity in the results—such as would have been obtained had the weighings been carried to decigrams only—might have been due simply to the fact that the balance, owing to its low sensitiveness, had failed to indicate really existing though small differences. Hence, in the case of the above experiment, all that can be said is that the two liquids most probably are the same, but that if they should be different their densities do not differ by more than ·03 at the outside, probably not by more than 0·003.

(2) Again, take Experiment XIX, p. 161, the object of which is to find out whether chemical change is or is not

accompanied by a change in weight. A chemical change, viz. the burning of phosphorus, is made to occur in a closed vessel, and the whole system is weighed before and after the reaction. The usual way of performing this experiment in elementary teaching is open to grave objections. It is a common practice merely to counterpoise the flask before the burning, and to show triumphantly that after the burning the same counterpoise again produces equilibrium. This is only permissible if it is distinctly stated that nothing but a very rough approximation is aimed at. As a matter of fact the extent of what is established in this way depends altogether on the sensitiveness of the balance used, and by deliberately making this very small the amount of care and of skill needed for successfully carrying out the above or any experiment similar in principle can be reduced at will. Hence, unless the position is frankly faced and the students are made to realise it, this experiment is likely to be very misleading as to what actually has been proved and what under the special conditions can be proved. A chain of reasoning on the following lines is necessary :—Owing to the limited and comparatively small supply of air contained in a flask of suitable size, the amount of phosphorus burnt is bound to be small; since ·02 gm. of phosphorus requires about 100 c.c. of air, with a flask of 300 c.c. capacity only about ·06 gm. of phosphorus could be burnt. Supposing now that the balance used requires an overweight of ·005 gm. to produce a definitely recognisable shifting of the pointer, it would follow that if a change of weight due to the burning of the phosphorus actually took place, but were smaller than ·005 gm.—1/12 of the weight of the phosphorus burnt—the fact of chemical transformation being accompanied by change in weight could not be established. Hence if, as is not at all unusual, this experiment is carried out under conditions such as those above described, no value whatever can be attached to the result.

But it might be said that it is neither necessary nor usual to work with so rough a balance, that as a matter of fact the instrument employed is generally sensitive to 1 mg., and that the amount of substance transformed may be increased by using a flask holding as much as a litre of air. If so, various further points would require consideration.

Even supposing that the students had been told, or had found for themselves, that with the balance used an overweight of 1 mg. produces a distinctly recognisable shifting of the equilibrium position of the pointer, and if it so happened that the weight of the flask and its contents had been found the same before and after the occurrence of the reaction, all that could be inferred from that one experiment (always supposing the absence of counterbalancing errors) would be that if a change of weight had occurred in this particular transformation it could not be greater than 1 mg.; moreover, since we do not know the amount of substance (phosphorus and portion of air) that had interacted, we could not evaluate the degree of approximation of our result to the exact relation which it had been the object of the experiment to illustrate.

$$A + B + C + D + \ldots\ldots \qquad = M + N + P + Q + \ldots.$$

sum of wts. before the reaction $\qquad$ sum of wts. after the reaction

But this is not all. Supposing that working with a balance sensitive to 1 mg. it were found that after the reaction the weight had changed by 3 mgs. (for the purposes of the present argument it is immaterial whether the change had been an increase or decrease), could this be taken as an indication of a real change in weight due to the chemical transformation? Surely not, unless we had previously established that the amount of change observed was of a different order of magnitude from that found when the same—or a similar—apparatus in which no chemical change had occurred was weighed several times in succession. Now as a matter of fact in this so-called 'blank' experiment, unless the precaution had been taken of counterpoising the vessel used by a flask of the same or nearly the same volume, differences amounting to several milligrams might and almost certainly would be registered. Archimedes' principle tells us that when a body is surrounded by air it is not the true weight that is indicated by the balance, but the true weight less the weight of the air displaced; but the weight of air so displaced depends on two factors, viz.

(i) the volume of the displacing body, and (ii) the density, *i.e.* the weight in grams per 1 c.c., of the air displaced. Obviously, then, it will be of great importance that in the phosphorus-burning experiment sufficient time should be allowed for the flask to cool after the phosphorus has been burnt; but even if for the sake of simplicity we assume that this was done, and that therefore the volume of the flask was exactly the same on the occasion of the two successive weighings, there still remains the effect of changes in the other factor, *i.e.* in the density of the air, a value which depends on temperature, pressure and amount of aqueous vapour present. Suppose that between the two weighings the temperature in the balance case had altered by $3°$ C., say from $17°$ C. to $20°$ C., then according to Charles' Law $V_2 = V_1 \cdot \frac{293}{290}$, which would correspond to an increase of volume and consequent decrease of density of about $1 °/_0$. But at ordinary temperature and pressure the weight of 1 c.c. of air is about ·0012 gram, so that for every 100 c.c. of air displaced the *change* in buoyancy would be about 1·2 mgs., increasing the weight of the flask by that amount. Hence, using a flask of 300 c.c. external volume, a temperature change of $3°$ C. in the air displaced would produce a change of 3·6 mgs. in the apparent weight of the flask; whilst with a flask of about 1000 c.c. external volume a change of temperature of $1°$ only would account for a change of 4 mgs.

This should show the difficulty of correct correlation between effect and cause, and the necessity for care and caution in the interpretation of quantitative results. To summarise: The quantitative aspect of the experiment in which phosphorus is burnt in a closed system has been dealt with somewhat in detail in order to show how on the one hand *maintenance of weight* might be due simply to the use of a not very sensitive balance; whilst on the other hand an observed *change of weight* might be due *entirely* to the experimental error, *i.e.* to the algebraic sum of a number of effects, all of which are produced by causes other than actual change of weight accompanying the chemical transformation. Hence it is not only useless, but also a violation of scientific principle, to use a very sensitive balance, say one which indicates an overweight of 0·1 mg., unless the magnitude of the experimental error is ascertained and taken into account.

This is a case of a students' experiment which in principle and practice follows the same lines as research work, though of course the limits of accuracy and the values of the experimental error are of a different order of magnitude. In an earlier part of this chapter, the determination of the density of hydrogen chloride was used for showing the fundamental *difference in technique* between students' work and standard work; a short account of the pertinent features of a recent research on the conservation of weight will now be given in order to bring out the *identity of the logical processes* guiding both the research work and the students' work, provided always of course that the latter is performed in a manner that makes it worth doing at all.

The question whether chemical transformation is or is not attended by change in weight was investigated by the German chemist Landolt in a series of researches extending over 17 years, from 1890 to 1907. Since Lavoisier made chemistry into a quantitative science, thousands of chemists have performed millions of analyses, in which it has always been found that the greater the accuracy of the work, the more nearly did the results approximate to 100 °/₀; thereby showing that if chemical transformation should be accompanied by changes of weight, such changes can only be very small. Hence it is obvious that in an enquiry such as that undertaken by Landolt, all conceivable sources of error must be eliminated or at least reduced; the accuracy of the weighings must be pushed to the uttermost limits of the attainable; and finally the residual experimental error must be evaluated.

The experiments extended to 15 reactions which occur in aqueous solution, *e.g.* silver sulphate and ferrous sulphate, iodine and sodium sulphite, etc., etc.[1] The two solutions were introduced separately into the two limbs of ∩-shaped vessels of Jena glass. Two such pieces of apparatus, *A* and *B*, were prepared, so nearly equal that errors due to change in the

---

[1] It was essential that the reactions chosen should occur without the evolution of gas and without the production of much heat.

buoyancy of the air were negligible. By this arrangement each experiment could be done in duplicate, the
difference in weight between the two vessels being determined: (i) in the original condition, (ii) after the reaction
produced by tilting the ⋂-tube and consequent mixing of the
contents of the two limbs had taken place in apparatus *A*,
(iii) after the reaction had taken place in apparatus *B*.
Though the reactions chosen for investigation were all of a
kind in which the concomitant heat effect is not great, yet
some heating of the glass vessels due to rise of temperature
in the solution was inevitable, and consequently sufficient
time had to be allowed to elapse between the successive
weighings in order to make sure that the original condition
had been re-established. All the earlier experiments were
vitiated by the error, only discovered later, that 10 days were
not sufficient for the re-establishment of the original temperature, but that as many as 20 days might have to elapse before
equilibrium, indicated by the weight becoming quite constant,
had been reached. Increase in the temperature of either
A or *B* alone produces (i) an increase of external volume,
hence an increase in the quantity of air displaced, and (ii) a
diminution in the film of water always condensed on the outer
surface of the glass, both of which changes in external conditions make themselves felt by a decrease in the weight
recorded. Of course these effects are almost infinitesimal, a
few thousandths or at most a few hundredths of a milligram;
but then, as has been pointed out above, it was known from
the outset that the effect looked for, if existing at all, was
very small. Then, after this and all other conceivable causes
of error had been considered and as far as possible guarded
against, it remained to determine experimentally the total
effect of all the remaining causes of error; this was done by
means of blank experiments in which the manipulation of the
apparatus was the same as in the reaction experiments, but
the substances used were such as did not react, *e.g.* water
and paraffin, iodine and sodium sulphate, silver sulphate and
ferric sulphate, etc., etc. In 19 such experiments 8 experiments gave a + result, the mean increase in weight being

+ 0·008 mg., whilst 11 experiments gave a − result, the mean decrease of weight being − 0·010 mg.; the *maximum* difference recorded in these blank experiments was 0·024 mg., and if therefore in the reaction experiments an alteration of weight exceeding 0·030 mg. had occurred, this could not have been due to errors of observation. Now as a matter of fact, in 48 reaction experiments 25 showed a decrease and 23 showed an increase in weight; in two cases only was the maximum experimental error of ± 0·03 mg. exceeded, whilst the *average* magnitude of the changes observed was 0·012 mg.

The final result is that in all the 15 decompositions involved it has not been possible to establish a change of weight. The observed deviations from absolute equality before and after the reaction are due to external physical causes and are not the result of the chemical reactions.

In Landolt's classical work, the question put to Nature was the same as in the students' experiment on the burning of phosphorus, namely, whether chemical transformation is or is not accompanied by change of weight; the principle of the experiments devised for the purpose of getting an unequivocal answer was also the same, viz. the carrying out of a chemical reaction in a closed system, the weight of which is determined before and after the reaction. In both cases the correct interpretation of the results obtained involves the taking into account of the probable experimental error. In Landolt's research the evaluation of the experimental error by means of blank experiments was a *sine qua non*, and necessitated a set of measurements quite as important and nearly as extensive as those carried out in the actual reaction experiments. In the case of the students' experiment, to have recourse to blank experiments would also have been the most satisfactory plan; but failing this, a good deal may be done by an approximate evaluation based on a rough calculation such as that given on p. 29.

3. Conditions which determine the degree of accuracy to be aimed at in any of these measurements.

No chain is stronger than its weakest link! If in the determination of a density [1], for which the data required are a volume V and a weight W, the volume cannot be measured with an accuracy greater than, say, 1 in 200, the final result will be uncertain to the same degree, *i.e.* 1 in 200, and to aim at a greater accuracy in the determination of W would only be waste of time. Thus in the experiment already referred to (p. 28), which has for its object the comparison of the densities of distilled water and of the liquid synthesised by the burning of hydrogen in the air, if the volume measurements, made by an ordinary 10 c.c. pipette, are affected by an error of ·03 c.c. (*ante*, p. 28) it would be absurd to push the accuracy of the weighings to milligrams; the weight of 10 c.c. of water being about 10 grams, weighing to 1 mg. would represent an accuracy of 1 in 10,000, which is about 30 times as great as that which under the given conditions it is possible to attain in the volume measurements. If it were desirable or imperative to ascertain the density with an accuracy of 1 in 10,000, this could not be achieved merely by weighing to milligrams, but some different and more accurate method of volume measurement would have to be resorted to. Now, though one is at a loss to account for it, it is a fact that the average student finds it difficult to realise, or at any rate in practice to allow for the relation between the individual errors of the various measurements involved and the error in the final result. Volumetric analysis especially seems to offer scope for grievous waste of time and reprehensible violation of scientific principle, such as when, the accuracy of the titrations being at most 1 in 500, the substance to be dissolved is weighed out with an accuracy of 1 in 10,000 or even 1 in 50,000. Thus it is not at all unusual to find in students' note-books, under the heading of "Preparation of standard

---

[1] Density = weight of unit volume = $\dfrac{W}{V}$.

solution of sulphuric acid," numbers something like the following:

Wt. of sodium carbonate in 1 litre of solution $= 53\cdot427$ gms.

Vol. of sulphuric acid solution required for the neutralisation of 25 c.c. of the sodium carbonate solution.................................... $= \begin{cases} 25\cdot40 \text{ c.c.} \\ 25\cdot50 \text{ ,,} \\ 25\cdot45 \text{ ,,} \end{cases}$ Mean $25\cdot45$ c.c.

The probable error of these special volume measurements is 5 in 2500, *i.e.* 1 in 500, and extensive experience of volumetric analysis has shown that this is the average maximum accuracy attainable; hence the sodium carbonate need not and should not have been weighed to more than 1 decigram, which is also about 1 in 500. But the not uncommon practice illustrated by the above-quoted numbers, in which sodium carbonate was weighed to an accuracy of 1 in 50,000, seems to be the outcome of an ethical factor (or aspiration), shown in the desire to attain to greater perfection by the fundamentally wrong process of pushing the accuracy of each measurement involved as far as the instrument available will allow, quite irrespective of what can be achieved in the other measurements required[1]. Some text books, especially recent ones on practical physical chemistry[2], give due consideration to this important subject of the proper relation between the accuracy of a number of connected measurements; but students to whom the principle involved is not self-evident should seek conviction through the consideration of a number of specific cases of which the following are examples:

(1) In the application of a temperature pressure correction to a gaseous volume, one is apt to find the calculation made with data such as the following:

Volume   167 c.c.

Temperature 15° C.

Barometer   757·45 mm.,

---

[1] This would seem to be an attitude of mind inherent in human nature, one noted and criticised by Scheele:

"People will often split a hair where it is not the least necessary."

(*Chemical Treatise on Air and Fire*, 1777.)

[2] Findlay, *Practical Physical Chemistry* (1906), pp. 15—17; Glazebrook and Shaw, *Practical Physics*, 1909, p. 36; Gray, *Practical Physical Chemistry*, p. 1.

that is, whilst the accuracy of the volume and temperature measurements are $1/167$ and $1/273 + 15$ respectively, the pressure is taken to $1/75745$, that is, to an accuracy about 450 times as great as that of the volume measurement.

(2) The data given on p. 36 are another case in point, and one which deserves actual calculation.

The volumes of a certain sulphuric acid solution required for the neutralisation of $\dfrac{53\cdot427 \times 25}{1000}$ grams of sodium carbonate were found in successive titrations to be $25\cdot40$, $25\cdot45$, $25\cdot5$ c.c. respectively; the empirically found equivalency factor between sodium carbonate and sulphuric acid represented by the quantitative aspect of the following equation :

$$H_2SO_4 + Na_2CO_3 = Na_2SO_4 + CO_2 + H_2O$$
$$\quad 98 \qquad 106$$

is $\dfrac{98}{106}$, and hence 25 c.c. of sodium carbonate solution containing $\dfrac{53\cdot427 \times 25}{1000}$ of the salt correspond to $\dfrac{53\cdot427 \times 25 \times 98}{1000 \times 106}$ of sulphuric acid, which according to the results of the titrations are contained in $25\cdot4$ c.c. or $25\cdot5$ c.c. of solution, giving for the strength of the acid (*i.e.* grams of H_2SO_4 in 1 litre of solution)

$$\dfrac{53\cdot427 \times 25 \times 98 \times 1000}{1000 \times 106 \times 25\cdot4} = 48\cdot617 \text{ gms.}$$

and

$$\dfrac{53\cdot427 \times 25 \times 98 \times 1000}{1000 \times 106 \times 25\cdot5} = 48\cdot426 \text{ gms. respectively.}$$

But these results being uncertain to the same extent as the volume measurements involved, the last two figures must in each case be rejected. The mean value, $48\cdot5$, is identical with that which would have been obtained if the weight of the Na_2CO_3 had been taken as $53\cdot4$, that is, reduced to a number of significant figures representing the same order of accuracy as the volume measurements made.

(3) In the experiment on p. 303 on the determination of the equivalent weight of magnesium by replacement of the hydrogen in dilute acid (sulphuric or hydrochloric), it is found that each milligram of metal used yields very nearly 1 c.c. of

hydrogen. If about 150 c.c. of hydrogen are collected and measured in a gas tube graduated in c.c., the utmost accuracy attainable in the volume measurement is estimation to one-half of a graduation, that is, to 1 in 300. From this it follows that: (i) the amount of metal used, which will be about 150 mgms., should be weighed accurately to $\frac{1}{2}$ mgm.; (ii) it is *not* a matter of indifference, as far as the numerical result goes, whether the standard of reference is hydrogen $= 1\cdot00$ or oxygen $= 8\cdot00$, in which latter case hydrogen becomes $1\cdot01$; (iii) it would be absurd to read the barometer to fractions of a mm., or the thermometer to fractions of a degree centigrade, or to use in the calculation involved

$$\text{Equivalent weight of Magnesium}_{O=8\cdot00} \Bigg\} = \frac{\text{Wt. magnesium} \times \text{equiv. wt. hydrogen}_{O=8\cdot00}}{\text{Vol. hydrogen} \times \dfrac{273\cdot p}{(273+t)\,760}} \times \frac{\text{density hydrogen}}{1000}.$$

the more accurate value $\cdot08995$ instead of the rounded-off number $\cdot0900$ for the density (weight in grams of 1 litre) of hydrogen.

4. Illustrative Experiment.

Experiment I.

To find the experimental error in the determination of the volume contained in or delivered from various types of measuring vessels in successive fillings.

The principle of the experiment is based on the fact that as long as the physical conditions remain the same there is strict proportionality between the weight and the volume of a substance, and the procedure consists in comparing volumes by ascertaining the corresponding weights of one and the same liquid held or delivered in successive fillings. Differences in temperature, which would of course vitiate the postulated strict proportionality between weight and volume, would exert a disturbing influence, since there is a considerable difference between the coefficient of expansion of glass ($0\cdot000024$ between $0°$ and $100°\,$C.) and that of water ($0\cdot000185$ at about $18°$) or of any other liquid. The assumption usually made, that the temperature has remained constant during the course of each set of measurements, is sufficiently near the truth in the case of ordinary students' work, but if really accurate measurements are to be made, the vessels and their contents must be brought to a definite and accurately known temperature; this is most easily achieved by immersion for a sufficient length of time in the water of a thermostat (an arrangement for the production and long-continued maintenance of any definite temperature required).

Various types of measuring vessels[1].

Instruments constructed to *deliver* a definite volume

Instruments constructed to *hold* a definite volume

Pipette   Measuring cylinder   Graduated flask   Specific gravity bottle (Pyknometer)

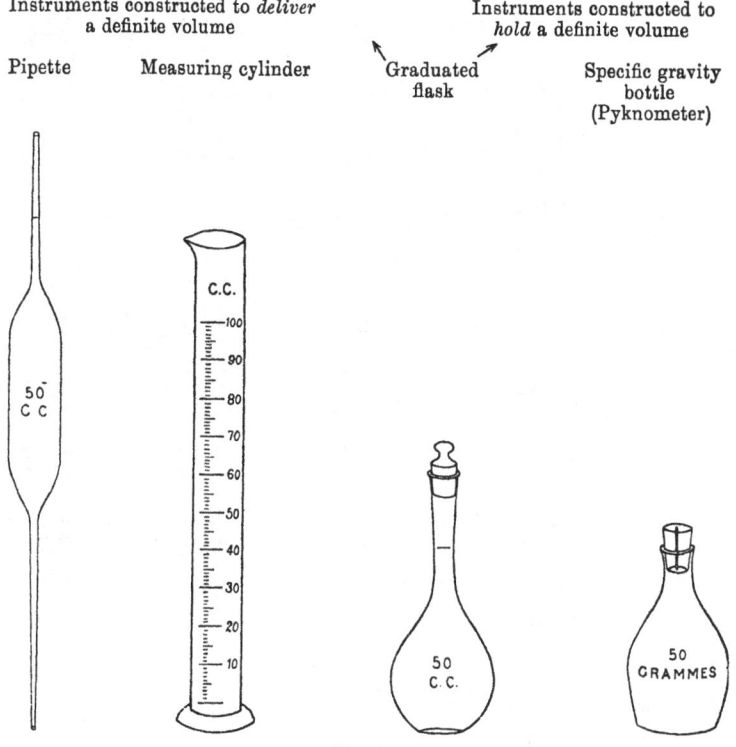

Fig. 4.

Graduated flasks belong to both types of instruments. Generally they are made to hold a stated volume and are used largely for the making up of standard solutions. But they can be made also for *delivering* a stated volume, or what is more usual, they are made available for either purpose by the provision of *two* marks on the neck. In the ordinary course of students' work it is not unusual to use a graduated flask for either purpose, though it may have one mark only and in most such cases very probably the experimental error of the other measurements involved (*e.g.* titrations, which generally are not accurate to more than 1/200) is such as to make it legitimate to neglect the error due to the liquid clinging to the walls of the flask when the volume delivered is treated as identical with the volume contained. But it is very important that in any case students using graduated flasks should realise this source of error, should know its approximate value, and should in each case when it is introduced consider whether it is or is not legitimate to ignore it.

[1] For the method of calibrating a burette, or graduated gas tube, see Clowes and Coleman, *Quantitative Analysis*, 9th edition, 1911, §§ 257 *et seq.*; Newth, *Manual of Chemical Analysis*, 1903, pp. 308 *et seq.*

Determination of the Experimental Error attaching to Volume Measurements made by means of various types of graduated vessels.

The table indicates the manner of making the measurements required for the purposes of the experiment, and the number of data wanted.

	I			II			III		
Sensitiveness of the balance used ...	·01 gram			·002 gram			·0002 gram		
	(i)	(ii)	(iii)	(i)	(ii)	(iii)	(i)	(ii)	(iii)
Pipette to deliver 50 c.c. Diam. at mark = 7 mm.	grs.	grs.	grs.	grs.	grs.	grs.	grs.	grs.	grs.
Wt. of empty beaker + cover glass	19·41	19·41	19·41	19·500	19·490	19·490	19·5050	19·5050	19·5045
Wt. of above + water delivered from pipette	69·29	69·27	69·29	69·390	69·350	69·360	69·4120	69·3730	69·3800
Wt. of water supposed to measure 50 c.c.	49·88	49·86	49·88	49·890	49·860	49·870	49·9070	49·8680	49·8755
Mean of (i), (ii), (iii)		49·87			49·873			49·8835	
Greatest difference between measurements		0·02			0·030			0·0390	
The above expressed as °/₀ of quantity measured		0·04			0·06			0·08	
Graduated Cylinder, 100 c.c. in ¹/₁ c.c. Diam. = 28 mm.									
Wt. of empty beaker + cover glass	34·45	34·45	34·45	34·247	34·245	34·245	34·2500	34·2505	34·2500
Wt. of above + water delivered from cylinder filled to the 50 c.c. mark	83·50	83·47	83·62	83·239	83·207	83·360	83·3360	83·3030	83·4575
Wt. of water supposed to measure 50 c.c.	49·05	49·02	49·17	48·992	48·962	49·115	49·0860	49·0525	49·2075
Mean of (i), (ii), (iii)		49·08			49·023			49·1153	
Greatest difference between measurements		0·15			0·153			0·1550	
The above expressed as °/₀ of quantity measured		0·3			0·3			0·3	
Specific Gravity Bottle, Capacity 50 c.c. Diam. of orifice = ·5 mm.									
Wt. of empty bottle	21·05	21·05	21·05	20·824	20·827	20·827	20·8295	20·8294	20·8292
Wt. of bottle filled with water	71·05	71·05	71·05	70·832	70·849	70·835	70·8458	70·8530	70·8594
Wt. of water supposed to measure 50 c.c.	50·00	50·00	50·00	50·008	50·022	50·008	50·0163	50·0236	50·0302
Mean of (i), (ii), (iii)		50·00			50·012			50·0233	
Greatest difference between measurements		0·00			0·014			0·0139	
The above expressed as °/₀ of quantity measured		0·0			0·028			0·028	
Graduated Flask, Capacity 50 c.c. Diam. at mark = 11 mm.									
Wt. of empty flask	20·43	20·47	20·50	20·520	20·520	20·520	20·5195	20·5190	20·5190
Wt. of flask filled with water	70·60	70·68	70·75	70·677	70·730	70·750	70·6920	70·7460	70·7610
Wt. of water supposed to measure 50 c.c.	50·17	50·21	50·25	50·157	50·210	50·230	50·1725	50·2270	50·2420
Mean of (i), (ii), (iii)		50·21			50·199			50·2138	
Greatest difference between measurements		0·08			0·073			0·0695	
The above expressed as °/₀ of quantity measured		0·16			0·14			0·14	

Inspection of the results set out in the table shows :

(i) The superiority of the pipette over the graduated cylinder, and of the specific gravity bottle over the graduated flask, due to the smaller diameter of the neck or the orifice by means of which the volume adjustment is made and the consequent diminution of the diameter of the layer of liquid above or below the level, representative of the exact, but (alas !) always ideal filling.

(ii) The sum of the various errors involved with measuring vessels of this capacity and construction is such as to make it unnecessary, in fact scientifically reprehensible, to do the weighings involved in the process of calibration with an accuracy, in any case greater than milligrams, whilst for the graduated cylinder, weighing to centigrams suffices.

APPENDIX
The determination of the density of a gas: Students' Experiment.

The apparatus used must include arrangements for the generation of certain gases, the displacement by these gases of the air contained in a flask or tube, and the absorption of the excess of gas used in this process.

When determining the density or the volumetric composition of gases such as hydrogen chloride, chlorine, carbonic acid, ammonia, sulphurous anhydride, nitric oxide, etc., which are easily absorbed by water or alkali or acid or by some specific solution (*e.g.* ferrous sulphate solution absorbs nitric oxide), the arrangement depicted in fig. 5 has been found to work very well. The apparatus, though fairly simplé, possesses two distinct advantages over the type usually employed : (i) It shows by the stoppage in the escape of air bubbles through F when the desired object of completely filling the vessel with the gas under investigation has been attained ; (ii) it disposes of the excess of this gas, making it possible to perform these experiments on the lecture table or the students' benches without having to use a draught cupboard.

Apparatus for filling flask with a gas soluble in water.

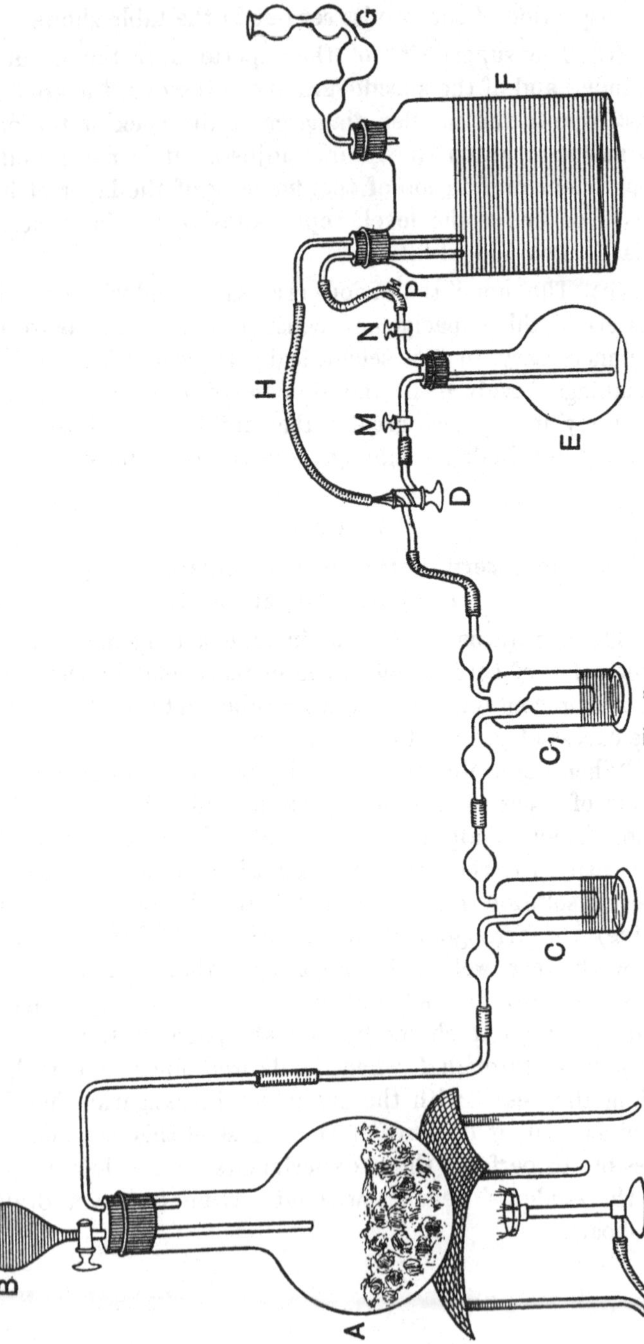

Fig. 5.

A. Round-bottomed flask of Jena glass, of capacity from $\frac{1}{2}$ to $1\frac{1}{2}$ litres, fitted with the dropping funnel B and a delivery tube, and holding the solid used in the generation of the gas required, *e.g.* lumps of rock salt (for hydrogen chloride), or of pyrolusite (for chlorine), or of marble (for carbon dioxide) ; ferrous sulphate and potassium nitrate (for nitric oxide).

B. Dropping funnel holding the liquid used in the reaction yielding the gas required, *e.g.* concentrated sulphuric acid (for hydrogen chloride and nitric oxide), dilute hydrochloric acid (for carbon dioxide), concentrated hydrochloric acid (for chlorine).

C and C_1. Gas wash bottles filled with suitable absorbents for the removal of impurities in the gas under investigation, *e.g.* in the case of chlorine, first water to remove the hydrogen chloride carried over with the chlorine from A, and then concentrated sulphuric acid to remove moisture ; in the case of the very hygroscopic hydrogen chloride, two lots of concentrated sulphuric acid to remove moisture.

D. Three-way stop-cock whereby to connect C_1 with F, either directly through H or indirectly through E.

E. Ordinary flat-bottomed flask (or Erlenmeyer flask) of about 250 c.c. capacity, with well-fitting rubber cork through which pass two glass tubes closed by the glass taps M and N.

F. Two-necked (or three-necked) Woulff bottle of at least $\frac{3}{4}$ litre capacity, communicating by two tubes with D and E respectively, and carrying the safety funnel G ; the vessel is about $\frac{2}{3}$ filled with a liquid which will completely absorb the gas under investigation, and the excess of which this part of the apparatus is intended to deal with, *e.g* water for hydrogen chloride, potash for chlorine. Whether the delivery tubes are made to dip under the surface of the liquid or are kept just above it, depends on the ease with which the gas is absorbed and the consequent danger of sucking back.

G. Safety funnel containing in its bend the same liquid as that in F, and intended to show when all the air has been expelled by the gas on its way from A into G.

H. India-rubber and glass tube connecting D with F.

M and N. Glass taps closing E.

P. Screw clip to prevent the escape of the gas investigated from the india-rubber tube when E is disconnected.

Procedure.

With D turned so that C_1 and F are connected through H, start the evolution of the gas in A by dropping liquid from the funnel B on to the solid, applying heat if needed and to the amount needed. Air bubbles will pass out at G, at first in quick succession, then more and more slowly until their stoppage shows that all the air has been swept out of those parts of the apparatus traversed by the gas, which now is completely absorbed in F. When this stage has been reached, open P and M and N and turn D so that there is communication between C_1 and F through E; the gas free from air now passes through E, sweeping out the air from it, the completion of which process is indicated as before by means of G. When we have evidence of E being full of gas, the various taps and clip are manipulated quickly in the following order : N is closed and then M, after which D is turned so that C_1 and F are again connected through H ; the next step is to stop the evolution of gas in A, or at any rate to make it as slow as possible, after which the screw clip P is closed and E is detached. The result of this way of manipulating is that the gas contained in E is at a pressure slightly higher than the atmospheric—N having been closed whilst the gas evolved in A had no other outlet—and hence, just before reading the thermometer, one or other of the taps M, N, should be opened for a second, when the pressure will adjust itself to that of the atmosphere, the value of which is obtained by reading the barometer. The gas contained in the ends of the tubes, beyond the taps M, N, should be replaced by air, which is most easily done by means of a glass tube drawn out to a long fine end and attached to hand bellows (fig. 6).

Bellows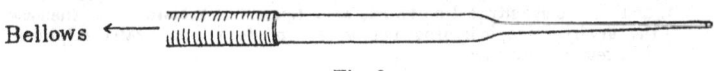

Fig. 6.

Whilst the process of filling would of course be the same for a graduated tube or any other vessel in which two suitable connections can be made for the inlet and outlet of gas, the actual measurements—other than temperature and pressure—which it is necessary to make will depend on the object of the experiment. If it is intended to determine a density, as in the case to which the figure specially refers, the flask must be weighed first when full of air (temperature t_a and pressure p_a being noted), and then full of the gas under investigation (the temperature and pressure being t_c and p_c). It will add greatly to the value of the work done if a further test, this time quantitative, is applied concerning the complete expulsion of air from E by connecting it with the gas supply for a second time and making the constancy of weight the criterion ; the second filling should follow quickly on the first, as of course otherwise we might have to make corrections for variations in temperature and pressure before the results of the two weighings could be compared. The only other measurement required is that of the volume of the gas, which is given by the volume of water filling the fitted-up flask to the taps M, N ; obviously it is essential that in making the volume measurement in the manner indicated we should be dealing with the volume actually corresponding to that occupied by the gas weighed, which postulates the cork being always pushed in as nearly as possible to exactly the same amount. Some variation in this is inevitable, but the error thus introduced need not be very large, it may have a + or − value, and repetition of the experiment would almost eliminate it. Its maximum value can easily be found experimentally by a series of weighings of the flask filled with water up to the taps, the cork having in each case been pushed in carefully to a mark made in the neck of the flask. To fill the flask for this purpose it is simplest to pour water right to the top of the quite empty flask ; then, when with the taps M and N open, the cork with its fittings is pushed in, water will fill the delivery tubes completely up to the taps, the excess flowing out at the open ends ; after closing the taps, the excess of water in the ends beyond the taps can be shaken out or soaked up with blotting paper. The volume of water is then measured, either somewhat roughly by being poured into a measuring cylinder, or more accurately by being weighed (do not commit the crime of weighing to more than decigrams !). Of course the error introduced into the volume measurement by the uncertainty in the pushing home of the cork may be obviated by arranging so as not to have to take the cork out at all in the course of any one experiment ; in this case the flask is filled by suction applied at the end nearest N, whilst M dips under water; in order to fill the flask completely in this manner, it is essential that the other end of N should be flush with the cork, otherwise it will be impossible to get rid of the air in the space indicated in fig. 7 by the dots.

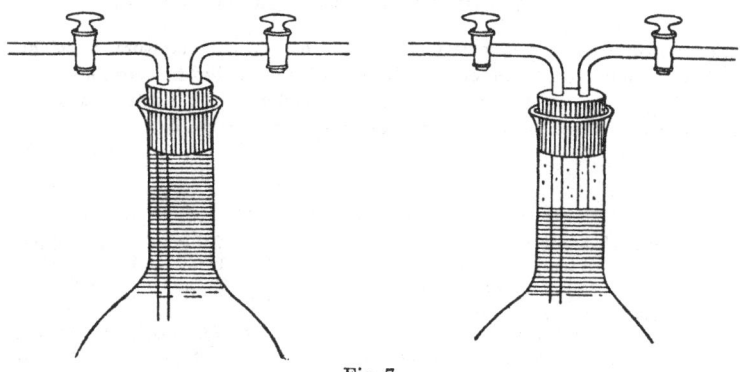

Fig. 7.

It should be remembered also that in any case, before the water is introduced, the gas filling the flask—which probably would be at least fairly soluble in water, perhaps very soluble —should be replaced by air. This is done to avoid the production of a gaseous solution which would not lend itself to the purpose of a volume measurement, in which we determine the weight of liquid filling the flask and assume that the specific volume (the volume occupied by unit weight, *i.e.* the reciprocal of the density) is that of water, which under the special conditions may be taken to be 1·000 c.c. per 1·000 gram.

Record of results.

The following tables give the results obtained by a class of students in the determination of the density of chlorine.

A. Specimens of note-book entries.

The weight of 1 litre of chlorine at 0° C. and 760 mm. pressure, found by determining the weight of a flask of known volume filled first with air at $t_a°$ and p_a mm., and then with chlorine at $t_c°$ and p_c mm., the density of air at 0° and 760 mm. being taken as 1·293 grams per litre.

	(i)	(ii)
Wt. of flask filled with air .... $= W$	97·281 gms.	115·731 gms.
at temp. .... t_a	13° C.	14° C.
and pressure .... p_a	763 mm.	756 mm.
Wt. of flask filled with chlorine :		
1st filling ⎫	97·773 gms.	116·172 gms.
2nd ,, ⎬ .... $= W_c$	97·773 ,,	116·182 ,,
3rd ,, ⎭		116·183 ,,
at temp. .... t_c	13° C.	14° C.
and pressure .... p_c	763 mm.	756 mm.
Vol. of gas held in flask (ascertained by use of water and measuring cylinder) $= V$ c.c.	274 c.c.	232 c.c.
$\therefore$ Wt. of air filling flask at $t_a°$ and p_a mm. $= ·001293 \times V \times \dfrac{273 \times p_a}{(273 + t_a)\,760} \dots = W_a$	·340 gms.	·284 gms.
$\therefore$ Wt. of vacuous flask .... $= W - W_a$	96·941 ,,	115·447 ,,
$\therefore$ Wt. of chlorine filling flask at $t_c°$ and p_c mm. .... $= W_c - W + W_a$	·832 ,,	·736 ,,
$\therefore$ Density of chlorine, *i.e.* wt. of 1000 c.c. at 0° and 760 mm. .... $= \dfrac{(W_c - W + W_a) \times 1000}{V \times \dfrac{273 \times p_c}{(273 + t_c)\,760}}$	3·17 gms. per litre	3·35 gms. per litre

B. Summary of results of class.

Summary of Results obtained by a class of 13 students

	Student A	3·23
	B	3·22
	C	3·11
	D	3·33
	E	3·28
	(F	2·76)
	G	3·23
	H	3·24
	I	3·21
	(J	3·00)
	K	3·30
	L	3·12
	M	3·24

Mean of 11 experiments = **3·228**

Calculated Value obtained by use of Avogadro's hypothesis.

$$\text{Density of chlorine} = \frac{\text{density oxygen} \times \text{molec. wt. chlorine } (O = 32\cdot00)}{\text{molec. wt. oxygen}}$$

which, since the atomicities of both gases are the same, becomes

$$\frac{\text{density oxygen} \times \text{atomic wt. chlorine } (O = 16\cdot00)}{\text{atomic wt. oxygen}} = \frac{1\cdot429 \times 35\cdot46}{16\cdot00} = \textbf{3·167}$$

Standard Values:	Gay-Lussac and Thénard	1801	3·19
	Bunsen		3·165
	Ludwig	1868	3·2075
	Jahn	1882	3·209
	Leduc	1897	3·220
	Moissan and Binet	1903	3·220

CHAPTER I

THE NATURE AND RECOGNITION OF CHEMICAL CHANGE

I. Desirability of describing and circumscribing subject-matter of chemistry.

It is always an important and generally a difficult problem to decide upon the plan to be followed in the presentation of a big province of knowledge to a special type of enquirers and learners. But whatever may be the outstanding feature of the method adopted, provided that the public addressed—whether in print or verbally—is composed of people who have reached some maturity of mind, the starting-point at any rate is nearly always the same: it consists in an *attempt* to *define the province* of the subject dealt with, *i.e.* to characterise the phenomena comprised in it. In the case of a subject which antecedent classification has already recognised as a concrete or experimental science, it aims at staking off the piece of ground which in the wide field of real occurrences, of so-called natural phenomena, has been selected for special exploration. This is a logically rational course, and hence one that should commend itself to all thoughtful students, provided always that we do not overlook the difficulties that beset it.

1. Difficulties inherent in the classification of the sciences.

There is the difficulty, inherent in all classification, of deciding on an intellectual basis;

There seems to be a peculiar fascination in attempting to classify the sciences, and many great intellects have puzzled over the problem.
(J. A. Thomson, *Introduction to Science.*)

Bacon, Comte, Herbert Spencer, Bain, Professor Karl Pearson, are the names quoted whenever this subject receives detailed treatment. But it is not the multiplicity of classifications that need trouble us.

The classification of the sciences is a matter of practical and intellectual convenience, but it is full of difficulties....It does not matter very much which classification is adopted ; the important thing is to have in mind some classification which one has made one's own. (*Ibid.*)

The real difficulty would seem to be of another nature, and to lie in the impossibility of applying hard and fast classification and definition to natural phenomena the very essence of which, when studied in their entirety and interdependence, is gradual transition by infinitesimally small differences, development by evolution. The boundary between allied sciences such as zoology and botany, physics and chemistry, is certainly not a line in the sense of the Euclidean definition, but a broad tract accommodating many organisms, many phenomena respectively. And even where we had been wont to accept, so implicitly as almost to postulate it, a clear line of demarcation, the growth of knowledge tends to blur it. Thus as regards the distinction between living and lifeless, animate and inanimate matter, the student is recommended to read in connection with this Prof. Schäfer's 1912 Presidential Address to the British Association. We have been in the habit of giving the name 'animate,' or possessed of life, to heterogeneous matter built up in a complex manner, and in its aggregate giving units endowed with special potentialities for spontaneous movement, assimilation and disassimilation, growth and decay, differentiation of parts and reproduction; and to consider inanimate or lifeless, matter devoid of these special properties. Prof. Schäfer has tried to show that, taking these properties one by one, each of them is also exhibited by some form of matter hitherto unhesitatingly classed as inanimate. Hence it is clear that when it comes to classification of closely allied branches of science, there is much opportunity for dialectic wrangling, of which those who are intent on drawing sharp boundary lines have always availed themselves fully.

2. Utility of classification.

But to those who approach the subject from the truly scientific, which is identical with the practically useful and

not with the barren metaphysical point of view, all classification is merely a matter of convenience in which we are
guided rather by the consideration whether it will take us a
long way towards the accomplishment of our aim, which is
a concise description of the knowable universe, than by the
insistence on not using any that does not go *all* the way.
After all, it is the special substance, the special phenomenon
studied that are of primary importance, and not the question
as to whether the organism rightly belongs to the class plants
or the class animals, or whether a definite change observed in
a certain substance should be classed as physical or chemical
(see *post*, p. 79). Take for instance the case of the distinction between animate and inanimate. We may assume
that Professor Schäfer, in spite of the fact that he has tried
to prove the absence of any sharp distinction between the
two, would not wish to see treated in the same text book, in
the same course of lectures, the properties of iron and the
embryology of the frog; that he would not quarrel with the
view which considers the substance iron and the organism frog
as exhibiting essential and easily apprehended differences,
which when we deal with other bodies make us unhesitatingly
class water and coal with iron, a butterfly and an amoeba with
the frog.

It may be taken, therefore, that in spite of the difficulty
arising from the fact that different branches of knowledge
almost imperceptibly merge into one another, it is always
possible, by choosing definite and easily apprehended examples, to discover some salient features which mark off into
distinct classes the majority of the cases met with, which
supply the base for some definition whereby the subject-
matter and scope of a special branch of knowledge may be
described and circumscribed. Such a procedure from its very
nature affords opportunity for the display of personal bias
and arbitrariness as to how much of the borderland is annexed
from or ceded to the realm on the other side. But from what
has been said already it should be evident that this is not a
matter of real practical importance; all the more as it is not
only according to the nature of the subject-matter that we

classify, but also according to the point of view from which we happen to consider it[1].

3. The place of chemistry in the scheme of classification of the sciences.

Having thus recognised the possibility and the practical utility of some *system* of classifying the sciences, it remains to find the place held within such a *system* by *chemistry*. The following may be accepted as a good working definition of the scope of science: 'The subject-matter of science includes all clearly defined facts of experience which are communicable and verifiable.' It may be useful to show by some examples the precise meaning to be attached to the terms 'communicable' and 'verifiable' used in the above connection, by indicating what sort of facts of experience do not fulfil these requirements and remain therefore outside the scope of science. We have all heard about the sea-serpent, and not a year passes without news of its having been seen, together with a more or less detailed description of its appearance; yet, it does not find a place in any text book of zoology, and will never do so unless a specimen is produced, whereby the experience of one can be *communicated* to all. Again, in the year 1903 much interest was aroused in the scientific world by the supposed discovery by Blondlot, a distinguished physicist and professor at Nancy, of a new kind of radiation, termed by him N-rays (from Nancy). These rays were said to be emitted by incandescent burners or by any material in a state of elastic strain, such as a file, wood or glass that had been compressed; they were described as waves of extremely small wave-length, which did not produce luminosity, nor phosphorescence, nor photographic effect, but which could be detected by their property of increasing the intensity of very feebly luminous objects, such as a faint electric spark or a very small feeble gas flame. English and German observers and the majority of the French physicists who occupied

---

[1] ...a frequent cause of confusion has been to attempt to map out territories as preserves of particular sciences...the chemist may give his life to the study of the odoriferous substances in flowers and yet never ask one biological question. (J. A. Thomson, *ibid.*)

themselves with the subject, when working under the conditions postulated, failed to reproduce any of the effects described by Blondlot and some of his French *confrères*; hence, since the experience was not *verifiable*, N-rays are not found in any text book of physics, and the importance of the episode, which 'will always form an interesting and instructive chapter in the history of human errors' is merely psychological, psychophysical and physiological.

The first step in the classification of the sciences is the division into concrete or experimental, and abstract sciences.

The concrete sciences, *e.g.* biology, physics, chemistry, deal with and describe the facts of experience, real existences and phenomena, as opposed to the abstract sciences (*e.g.* mathematics and logic), which are methodological, *i.e.* deal with modes and methods of such descriptions, and supply the intellectual instruments required for the purpose of these descriptions.

The concrete or experimental sciences comprise the two classes termed 'biological' (*e.g.* botany, zoology) and 'physical' (*e.g.* mineralogy, physics), which deal with animate and inanimate matter respectively. The attributes summed up in the term 'animate,' possessed of life, have already been enumerated (*ante*, p. 49). When the inanimate matter investigated in the physical sciences is considered irrespective of its size, shape, occurrence in nature and origin, we deal with what are termed 'substances,' and the study of the properties possessed by and the changes undergone by all the different substances met with in nature and prepared artificially is the province of the two sciences physics and chemistry. Moreover in these two sciences, especially in chemistry, there is a tendency to select for separate consideration the homogeneous[1] constituents of the heterogeneous[1] agglomerations found native or produced industrially, *e.g.* to

---

[1] All matter may be classified as being either *homogeneous* ($\delta\mu\delta s$ = the same, and $\gamma\epsilon\nu os$ = kind), when the physical properties measured at any point are the same as those measured in the same direction at any other point, or *heterogeneous* ($\xi\tau\epsilon\rho os$ = other), when this is not the case. Ice in contact with liquid water is a heterogeneous system; so is a piece of granite, in which the constituent particles of quartz, felspar and mica may be detected by the naked eye. Solutions, gaseous mixtures, chemical elements, and compounds in the same state of aggregation and in the absence of isomeric modifications, are instances of homogeneous systems.

investigate separately the quartz, mica and felspar of which granite is made rather than the granite itself; to deal with water, chalk, and carbon dioxide, and not with the river or spring water which by boiling can be separated into these substances.

II. The nature of chemical change.

The next question is that of the differentiation between the changes in matter studied under physics and chemistry respectively. In accordance with what has been said before, it will be best to deduce these inductively from the consideration of a number of simple typical cases, not near the boundary line but on the contrary showing very clearly an essential difference.

1. **Typical cases to show the existence of two kinds of change that substances can undergo.**

    Experiment II. Investigate the effect of heat on :
    (i) Platinum.
    (ii) Zinc oxide.
    (iii) Nitrate of lead or carbonate of manganese.
    (iv) Copper-calcium acetate or lead-potassium iodide.

All these are cases in which experience has shown that the change observed is independent of the presence of air, *i.e.* that if the heating were done in an evacuated vessel the result would be precisely the same as that observed when, as is proposed here, the substances are heated in contact with air.

Iodine, lead peroxide, thermite (mixture of aluminium powder and ferric oxide in the ratio 1 to 3 parts by weight) are other suitable substances for use in addition to or instead of those given above.

The principle of this and the following experiment consists in raising the temperature of the substance investigated by suitable means, noting any change occurring, re-establishing the original conditions by cooling to the temperature of the surrounding air and comparing the resulting substance or substances with the original one.

(i) Use the metal platinum in the form of wire or of foil or of powder deposited on asbestos fibre[1], and heat directly by holding in the Bunsen flame.

---

[1] This so-called 'platinised asbestos' is easily made by dipping an asbestos fibre into a solution of platinum chloride and then igniting it in the Bunsen flame.

(ii)　Use a specimen of zinc oxide in the form of powder which has been dried by keeping for some time in a desiccator. Heat a little in a small crucible or dish, or dry test-tube[1] or ignition tube of hard glass, preferably the latter.

(iii)　In the heating of lead nitrate (powder the crystals before use) and manganous carbonate, proceed as under (ii), heating until no further change seems to occur. In the case of the manganous carbonate, demonstrate the evolution of a colourless gas which extinguishes a glowing splint and which turns milky a drop of lime water held at the end of a rod[2].

(iv)　Heat the beautiful blue crystals of copper-calcium acetate to a

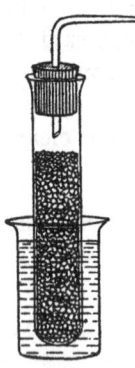

Fig. 8.

temperature below 100°C., under such conditions that the water liberated at that temperature cannot escape readily, and that in cooling it can be reabsorbed. A simple plan is to fill a clean dry test-tube with the crystals, closing it by a cork through which passes a tube drawn out to a fine point, and to effect the heating by immersion of the *lower* portion of this tube in a beaker of hot water (fig. 8); as soon as a colour change is noticed, the tube should be cooled by holding it under the tap. The double acetate can be easily prepared by dissolving 25 grams of calcium acetate in 150 c.c. of water, adding 7 grams of finely powdered copper acetate in small quantities at a time to the boiling solution, filtering and allowing to stand over night to crystallise.

Lead potassium iodide can be prepared and experimented with according to the following directions:

A hot solution of 4 grams of lead nitrate in 15 c.c. of water is mixed with a hot solution of 15 grams of potassium iodide in 15 c.c. of water....On cooling to room temperature, the [deep yellow] crystals of lead iodide disappear, and a very pale yellow, felted mass of crystal needles is produced. On heating, the [pale yellow] crystals...disappear, with the re-formation of the [deep yellow] lead iodide ; on cooling [the change is reversed].

(Biltz, *Laboratory Methods of Inorganic Chemistry*.)

Experiment III. Investigate the effect of heat on the following substances :

(i)　Magnesium.

(ii)　Iron.

(iii)　Cuprous oxide.

---

[1] An ordinary test-tube of thin glass will of course collapse if heated in the Bunsen flame and become unsuitable for further use.

[2] A better way of carrying out this test consists in pouring the gas, which is heavier than air, into a small test-tube which contains a *very little* lime water, leaving the larger part of the tube empty for the reception of the gas to be introduced into it, and then shaking up.

These are all substances in which experience has shown that in the change which is observed to occur when the heating is done in the presence of air, the air is an essential factor, since when the heating is done in an evacuated vessel no such change occurs.

(i) Use the magnesium in the form of wire or ribbon or wool[1], preferably the latter, and try to obtain a product quite white and not mixed with any unchanged magnesium. For this it is necessary to burn the magnesium, holding it with crucible tongs or hanging over a glass rod, and letting the completely burnt mass drop into a porcelain dish. Note that if burned when supported on porcelain or glass, the white powder produced is mixed with a black mass, and that if the burning is done in an insufficient supply of air (as may happen when the magnesium is packed tight in a crucible or wrapped in a fairly tight bundle) a yellow core[2] is formed.

(ii) The iron can be used in the form of powdery *reduced* iron, put in a little heap on a strip of bent-up fine iron gauze supported on an

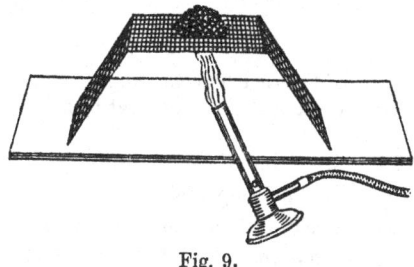

Fig. 9.

asbestos sheet and heated by a Bunsen burner (fig. 9); or as steel wool[1] held in the flame with crucible tongs; or as the so-called *pyrophoric* iron kept in a sealed or corked-up tube out of contact with air, which when shaken out of the tube catches fire spontaneously, the burnt substance being allowed to fall into a dish or on to a sheet of asbestos. The pyrophoric iron can be bought in sealed tubes, or can be easily prepared as wanted by heating ferrous oxalate (interesting as a *yellow* ferrous salt) contained in a thin-walled test-tube which is sealed up whilst the evolution of gas is still going on. If the substance is wanted for immediate use it suffices to close

---

[1] An article in the *Zs. f. d. physik. u. chem. Unterricht*, 1913, p. 155, advocates the use of metal wools in a number of students' experiments.

[2] The black substance is magnesium silicide, due to the reduction by the magnesium of some of the silica of the crucible to silicon, which then combines with a further amount of magnesium; the yellow substance is magnesium nitride, due to the combination of the magnesium with the nitrogen of the air. The great ease with which these substances are formed makes magnesium quite unsuitable for use in experiments on the composition of air.

the tube with a well-fitting cork when the colour of the substance is a uniform black and then to let it cool. (Fig. 10.)

Manganous oxide, which can be conveniently prepared by heating manganous oxalate in a stream of hydrogen, may if desired be substituted for pyrophoric iron.

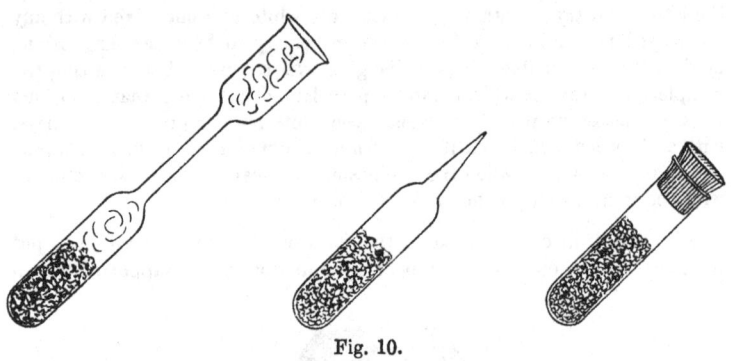

Fig. 10.

(iii) Use cuprous oxide dried by being kept in a desiccator. Spread a thin layer in an open dish and heat by a Bunsen burner.

Experiment IV. Investigate the effect on iodine of

    (i) Carbon bisulphide.

    (ii) A solution of ammonium sulphite,

in each case evaporating to dryness the solution obtained.

(i) Treat a small crystal of iodine with about 5 c.c. of carbon bisulphide in a test-tube, shake up and place the solution on a fair-sized watch-glass (6 to 7 cm. diameter), evaporate by blowing air over the surface by means of a hand bellows or any other suitable method which the means at your disposal enable you to use (*e.g.* suction applied by means of a glass tube connected with a water exhaustion pump).

(ii) Prepare an alkaline solution of ammonium sulphite from fairly dilute solutions (about $N/5$) of ammonia and sulphurous acid by the following method. Find by titration, using methyl orange as indicator[1], the volume m of ammonia solution required to neutralise n c.c. (say 10 c.c.) of the acid, and prepare the alkaline sulphite solution required by adding to 250 c.c. of the acid $\dfrac{2 \times 250 \times m}{n}$ of the ammonia solution, that is, a quantity twice that required for neutralisation as found by titration. Suspend a small amount of powdered iodine (about 1 gram) in water, and cautiously add the alkaline

---

[1] The pink colour of the methyl orange in the solution of sulphurous acid changes to yellow when the acid is all converted to ammonium hydrogen sulphite.

sulphite solution until all the colour due to the iodine is just destroyed and an absolutely colourless liquid has been obtained. If too much sulphite should have been added, restore the brown or yellow colour by adding a tiny speck of iodine, and then find the exact end point by the cautious addition of more sulphite solution[1]. Evaporate to dryness on the water bath, keeping the dish protected by putting over it a big funnel supported by corks (fig. 11).

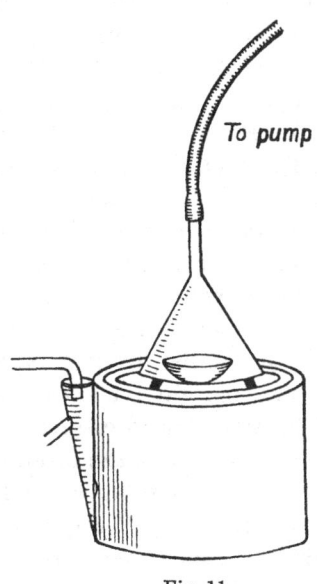

To pump

Fig. 11.

A review of the results obtained in these three experiments brings out the fact that they are of two essentially different types. Comparison of the original substance with the product obtained when the original conditions have been re-established, shows quite unmistakably and conclusively that in Exps. II (iii), III (i), (ii), and (iii) and in IV (ii) a substance different from the original has been produced, *i.e.* that a *permanent change* has been effected. The mere appearance indicates this, and further tests, in which the solvent effect of water and various acids may be tried, corroborate it. On the other hand, as far as appearance goes, in Exps. II (i), (ii), (iv) and in IV (i) the product shows no difference from the original substance, and as far as the evidence goes, any change observed under the altered conditions has been temporary only, re-establishment of the original conditions reproducing the original properties. Here, of course, our evidence for the identity of the two substances and the consequent non-occurrence of a permanent change is far from conclusive, being based only on appearance. This point will be dealt with fully later on. But for

---

[1] This experiment is of historical interest, because it reproduces the method used by Stas for the change of iodine to iodide in his complete gravimetric synthesis of silver iodide. See *post*, pp. 217 *et seq.*; also Freund, *The Study of Chemical Composition*, p. 66.

the purposes of the present argument, irrespective of whether there has or has not been further evidence for the identity of the two substances, the position can be legitimised by the continual use of the qualifying sentence ' as far as the evidence available goes.'

Moreover, whilst the temporary changes can be repeated at will an indefinite number of times, in the case of the other type—provided the change of conditions had been maintained long enough to make the change complete—the phenomena observed during the change cannot be repeated. Thus the change of yellow to orange and back again to yellow in the double iodide of lead and potassium can be made to recur as often as we wish by just plunging the tube into beakers of hot and cold water successively; on the other hand, when in the heating of lead nitrate the white crystals yield a red gas and a yellow solid, the evolution of gas soon comes to an end, and the yellow solid, when heated again after it has been allowed to cool, either apparently remains unchanged or fuses without a recurrence of the phenomena first observed in the heating of the nitrate.

2. Current definitions which bring out the accepted division between physics and chemistry.

This classification of the changes that matter can undergo, necessitated by the results of the preceding experiments, supplies the basis for the accepted division between physics and chemistry, and finds expression in current definitions such as the following :

Physics deals with the temporary, chemistry with the permanent changes that matter can undergo.

Physics investigates the properties common to all kinds of matter and differing only in degree, such as density, power of conducting an electric current, etc.; chemistry deals with the properties which belong to certain kinds of matter and not to others, which characterise one kind of matter and differentiate it from all other matter, such as the property possessed by water of yielding under suitable treatment the two gases hydrogen and oxygen in a fixed ratio.

Chemistry is the science dealing with the study of all the homogeneous kinds of matter met with in Nature, and with the permanent changes these can undergo when transformed into other kinds of matter.

The common feature of all these definitions of chemistry is evidently the insistence on the production of a *permanent change*, the formation of a *different kind of matter*; but 'permanent change' and 'production of a different kind of matter' are really only different names for one and the same thing, since it is the production of a different kind of matter that constitutes the permanent change, and *vice versa*.

III. Criteria for the occurrence of a chemical change.

In accepting the above definitions as merely verbal variants of one and the same view concerning the essential nature of chemical change, we are confronted with the problem of the deductive application of the principle involved, with the actual determination in practice of whether a change observed is chemical or physical, whether a new kind of matter has or has not been formed.

1. Let us begin by considering the problem in its theoretical aspect, so as to realise clearly the nature of the method that has to be followed. The question we have to answer is whether the substance resulting from the change is *identical* with the original or not. What does this mean precisely, and how are we to decide?

It has been found as the result of innumerable investigations that different specimens of the same substance have exactly the same properties, and that if these properties are measurable (*e.g.* density, heat capacity, electrical and thermal conductivity, [quantitative composition]), their values are not merely similar, but that within the limit of experimental error they are identical. (Ostwald.)

This is a statement so generally accepted that it is apt to be looked upon as a truism, but there is no *a priori* reason for so regarding it, and many an investigation has been undertaken with the object of putting it to an experimental test. Thus Stas synthesised silver chloride and analysed ammonium chloride with the express object of ascertaining whether the ratio between the components was always the same or not, however widely different might have been the modes of preparation of the various specimens used. But if we accept the above statement, not on *a priori* grounds but as an empirical generalisation, it follows that :

(*a*) *One* definitely established difference between the original substance and that resulting from the change is sufficient proof of the occurrence of a permanent or chemical change[1]; *e.g.* whilst iron dissolves easily in dilute acid with the evolution of a combustible gas, the burnt iron (which moreover differs also in colour) dissolves much less easily and does so without effervescence. The change may consist in the *appearance* or the *disappearance* of a *qualitative property* (such as solubility in water or in a special acid), or in a *variation* of the value of a *quantitative property* (such as the value of the coefficient of solubility[2]).

(*b*) Absence of permanent change involves the identity in *all* properties of the original substance with that resulting from the change. Hence experimental work which merely shows identity in *one* or *two* properties (*e.g.* colour, behaviour towards acids) fails to establish complete identity. It would seem therefore as if classification of an observed change as temporary, and hence physical, must always be provisional, as every substance has almost innumerable properties, and it obviously would be impossible to investigate them all. But here we are helped by an empirical law, *i.e.* a generalisation from many actual observations, which states that

when two substances agree *entirely* in a few properties they agree also with regard to all other properties   (Ostwald.)

The term 'few' is rather vague; does it mean two or three or twenty? but 'entirely' saves the situation, since it carries with it a quantitative meaning.

It has been found that when properties capable of being measured in terms of some suitable unit, so-called *quantitative properties*, are considered, each substance possesses these to a specific amount characteristic of itself, and that even where at first the agreement in properties between two given substances had seemed so close as to be compatible with identity, increased accuracy in the measurements reveals undoubted

---

[1] A very clear pronouncement on this point, made as far back as 1777 by Scheele, is contained in the quotation on p. 77.   [2] See *post*, p. 79.

differences[1]. The manner in which the results of qualitative experiments are amplified and modified through the determination of the quantitative value of the properties involved, may be illustrated by the following example:

(1) The white solid sodium bicarbonate, when heated, leaves another white solid; the original substance and the product obtained on heating are not only both white but are both soluble in water; both effervesce with acids, yielding carbonic acid gas; both colour the Bunsen flame brilliantly yellow. Are the two identical? The required agreement in a few properties seems to be supplied, but is the agreement of a kind justifying the use of the term 'entirely'? The following table, giving the values for certain quantitative properties, supplies the answer:

	Sodium bicarbonate	Product of heating sodium bicarbonate
Colour, state of aggregation	White solid	White solid
Action of water	Soluble	Soluble
100 parts of water at 20° dissolve	9·6	21·5
Action of dilute acid	Effervescence	Effervescence
Amount of carbon dioxide evolved from 100 gms.	52·4	41·5

(2) An even more striking example, well suited to show the desirability of a provisional character in most of our empirical classifications, is afforded by the case of the substances called dextro- and laevo-tartaric acid already referred to (p. 4). Most searching and most accurate study, qualitative and quantitative, yields identical results as regards a number of properties very large and very varied. The two substances possess the same melting point, the same solubility in various solvents, the same electrical conductivity and rate

[1] Cobalt and nickel, elements which had been recognised since the time of their discovery by obvious differences as distinct substances, at one time were supposed to have the same atomic weight, the values given in Lothar Meyer's *Modern Theories of Chemistry*, ed. 1888, being 58·6 for both ; but increased skill in the preparation of the pure substances and increased accuracy in the measurements involved have yielded the present standard values, which are $Co = 59·0$, $Ni = 58·7$. Conversely, a definitely established difference in the value of a quantitative property, however small, *proves* a difference between the two substances ; the search for the cause of this difference in the case of nitrogen obtained from the air and from chemical compounds, which were found by Lord Rayleigh to differ slightly in density, the values being 1·2571 and 1·2507 grams per litre respectively, led to the discovery of argon.

of diffusion ; their metallic salts have the same composition, crystallise with the same number of molecules of water of crystallisation, have the same solubility; their esters[1] fuse and boil at the same temperatures. Would it not seem, then, that the evidence for the identity of the two substances is absolutely conclusive? and yet, when the investigation is pushed further still, there are found differences so distinct as to make the opposite inference inevitable. Whilst the one substance rotates the plane of polarisation of light to the right, the other does so to exactly the same amount to the left; whilst the acid which is dextro-rotatory forms with cinchonine (a base allied to quinine, which itself rotates the plane of polarisation) an acid salt crystallising with 4 molecules of water and very easily soluble in alcohol, the corresponding salt of the laevo-acid crystallises with 1 molecule of water and is but little soluble in alcohol; the dextro-acid is very much more readily attacked by micro-organisms active in the process of fermentation (*e.g. Penicillium glaucum*) than is the case with the laevo-variety. The difference in crystalline form, as shown by the exhibition of opposite hemihedrism, which originally led to Pasteur's discovery of the laevo-acid, has already been mentioned (*ante*, p. 4). If we remember that it was only in 1815 that the phenomenon of optical rotation was discovered by Biot, and consider how recondite are the other properties enumerated above (all discovered by Pasteur[2]) which differentiate the two substances, we may well wonder what the future may still have in store for us in the matter of finding difference where hitherto we have assumed identity.

2.   After this discussion of the theoretical nature of the problem that has to be solved in classing a change as physical

---

[1] ' Esters ' are compounds obtained by the interaction between alcohols and acids, when water is eliminated and substances are produced which are analogous to metallic salts.

KOH	+	HCl	=	KCl	+	H_2O
potass. hydrate		hydrogen chloride		potass. chloride		water
C_2H_5OH	+	$H . C_2H_3O_2$	=	$C_2H_5 . C_2H_3O_2$	+	H_2O
alcohol		acetic acid		ethyl acetate		water
(ethyl hydrate).						

[2] Alembic Club Reprints, No. 14, Pasteur, *Researches on Molecular Asymmetry*.

or chemical, it remains to show in what the actual work to be done consists. No definite rules for procedure can be given for the experimental work required in the comparative investigation of two substances. Whilst *chance trial* often has to be the dominant feature in what is done, full scope is of course given for the experience which has shown that trial along certain definite lines (*e.g.* determination of the melting point, the boiling point and the density; the action of water, of dilute and of strong acids, of alkalis, of oxidising and of reducing agents) is likely to lead to some success. Obviously, therefore, the experience and instinct (itself a result of experience) of the experimenter are all-important, and students who carry out an experiment illustrative of the difference between physical and chemical change according to the directions given in a text book should realise that they come in for the result of a vastly greater amount of previous work than appears at first sight; that what they are directed to do represents but that small residual fraction of the various devices tried which has yielded positive results; and that if a student were set to work on an unknown substance without directions as to the choice of the special properties to be investigated, a complete record of his work would almost certainly contain a considerable number of negative results, though of course it might so happen that at a quite early stage the establishment of some definite qualitative or quantitative difference, or of some quantitative identity[1] decided the case. But however great may be the diversity of the various devices adopted and of the tests applied, three distinct steps may be recognised in the sequence of the work.

(*a*) The production of a change in conditions, whereby opportunity is afforded for the occurrence of a change, permanent or temporary, in the substance A which is being investigated. This may consist in raising or lowering the temperature; or in making A interact in solution with another

---

[1] In research dealing with organic substances (*i.e.* carbon compounds) the determination of the melting point of a substance which has been put through a series of continued crystallisations or other changes is taken as a most important criterion concerning the occurrence or non-occurrence of a change.

substance, etc., etc. Any change occurring, such as fusion, ebullition, evolution of a gas, precipitation of a solid, is carefully watched, and the operation of heating, or of addition of another solution, or whatever else the change in conditions may consist in, is continued until such change as occurs is completed, a point which is decided either by appearance (*e.g.* cessation of the appearance of a red gas in the heating of lead nitrate, Exp. II (iii)), or by some special test applied for the purpose (*e.g.* absence of hydrogen evolution on the treatment with acid of a specimen of iron that is being 'burnt,' Exp. III (ii)). The original conditions are then re-established, *e.g.* heating or cooling is stopped, a solvent used is removed by evaporation (Exp. IV (i) and (ii)).

(*b*) Suitable arrangements must be made for collecting the product of the reaction in sufficient amount for the subsequent investigation of its properties. The means adopted for this end must of course depend on the special case. If the change occurring involves the production of a vapour, this must be condensed and the liquid collected ; if a gas, this must be collected, and the mode of collection will have to vary according as to whether the gas is soluble or insoluble in water, whether it is heavier or lighter than air. If a precipitate is formed by the interaction between two solutions, this must by filtration be separated from the liquid and as much as possible freed from the latter, etc., etc.

(*c*) The properties of the original substance (or substances) are compared with those of the substance (or substances) resulting from the change, the properties investigated being (*a*) qualitative or (*β*) qualitative *and* quantitative ; the wider the scope of the comparison the better, though, as has been said already, it may so happen that at a quite early stage material for a definite conclusion is found. It may be well to repeat what has already been emphasised, that no rule can be laid down with regard to the choice of properties to be investigated. In the case of a students' experiment it is highly desirable to record the results in tabular form ; a way of doing this is suggested below. Obviously it will be possible for the student to arrange the order of the work in two ways, either trying each test successively with A and A', the original substance and that resulting from the change (filling in the columns of the table horizontally), or dealing exhaustively first with A and then with A' (filling in the columns vertically). There is no *a priori* theoretical preference for either course, but in practice it will probably in most cases mean a saving of time to follow the first plan, which of course assumes that a sufficient supply of A' has been made available before beginning. If however A' should take long to prepare, and the process should be of a kind which does not require much attention, it might be carried on simultaneously with the complete study of A. The justification for saying so much on this point is that it clearly brings out the necessity for preceding the actual experimental work by a critical examination of the problem to be dealt with and by drawing up a definite plan of action.

3. **Examples of the manner of dealing with such problems.**

Experiment V. The investigation of the effect of heat on *ammonium chloride* (sal ammoniac[1]), with the object of determining whether the change produced is physical or chemical, introducing incidentally :

(1) the technique of sublimation,

(2) the technique of the volumetric estimation of chlorides by means of silver nitrate in neutral solution, with potassium chromate as indicator.

(a) Heat a little pure dry ammonium chloride in a long narrow dry test-tube ; note the production of white fumes, which deposit a white solid in the cold part of the tube ; repeat the process by heating the lowest part of the tube at which condensation has occurred. Note carefully all the phases of the change, the characteristic of which is change of the solid to gas and back again to solid without the intervention of the liquid state ; this is named sublimation[2] and is exhibited also by other well-known substances, such as iodine, camphor, benzoic acid.

[1] Sal ammoniac was brought into Europe from Egypt, where it was prepared from the soot obtained by burning camels' dung. The original name...*sal ammoniacum*...served originally among the Alexandrian alchemists to describe the common salt (chloride of sodium) and native sodium carbonate, which was found in the Libyan desert, in the neighbourhood of the ruins of the temple of Jupiter Ammon. (Roscoe and Schorlemmer, *Treatise on Chemistry*, 1894, I, p. 453.)

[2] Sublimation is one of the methods used in the preparation of pure substances, *i.e.* in the separation of a substance which possesses this property from

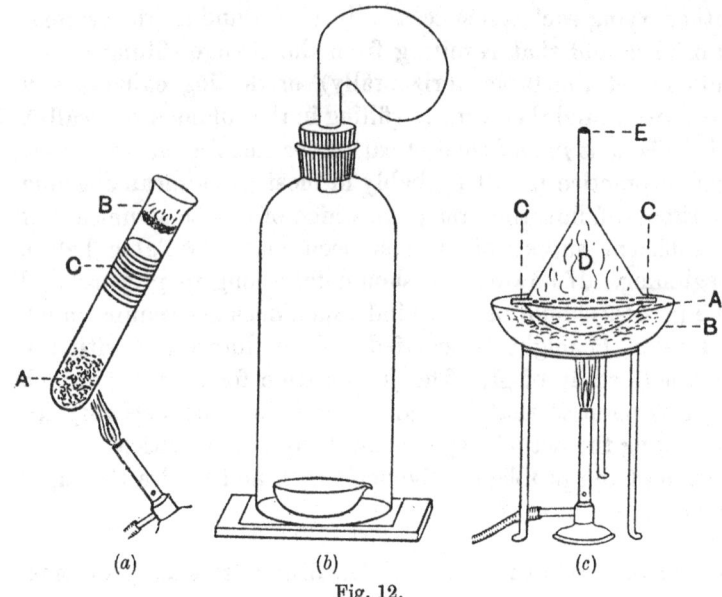

(a)   (b)   (c)

Fig. 12.

Fig. (a).

A. Dry ammonium chloride, heated by small Bunsen burner.

B. Plug of ignited asbestos.

C. Sublimate deposited in form of ring, which is most easily got out by breaking the tube in a mortar and picking off the glass.

Fig. (b).

The apparatus consists of a narrow bell-jar, fitted with a cork through which passes the delivery tube of a small glass retort, the tube being cut off fairly short. The substance to be sublimed is placed in the retort, which is then connected with the bell-jar, and the whole inverted over a watch-glass or small porcelain dish resting on a few thicknesses of moist filter paper laid on a tile.

On gently heating the retort bulb the substance sublimes and collects in the dish. Slow heating is necessary, otherwise the substance may be deposited on the sides of the jar.

By placing the jar on a ground glass plate and connecting with a pump by means of an extra tube through the cork, sublimations may be carried out *in vacuo*. *Chem. News*, 1911, **103**, p. 138.

Fig. (c).

A. Dish holding substance to be sublimed.

B. Sand bath in which the dish A is buried nearly to its edge, and which is heated in the ordinary way by a Bunsen burner.

C. Sheet of thick perforated paper laid across the dish; the vapour rises through the holes and the condensed solid falls back on to the paper.

D. Funnel closed by asbestos plug E, acting as condensation chamber.

solid impurities which do not volatilise thus and are left behind as a residue. The preparation of benzoic acid from gum benzoin is an experiment that might with advantage be done at this point.

(*b*) The collection of a sufficient amount of the sublimate for the purpose of the work in (*c*) affords opportunity for becoming acquainted with the process of sublimation, which is of considerable importance in the technique of the laboratory. A variety of devices can be employed, each of which has its own specific advantages. The figures *a*, *b*, and *c*, fig. 12, represent some of the simpler arrangements, and it is recommended that their efficiency should be tested by letting different members of the class find what weight of sublimate can be obtained with each of them in a given time.

(*c*) Comparison of the properties of the original ammonium chloride with those of the sublimate obtained from it.

(*a*) Qualitative properties. The following table summarises the characteristic properties which have been found to lend themselves well to the purpose of this comparison.

	Ammonium chloride	White solid obtained by the solidification and deposit in cold part of apparatus of the vapours obtained on heating ammonium chloride
Appearance: Colour State of aggregation		
Solubility in water		
Action of heat on the solid		
Action of conc. sulphuric acid on the solid		
Effect of boiling solution with potash		
Effect of the addition to the solution of silver nitrate solution		
Inference as to the nature of the change that had occurred		

(*β*) Quantitative properties. In accordance with what has been said before as to the measurement of a quantitative property being required for the conclusive establishment of *entire* agreement between the properties of two substances, the comparative study of the qualitative properties of ammonium chloride and the sublimate obtained from it requires to be supplemented by the measurement of some one quantitative property of the two substances, which might be that of the coefficient of solubility, or of the composition by weight. The experiment about to be dealt with refers to the composition. One of the qualitative properties found to be possessed in common by the ammonium chloride and its sublimate is that of the production of a white curdy precipitate on the addition of a solution of silver nitrate. The end point of this reaction can be found with great accuracy,

Analysis of A (ammonium chloride) and A′ (the sublimate obtained from A) by the determination of the ratio of the weights W of A (or A′) to W′, that of the silver nitrate required for the complete precipitation of the chloride.

		A Original Ammonium Chloride		A′ Sublimate	
		(1)	(2)	(1)	(2)
Weighing tube and contents .... (i) $=W_1$		7·620	7·789	7·435	7·896
Weighing tube and contents after shaking out some substance................................ (ii) $=W_2$		6·785	6·802	6·593	6·927
∴ Wt. of chloride used for making solution $=W_1-W_2...=W$		·835	·987	·842	·969
Volume* to which solution of W is made up ....$=V$		100 c.c.	100 c.c.	100 c.c.	100 c.c.
∴ Wt.* contained in 1 c.c. ....$=\dfrac{W}{V}$		·00835	·00987	·00842	·00969
Aliquot part of V taken for titration.....................$=v$		5 c.c.	5 c.c.	5 c.c.	5 c.c.
Volume of silver nitrate solution required for the complete precipitation of all the chloride, potassium chromate being used as indicator	(i) (ii) (iii)	13·05 12·90	15·60 15·55 15·65	13·30 13·20	15·1 15·1 14·95
Mean of (i), (ii) and (iii) ....$=m$		12·97	15·60	13·25	15·05
∴ Vol. of the special silver nitrate solution used required for the precipitation of 1 gram of chloride $\Big\} =\dfrac{mV}{vW}$		310·7	316·1	314·7	310·6
Strength of the silver nitrate solution used, expressed in grams per c.c. ....$\Big\} =w$		·01013	·01013	·01013	·01013
Wt. of silver nitrate used $=mw$ ....$=W'$					
∴ Wt.† of silver nitrate required for the precipitation of 1 gram of chloride ....$\Big\} =\dfrac{W'}{W}=\dfrac{mVw}{vW}$		3·15	3·20	3·19	3·15
Mean of $\dfrac{mVw}{vw}$ for (1) and (2)		3·17		3·17	

* † For footnotes see pp. 69 and 70.

owing to the fact that in neutral solution silver nitrate gives with potassium chromate a reddish brown precipitate, which however is not formed so long as any ammonium chloride (or any other soluble chloride) is available in the solution for the formation of the white curdy silver chloride. The plan followed is to add a drop of potassium chromate to a solution containing a known weight of the ammonium chloride or of its sublimate, and then to run in cautiously from a burette a solution of silver nitrate of known strength until a faint pink tinge appears, indicative of the formation of silver chromate, and hence of the presence in very small excess of the silver nitrate. This process, which is used in the volumetric estimation of soluble chlorides and soluble silver salts respectively, is described in detail in any text book of quantitative analysis, *e.g.* Clowes and Coleman, *Quantitative Analysis*, 10th ed., 1914, pp. 192 *et seq.* The results of an experiment are given on p. 68. The table on p. 71 summarises the results obtained by a class of thirteen students, and gives the values found by various investigators.

The final inference from the result of the qualitative and quantitative experiments above described obviously is, that the original ammonium chloride and the sublimate are possessed

* (p. 68) There is the alternative of either working with a larger amount W, making the solution of it up to a definite and known volume V, and using for the titration with the silver nitrate an aliquot part v of it; or using a smaller amount of W and using the whole volume of solution, the value of which need not be known, for the titration. The second plan has the obvious advantage that the factor V/v disappears from the calculation, and with it any of the errors incidental to these two measurements; on the other hand, in the case of students requiring practice in the determination of the end part of the reaction, there is a decided advantage in the possibility of a number of determinations of m, the volume of the silver nitrate solution required for the completion of the reaction, making it possible to reject individual titrations which had been 'overdone' without thereby losing the whole of the work.

† Obviously as far as the special purpose of this experiment goes, which is one of comparison and in which it is assumed that the *same* silver nitrate solution is used throughout, the statement of results in terms of *weight* of silver nitrate $\dfrac{mVw}{Wv}$ presents no advantage over that in terms of *volume* of solution $\dfrac{mV}{Wv}$, since in this case *volume* of solution and *weight* of silver nitrate are strictly proportional. The ratio between the two values of $\dfrac{mV}{Wv}$ for A and A' is not affected by multiplication of each of them by the constant w, but there are certain advantages to be gained by knowing the concentration of the solution, viz. :

(1) If in the course of the work it should for any reason become desirable or necessary to change the silver nitrate solution used, this change would not invalidate the work already done.

(2) It becomes possible to compare the value obtained with the *standard value*, a matter which is always of interest to the students, and rightly so. In connection with this it must be noted, though, that the use of this value w, termed an *antecedent datum*, introduces a *constant error*, which represents the

of the same properties, and hence that heating in this case does not produce a chemical change. It is however most important to realise that this inference refers only to the relation between the original ammonium chloride and the sublimate, and does not even touch the quite distinct question of the nature of the substance whilst in the gaseous state, leaving it an open question whether the substance then was ammonium chloride in a different state of aggregation or whether a different kind of matter had been produced, which reproduced the original ammonium chloride as soon as the physical conditions had been made the same again.

Experiment VI. The investigation of the effect of heat on *calcium nitrate*, with the object of determining whether the change produced is chemical or physical, introducing incidentally the technique of collecting a gas and its separation into portions soluble and insoluble in water.

sum of the errors due to (i) the errors in weighing and in measuring incidental to the preparation of the silver nitrate solution, (ii) the presence of impurities in the silver nitrate used. Usually it may be taken that the error in w is likely to be smaller than that attaching to the measurement of W, V, v, and m (this last especially) in any *one* of the students' experiments. But when the class is a fairly large one, and the mean of all the results is calculated, there is the likelihood of a compensation and consequent great reduction in the total value of this so-called *experimental error*, whereby the constant error introduced through w becomes much more prominent. That this is so is shown by an inspection of the next table, from which it appears that whilst the mean of the values obtained for A and A' is practically identical, differing only by 3/3000, these means differ from the standard value by about 16/3000, an amount five times as great, and probably due chiefly to the constant error in the value of w. This makes it desirable to examine further the probable causes of the discrepancy between the mean of the values obtained in the students' experiments for $\dfrac{mVw}{Wv}$ and the standard value for the ratio ammonium chloride : silver nitrate.

Obviously the true value of w might be greater or smaller than that used in the calculation, with the result that the value for $\dfrac{mVw}{Wv}$ might be affected either way. Then there is the error due to such impurities in the original ammonium chloride as were not removed in the process of sublimation, and which, through W being taken as larger than it is in reality, would make the result too small. Finally, there is the constant error inherent in the method, whereby m, and hence the final result, is too large. This is due to the fact that since the silver nitrate solution is run in until a pink tinge appears, it must always be used in some excess, though small. It is suggested that, assuming the excess to be one drop only, which when working with an ordinary burette corresponds to about 1/20 c.c., the student should calculate the percentage effect on the final result with the values for V and v, W and w used in their own experiment.

Weight of silver nitrate required for the complete precipitation of 1 gram of ammonium chloride.

Summary of results obtained by a class of 13 students		
	Amm. chloride	Sublimate
Student A	3·16	3·17
B	3·20	3·19
C	3·14	3·13
D	3·14	(3·70)
E	3·25	3·23
F	3·29	(3·35)
G	(3·66)	(3·74)
H	3·29	3·26
I	3·15	3·22
J	(2·85)	3·15
K	3·16	3·17
L	3·05	(3·50)
M	3·28	3·23
Mean	**3·192** (11 exps.)	**3·195** (9 exps.)

Standard Values *

Stas, 1860   **3·1758**

Of greater value even than the direct determination, because of the increased accuracy of the latest measurements, are the indirect determinations calculated from the actual measurement of the ratios amm. chloride : silver and silver : silver nitrate.

	Amm. chloride : silver $= m$	Silver : silver nitrate $= n$	Amm. chloride : silver nitrate $= m \times n$
Penny, 1839		1·57442	
Pelouse, 1845	2·0188		
Marignac	2·0179	1·57424	3·1766
Stas	2·0163	1·57486	3·1754
Scott	2·0167		
Hardin, 1896		1·57484	
Richards and Forbes, 1907		1·57479	
	2·0174	**1·57463**	**3·1760**

In the calculation of the mean, some of the results have been omitted. These have been indicated by being enclosed in brackets.

For a critical examination of the discrepancy between the standard value and that obtained in the students' experiment, see footnote † 2 on page 69.

* For footnote see p. 72.

(*a*) Heat a little finely powdered calcium nitrate in a small ignition tube until the evolution of coloured gas has ceased. Note carefully the successive phases of the process: the liquefaction of the solid; the evolution of what in appearance suggests aqueous vapour; the solidification of the mass; the second liquefaction; the evolution of red gas; the second solidification. Also heat a very little of the powder placed on the edge of a piece of platinum foil, and note the bright glow of the solid when the final stage has been reached.

(*b*) As far as appearance goes, the original white crystals have been decomposed by heat into a white solid remaining behind in the decomposition vessel and a deep red gas, for the collection of which suitable arrangements have to be devised. It will probably result in a saving of time if the solid and the gas required for the investigation of the properties of the substances produced are prepared in separate experiments. In the case of the solid it is essential that it should not be mixed with any unchanged calcium nitrate, a result which will be accomplished more rapidly if only a

---

* (p. 71) At the risk of appearing hypercritical, this opportunity cannot be missed for entering a protest against the frequent wrong use of the term ' theoretical value.' In the above and many other cases—such as the comparison of the experimental results of a quantitative analysis with those calculated from the formula—' theoretical ' should be replaced by ' standard ' or ' calculated ' as the case may be. If in a complete gravimetric synthesis we actually measure A and B, the weights of the constituents, and AB, the weight of the compound formed, and compare AB with the value $A + B$, the latter *is* the *theoretical value*, calculated on the theory that conservation of mass (Ch. V) holds as an exact law. But if we *calculate* composition from the formula, the values used for the combining weights (identical with or simply related to the atomic weights) are the determining factor, and in the present state of our knowledge there is nothing ' theoretical ' about these, nothing to fix in terms of a law or generalisation what these values should be. It is quite conceivable that the future may bring the discovery of the much sought ' function ' of the periodic law, which would enable us to calculate the exact value of each individual atomic weight in terms of some one fundamental unit (*Study of Chemical Composition*, pp. 500 *et seq.*). But the combining weights are empirical constants, about which there is no permanency or finality, such as there would be about a ' theoretical ' result. What it comes to is that from the determinations that have been made (and still are being made) for many years in many lands by many men, we—or more precisely the International Committee on Atomic Weights, that is, the men whom chemists all over the world have chosen to act for them in this matter—have selected those results which on searching critical examination seem the most trustworthy, *i.e.* appear likely to be nearer than any others to that true value (*Study of Chemical Composition*, p. 192), which we do not know and may never know: these are the values that are made the basis of the calculations whose result serves us for comparison with results of analysis. But re-determinations being the order of the day, it follows that the calculated values all vary according to the data used, though the variations may be but slight and have a tendency to get smaller as the development of experimental methods brings them nearer to the true value.

small amount of solid is used and the heating is done under conditions conducive to the rapid removal of the vapours and gases evolved. On the other hand, as regards the collection of the red gas evolved, it is desirable to obtain a considerable volume, so as to be able to postpone the collecting until all the air has been driven out of the apparatus; and for this it is necessary to use a fairly large amount of the crystals, which however need not be heated sufficiently long to ensure complete decomposition. Hence it is recommended to prepare a specimen of solid by heating a few crystals

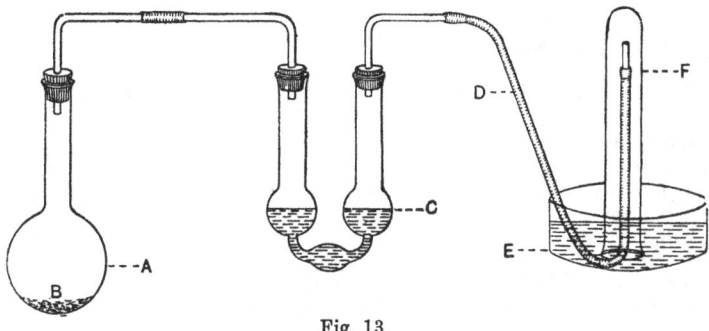

Fig. 13.

A. Long-necked flask of Jena glass of about 100 c.c. capacity, used in preference to hard glass tube because reducing the difficulty incidental to the probable frothing of the fused mass.

B. Powdered calcium nitrate, about 10 grams.

C. Bulbed U-tube holding some water for the absorption of the portion of the gas soluble in water.

D. Long rubber tube with short piece of glass tube at the end for pushing up to the top of the collecting tube F when the evolution of the gas has become vigorous.

E. Small trough or glass or porcelain dish filled with water.

F. Test-tube (or small graduated gas tube) for collection of the portion of gas not soluble in water.

The arrangement of C, D and F is devised with the special object of preventing the sucking back of liquid into A.

very strongly and for a considerable time (15 to 20 minutes) in a small open crucible, and to heat a larger quantity in a suitable apparatus until a sufficient volume of gas has been collected. The first stage, comprised between the first liquefaction and solidification, and consisting apparently in the removal of water, can with advantage be carried out in an open dish, after which the powdered solid is introduced for further heating into the flask of the apparatus (fig. 13).

(c) The qualitative properties suitable for a comparison between the original calcium nitrate and the substances into which it is broken up by the action of heat are indicated in the following scheme :

	Calcium Nitrate	Substances obtained by the heating of calcium nitrate		
		Solid left in decomposing vessel	Aqueous solution of red gas	Colourless gas collected over water
Appearance: State of aggregation Colour				
Effect of heat on solid				
Introduction of glowing splint into gas				
Action of water				
Action on moist neutral litmus paper				
Addition of solution of ferrous sulphate, followed by that of conc. sulphuric acid				
Inference as to the nature of the change that had occurred				

This is a case in which mere inspection of what happens on heating is sufficient to demonstrate the occurrence of a chemical change ; but the experiment should be well worth doing in the somewhat elaborate form described, because of the opportunity it affords for increasing the students' knowledge of a variety of qualitative tests and of the technique required in the collecting of a gas.

The students performing this experiment will probably assume tacitly that here (as also in the preceding experiment) the change occurring is independent of the presence of air, which really is the case. It is suggested that an attempt should be made to specify any points in the occurrences observed which afford experimental evidence for this view, and further to consider what form the experiment would have to be given if it were required to prove that the presence of air is immaterial.

Experiment VII. The investigation of the effect on iron of

(i) a magnet,

(ii) dilute sulphuric acid,

with the object of determining in each case whether the change produced is chemical or physical, and involving incidentally the technique of crystallisation.

(i) Using clean iron filings, or small clean nails, or steel wool, show that the process of attraction by means of a magnet can be reproduced at will any number of times ; as far as appearance goes there is absolutely no difference between the iron before and after it had been acted on by the magnet. If time allows and circumstances make it desirable to use this opportunity for a study of the chief physical and chemical properties of iron, the students might be made to supply the proof of the identity of the substances, or the absence of any permanent change, by demonstrating a certain number, at any rate, of the properties given under the heading " Iron—The Element " in any text book of descriptive chemistry (*e.g.* Newth, *Inorganic Chemistry*, 1907, pp. 676, 677 ; Ostwald, *Principles of Inorganic Chemistry*, 4th ed., 1914, pp. 593 *et seq.* ; Molinari, *Treatise on General and Industrial Inorganic Chemistry*, p. 627). A determination of the density of the solid, done by different members of the class using different methods, would be instructive[1].

(ii) (*a*) Treat some clean iron filings or steel wool (preferably the latter) with dilute sulphuric acid; note the evolution of a gas which is inflammable and lighter than air, and the production of a green liquid ; show that if left long enough and provided with sufficient acid, the whole of the iron dissolves (it may be necessary to add a few drops of concentrated acid if effervescence stops while there is still some iron present), and that the green solution, if left in a desiccator, will after a time deposit pale green transparent 'crystals.'

(*b*) In order to produce the crystals in quantity and as well-developed fair-sized individuals, it is desirable to obtain a strong solution and to allow the evaporation, which leads to the production of a super-saturated solution and the consequent precipitation of some of the solute in crystalline form, to occur slowly. Use a considerable amount of the iron, heat in a dish with the dilute acid until all effervescence has ceased (when there should still be present undissolved iron), add a few more drops of acid and separate the clear liquid by pouring whilst still hot through a folded filter paper into a rather narrow beaker. Prevent rapid cooling by surrounding the vessel with hot sand, cover with a watch-glass and allow to stand until the solid separates out ; drain this on a filter plate, dry on a porous plate.

---

[1] For the determination of the specific gravity of solids by the hydrostatic balance, see Glazebrook, *Hydrostatics*, 1904, § 57; by Nicholson's hydrometer, § 59; by Jolly's balance, § 60; by the specific gravity bottle, § 62; by calculation of the volume from measurement in the case of a regular solid, Perkin and Lean, *Introduction to Chemistry and Physics*, 1909, I, pp. 54—56; by determination of the volume with a measuring jar, p. 56.

(c) Mere inspection shows the fundamental difference between the original substances, viz. the solid metal iron and the liquid acid, and the substances produced by their interaction, viz. the inflammable gas and the green transparent individual solid bodies bounded by plane surfaces. The properties of the iron used have already been studied in (i); those of the sulphuric acid solution may be demonstrated in the same way, *i.e.* by following the account as given in any text book of descriptive chemistry. As regards the gas evolved, which is hydrogen, the two chief characteristics, viz. inflammability and great lightness, can be very easily demonstrated. There remains for consideration the green solid, which can be shown to be easily soluble in water, to turn brown on exposure to air and moisture, and to have a number of other characteristic properties which are given in the text books under the heading of "ferrous salts." The chief point to which it is wished to draw the students' attention is that of the external form (fig. 14). If the deposition of solid has been sufficiently slow (slow cooling and slow evaporation of a solution initially not too strong) it should be possible to find three or four individuals sufficiently large and sufficiently well developed to show the external characteristics of the crystalline state. These are plane surfaces inclined at definite angles, constant and characteristic for each

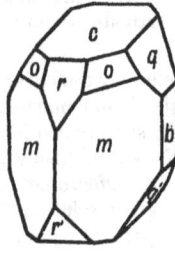 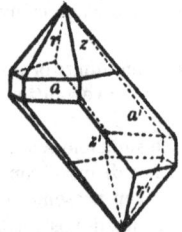 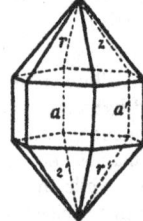

Fig. 14.   Fig. 15 a.   Fig. 15 b.

substance. Place the crystals on paper with similar faces in corresponding positions, and compare the inclination of these by drawing straight lines along the edges until they meet, and then measuring the angles thus included.

The faces of crystals may vary in shape and size, but for the same substance they are always inclined to each other at angles characteristic of that substance. The variation (fig. 15 a), which is termed distortion, from the ideal crystal as depicted in fig. 15 b, represents an unequal development of quartz due to the conditions under which the crystals had been formed.

(Freund, *Study of Chemical Composition*, pp. 387, 388.)

IV. Historically interesting and debatable cases.

The number of experiments illustrative of the nature of changes designated as chemical, and of the method that is followed in establishing the identity or diversity of substances need not be further increased, but it might be of advantage before leaving the subject altogether to draw attention to two more points.

1. **Classical researches in which the point at issue was the establishment of the occurrence or non-occurrence of chemical change.**

(i) Scheele, in the record of his classical researches on air and fire (1777 ; see *post*, p. 144), enunciates his belief, based on experimental evidence, in the existence of different kinds of air, 'air' being used by him as the generic term for substances which retain their elastic fluidity, *i.e.* remain gaseous, even in the greatest cold[1]; his criteria for recognising a gaseous substance as a peculiar kind of air are stated with a precision which makes the attraction of a verbal quotation irresistible. Following on a summary of what he considers the five[2] general properties of ordinary air, he concludes :

Consequently, when I have a fluid resembling air in its external appearance and find that it has not the properties mentioned, even when only one of them is wanting, I feel convinced that it is not ordinary air.

The application of this principle to a large mass of experimental material relating to the part taken by air in a number of changes, prominent amongst which is combustion, led him to the recognition that the air does not act as a mere recipient for foreign material, but that it is a complex substance made up of two constituents, into which it is separated by the burning substance, the one constituent named by him 'fireair' being absorbed and the other left behind (*post*, p. 144).

(ii) The view held at the beginning of the nineteenth century concerning the composition of the substances now named *hydrochloric acid gas* and *chlorine*, then named *muriatic acid gas* and *oxymuriatic acid gas*, was that whilst the first was a compound of some unknown element with water, the second contained this element united with oxygen. In

---

[1] This of course covers only the temperature range then known, which can barely have extended as far as $-50°C$.

[2] Supports combustion ; combustion reduces the volume to between 1/3 and 1/4 of the original ; not soluble in water ; supports animal life ; supports vegetable life.

the series of researches extending from 1808 to 1819, in which Sir Humphry Davy proved the elementary nature of chlorine, one important experiment took the form of showing that so-called 'oxymuriatic acid' when ignited with charcoal was left quite unchanged. Owing to the great affinity of oxygen for charcoal, it was legitimate to assume that had this element been present in the chlorine it would have been removed by combustion with the charcoal, just as happens in the case of a metal calx (oxide) :

> One of the most singular facts that I have observed in this subject is, that charcoal, even when ignited to whiteness in oxymuriatic or muriatic acid gases by the voltaic battery, effects no change in them.

(iii) Comparatively recently the discovery of argon, soon followed by that of helium, neon, xenon and krypton, enlarged the scope of chemistry by the introduction of a number of new substances characterised by their *absolute* inertness. Argon was the first of these for which it was proved that all the various changes of condition which experience had shown to be conducive to the production of a chemical change had no effect, and that the gas left at the end was identical in amount and in properties with the original argon (see *post*, p. 126).

2. Then there are the debateable cases, *i.e.* instances of changes, about the classification of which as physical or chemical there is difficulty and difference of individual opinion.

Solidification and volatilisation, *e.g.* change of liquid water to ice and to steam respectively, the production of allotropic modifications, *e.g.* change of yellow phosphorus to red, and solution, are examples. As regards solution, which deserves to be considered in some detail, how does the case really stand ? Solutions are homogeneous[1] substances, gaseous[2], liquid[3]

---

[1] See *ante*, p. 52 (footnote).

[2] Produced by the solution of a gas in another gas, *e.g.* carbonic acid or water vapour in the air ; all gases are completely miscible, *i.e.* the solubility is unlimited.

[3] The most common case, and the one that is meant when the term is used without further specification. They are produced by the solution in a liquid of (i) a solid, *e.g.* salt in water, sulphur in carbon bisulphide, carbon in molten iron ; (ii) a liquid, *e.g.* alcohol in water, and *vice versa* ; (iii) a gas, *e.g.* air or carbonic acid in natural waters.

or solid[1], made up of two (or more) components A, B, etc., which are termed the solute and the solvent[2] respectively; the ratio of A to B can vary continuously, either from $0 \times A$ and $100 \times B$ to $100 \times A$ and $0 \times B$, when the case is one of complete miscibility, such as is exhibited by alcohol and water; or from $0 \times A$ and $100 \times B$ to $m \times A$ and $100 \times B$, when we are dealing with limited miscibility, such as is exhibited by the solution of salt in water, in which case at, say, $20°C$. 100 grams of water dissolve at most 36 grams of salt[3]; moreover the properties of solutions, though not additively those of the components, are a continuous function of the composition (see fig. 16).

There has been a great deal of useless and time-wasting argument as to whether the change produced in solution should be classed as physical or chemical; the question when put thus is really quite independent of any hypotheses and theories concerning the state of the solute in the solution, that is, of the condition of its constituent particles and of their relation to those of the solvent. In the present instance it is merely a question of adjustment between definition and actual phenomena: if, as is the case, we have by our own free choice decided upon a definition of chemical change which

---

[1] Substances which combine with the solid state the essential characteristic of solutions, viz. continuous change in composition and in properties. Though cases are known in which the solute is a liquid (*e.g.* the zeolites, natural hydrated silicates) or a gas (*e.g.* solution of hydrogen in palladium or in iron), the most important and best investigated cases are those of the solution of a solid in another solid, such as is produced in the crystallisation of isomorphous substances, *e.g.* potassium permanganate and potassium perchlorate, which exhibit continuous alteration in composition and in properties, the colour of the mixed crystals showing all shades intermediate between the pure white of the perchlorate and the deep purple of the permanganate.

[2] The substance dissolved, which may be a gas, a liquid or a solid, is termed the solute, the dissolving substance the solvent. Of course in the case of the solution of a liquid in a liquid the distribution of the names solute and solvent is an arbitrary process, but practical convenience, which in this as in many other cases is the determining factor, has decided that in the case of limited solubility the quantity present to a smaller amount shall be designated as the solute; *e.g.* in shaking up ether with water two layers of liquid are formed; in the upper layer, which is a solution of water in ether, water is the solute and ether the solvent, whilst in the lower layer the opposite relation holds.

[3] m is called the coefficient of solubility.

hinges on the formation of a new kind of matter possessed of new properties, then, applying these criteria, we have no choice but to class the process of solution as a chemical change. Taking as an example the solution called sulphuric acid, formed from water and hydrogen sulphate, we find that

Curve showing the specific gravity of an aqueous solution of hydrogen sulphate (sulphuric acid) as a function of the composition.

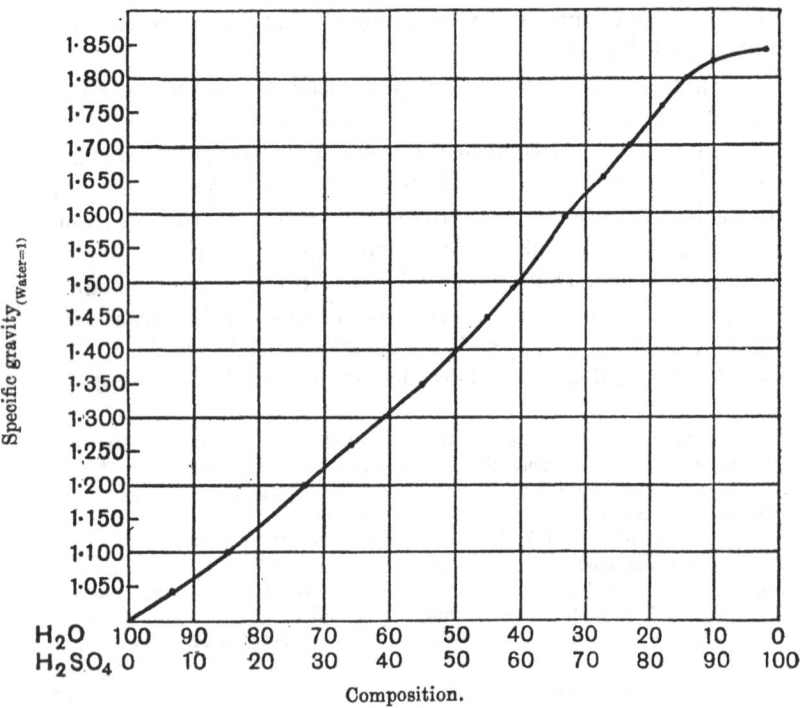

Composition.

Fig. 16.

its density (see fig. 16), boiling point and freezing point and a number of other physical and chemical properties differ from those of the components and are not even always intermediate between these: thus, while neither cold water nor perfectly pure hydrogen sulphate has any action on litmus or zinc, the solution sulphuric acid turns blue litmus red and dissolves zinc with the evolution of hydrogen: surely conclusive evidence for the formation of a new substance. But if our views concerning the nature of solution, resulting from a study of the characteristic properties exhibited by solutions, should make us feel reluctant to class the kind of change occurring in the process of solution with changes such as occur in analysis (*e.g.* the breaking up of calcium nitrate under the action of heat) or synthesis (*e.g.* the burning of iron in oxygen or air) or double decomposition (*e.g.* the formation of silver chloride from ammonium chloride and silver nitrate), the only way out of the difficulty is to change the basis of the apparently unsatisfactory classification, and to find a basis for another classification, another definition.

CHAPTER II

THE CLASSIFICATION OF SUBSTANCES INTO COMPLEX AND SIMPLE (ELEMENTS)

I. The importance of quantitative methods.

It has been asserted so often as to reduce it almost to a platitude that chemistry first became a science when towards the end of the eighteenth century Lavoisier took it into the realm of the quantitative. Like every pronouncement which tries to arrest attention by picturesque conciseness, there is a good deal of exaggeration in this assertion. On the one hand, Black[1] in the middle of the eighteenth century had been a forerunner of Lavoisier in the use of quantitative methods, and on the other hand Boyle in the second half of the seventeenth century and Davy in the first two decades of the nineteenth century had shown that work pre-eminently qualitative[2] may be as truly scientific in conception and execution, as fundamentally important and as fruitful as any that is pre-eminently quantitative. But the determination not to accept without some reservation the current apotheosis of Lavoisier and his method need not and should not interfere with our belief in the importance of quantitative methods

[1] *Experiments upon Magnesia Alba*, 1755 (Alembic Club Reprints, No. 1); *Study of Chemical Composition*, p. 59.

[2] See Robert Boyle's refutation in the *Sceptical Chymist* (1661) of the Aristotelian doctrine of Four Elements and the Paracelsian doctrine of Three Principles (*Study of Chemical Composition*, pp. 274 *et seq.*) and his introduction of the sound empirical conception of element; Sir Humphry Davy's recognition of the elementary nature of chlorine, 1809—1818 (Alembic Club Reprints, No. 9), the isolation of the alkali metals, 1807—1808 (Alembic Club Reprints, No. 6), and proof of the composition of water, 1806 (*Study of Chemical Composition*, pp. 10 *et seq.*).

and the recognition that the time of their introduction into any science marks an important stage in its development: "all science begins with measurement." If therefore we consider it a generally recognised principle that in the solution of any and every problem the use of quantitative methods should be resorted to, it follows that the study of the characteristics of chemical change, which so far as we have pursued it in the preceding chapter has been mainly of a qualitative nature, should now be taken into the realm of the quantitative. What was done in all the cases investigated in the previous chapter was to examine the properties of a substance *before* and *after* it had been subjected to a change of conditions; and it was pointed out that conclusive proof of the absence of chemical change required the inclusion of some measureable properties, such as boiling point and freezing point, density, electrical conductivity, or any other of the many so-called 'physical properties' whose characteristic it is that they are possessed by all substances, but to a different degree. On the other hand, we found that the appearance or disappearance of one or more qualitative properties (*ante*, p. 60) is sufficient to establish the occurrence of a chemical change. But accepting the sufficiency of qualitative tests in this case, there still remains the importance which attaches to the concomitant phenomenon of a change in the values of the quantitative properties characteristic of the substance investigated ; the change may be so great that it is easily and quickly recognised, or it may be so small that it requires great refinement in the methods of measurement to detect it at all, but it always exists.

II.   The study of the changes in weight which accompany chemical change.

To the chemist the quantitative properties of greatest interest and importance are those referring to the total mass and the composition of the substances dealt with, and the subject-matter of this chapter is the investigation of the changes in mass that accompany chemical changes when considered from the point of view of *one* of the participating and resulting substances[1].

It has been found as the result of innumerable experiments that in the largely preponderating number of cases every qualitatively established chemical change is accompanied by a change in weight of the substance dealt with. The resulting substance, if collected without loss, weighs either more or less than the original substance except in a comparatively small number of cases in which the weight remains unaltered. It is not proposed to consider further at this stage these exceptional cases, beyond saying that they nearly all belong to the province of allotropic modifications and so-called 'intramolecular transformations,' that a remarkable instance of the latter was Wöhler's production of urea from ammonium cyanate[2] in 1828, and that they have played an important part in the development of the theory of isomerism.

Turning to the illustration, by means of students' experiments, of the results obtained in the study of the weight changes accompanying the production of a new kind (or kinds) of matter, there are some points that require preliminary discussion.

Each experiment is really made up of two parts:

(*a*) The qualitative proof that in the case dealt with a chemical change has occurred. This point has been fully dealt with in the preceding chapter, but it may not be amiss to remind the student that if the change consists in the disappearance of a particular property, the change must be made complete, *i.e.* the whole of the material worked with must be affected.

(*b*) The investigation of the concomitant change in mass by weighing the substance contained in a suitable vessel

---

[1] The law of conservation of mass, that fundamental principle on which all modern chemistry rests, is the outcome of the consideration, not of one isolated substance but of the sum of the weights of all the substances participating in, and all those resulting from, the chemical change (*ante*, p. 32 and ch. v).

[2] Perkin and Kipping, *Organic Chemistry*, 1909, pp. 2, 301 ; Schorlemmer, *Rise and Development of Organic Chemistry*, 1894, pp. 21, 22.

before and after the reaction. The vessel chosen must be such as to justify the assumption that its own weight has not been altered by the changed conditions (*e.g.* heating, action of acids, etc.); thus a copper crucible, which on being heated in air becomes coated with a layer of black oxide, would not be suitable. Obviously in this part of the experiment it is not essential that the change should have been made complete; a change in part of the substance worked with will be sufficient for showing whether the weight has increased or decreased (or remained the same), if that is all that is wanted. Of course in most cases the investigator will not be content with a mere indication of the *direction of the change*, but will want to obtain that further information concerning the specific nature of the change which requires a knowledge of the *amount of change*, positive or negative, per unit weight of original substance, and which can only be ascertained by making the change complete.

It follows from the difference of conditions to be fulfilled in (*a*) and (*b*) that in the majority of cases the most time-saving procedure will be to perform the qualitative and quantitative parts of the experiment separately, using for each the amount of material most suitable for obtaining the desired effect. Moreover in (*b*) it may be a matter of the utmost importance to devise conditions which will minimise the introduction of experimental error due to: (i) mechanical loss of material, *e.g.* white fumes evolved in the burning of magnesium (Exp. VIII (ii)); particles of solid carried off by the escaping oxygen in the heating of potassium chlorate (Exp. IX); loss of some of the precipitated silver chloride in the transference from one vessel to another (Exp. X); (ii) the introduction of an impurity, *i.e.* extraneous matter, the weight of which will increase that of the changed substance, obscuring the effect investigated. Thus in weighing the amount of silver chloride obtained from a known weight of metallic silver, it is essential that the chloride weighed should have been thoroughly dried (Exp. X), so as to ensure that the increase in weight observed has not been wholly or partially due to this addition of water, etc., etc.

The following experiments can be adapted to the varying needs of students at different stages. The qualitative part is specially suitable for those who require more practice in the work to which they have been introduced in the previous chapter, and it need not necessarily be done at all by those who wish to devote all their time to practice in quantitative work. Again, when it comes to the quantitative part, it will depend on the student's attainments whether the experiment is restricted to just proving that there has been an increase or decrease in the weight of the substance used, or whether the object is to ascertain with all possible accuracy the amount of the completed change for a known weight of material. The experiments in which the weight change occurring is to be measured in this accurate manner (viz. decomposition by heat of potassium chlorate or perchlorate, and production of silver chloride from metallic silver) have been specially chosen because the data obtained in this analysis and synthesis are of great importance in the determination of combining weights, and will be used as antecedent data in subsequent experiments (*post*, pp. 102, 265).

Experiment VIII. The heating in air of
- (i) iron,
- (ii) magnesium,
- (iii) copper.

(i) (a) The effect of heating iron in air has been investigated already (Exp. III). In order to complete the reaction it is desirable to work with a small amount, to use the iron in a state of fine division, and to expose its surface as much as possible to the action of the air by grinding up the mass and heating again several times.

(b) The proof that the burnt iron weighs more than the metal can be furnished by using a variety of experimental devices: the iron can be heated in an open crucible or held by a horseshoe magnet suspended from the hook of a balance pan (fig. 17). Of course the finer the state of division of the iron, the greater the effect and the less heating

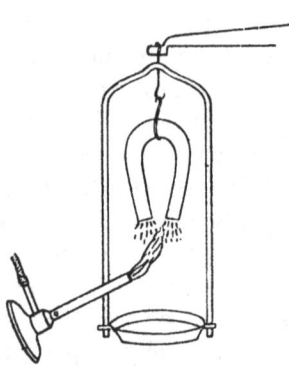

Fig. 17.

required; no difficulty arises from mechanical loss, as neither the metal nor the burnt iron is volatile. But it is essential that the material should be free from the grease often used to keep iron from rusting, and that it should be dry; the powdery reduced iron always holds some moisture, which must be removed before use by heating in a stream of pure dry hydrogen, cooling in the hydrogen and keeping in a desiccator until required. The experiment in this form does not lend itself well to finding the actual increase in weight per unit weight of metal, and hence there is no point in even trying to complete the reaction, or to ascertain the weight of iron used, which should be considerable in order to get a large effect.

Record of results.

Scheme for note-book entry.

Properties of:	The original substance Iron	The substance left after the heating of the original iron Burnt iron

(a)   Action of heat
     Appearance:
       Colour
     Action of:
       dilute hydrochloric acid
       conc. hydrochloric acid
       solution of ammonium
         (or potassium) thio-
         cyanate on the solu-
         tions in hydrochloric
         acid		

(b)   Wt. of crucible + iron   before heating = 20·46 grams
       ,,   + unchanged iron + burnt iron after heating   = 20·68   ,,
                   Difference   = 0·22   ,,

(ii)   (a)   Here again the qualitative experiment of heating magnesium in air has been described already, and nothing need be added.

(b)   In performing the quantitative part of the experiment, special precautions must be taken to prevent the loss of material in the form of dense white fumes. It is obvious, therefore, that if any increase in weight at all is observed, this increase represents only a part of the actual increase, owing to the probability of some mechanical loss having occurred. The great difficulty of preventing mechanical loss and the complications due to the simultaneous occurrence of a number of different reactions (formation of silicide and nitride, see *ante*, p. 55) make this experiment unsuitable for the actual determination of the ratio magnesium : burnt magnesium (magnesium oxide).

The heating may be done in a crucible with a plug of ignited asbestos covering the magnesium (to retain the solid otherwise carried off in the form of white fumes), and with the lid slightly lifted to allow the access of the air necessary for combustion.

A. Magnesium wire or ribbon or filings.

B. Loose plug of ignited asbestos.

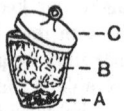

C. Lid of crucible put on slanting, so as to leave an opening.

Fig. 18.

A more elaborate contrivance is depicted in fig. 19; the object of this special arrangement of apparatus is to produce conditions in which it is possible : (i) to weigh the combustible and the burnt substance in a vessel so arranged that no change can occur during the process of weighing ; (ii) to supply the air necessary for combustion at an easily adjustable rate— the aspirator could of course be replaced by a water suction pump, or by an arrangement similar to that shown in fig. 36 A, p. 153 ; (iii) to retain by means of the asbestos plugs any solid produced by the condensation of fumes.

Record of results.

Specimen of note-book entry.

		(i)	(ii)
		grams	grams
Wt. of glass tube + asbestos plug +	magnesium	= 42·942	35·558
,, ,, +	burnt + unchanged	= 43·074	35·774
	magnesium magnesium		
	Difference	= 0·132	0·216

(iii) (*a*) For the qualitative test, detach some of the black scales formed on the surface of the copper ; or some of the turnings changed only superficially may be used.

(*b*) For the quantitative test, the copper should be used in the form of fine turnings; in order to produce an effect as great as possible, it is necessary to heat in a current of air, using the apparatus depicted in fig. 19. To complete the change is almost impossible, as the layer of burnt copper formed on the surface protects the core from being acted upon[1].

[1] Examine the copper oxide made from wire for organic analysis, by breaking a wire and noting the core of red copper.

**Apparatus for heating a substance
in a stream of dry air.**

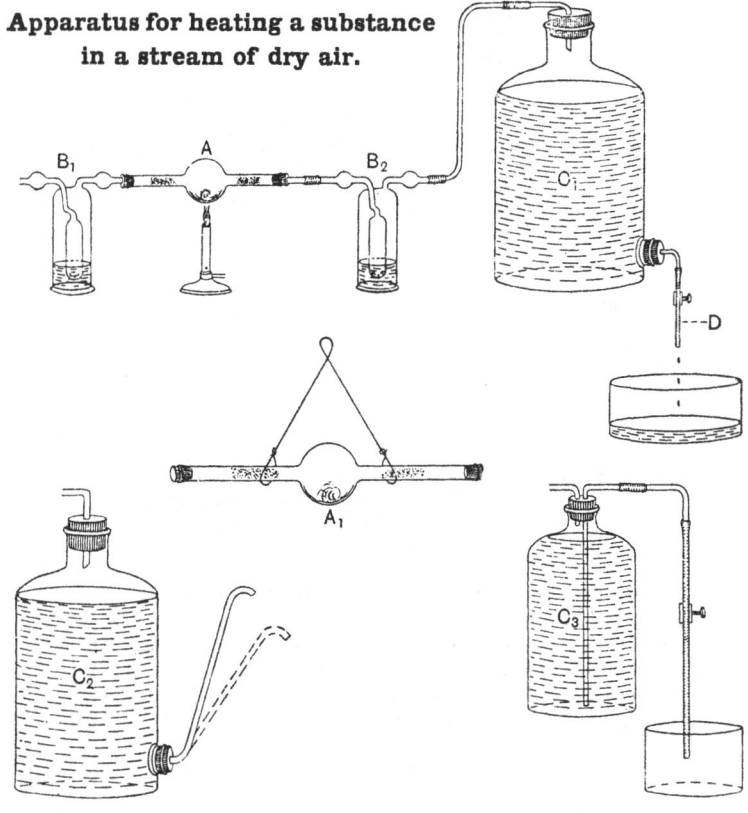

Fig. 19.

A. Hard glass tube holding substance to be heated (magnesium or copper or
    phosphorus), with plugs of ignited asbestos at both sides to keep the sub-
    stance in place and to retain products of combustion.

A_1. Same tube (slightly enlarged) prepared for weighing, with contents as in A,
    but closed with corks and fitted with suspending wire.

B_1. Wash bottle with concentrated sulphuric acid, to dry the ingoing air and to
    show the rate at which it passes.

B_2. Wash bottle with concentrated sulphuric acid, to prevent moisture from C_1
    passing back into A. A calcium chloride tube might be substituted here.

C_1. Aspirator filled with water and fitted with rubber tube closed by clip.

D. Rubber tube and screw clip, for producing and regulating the flow of air
    through the whole apparatus.

C_2, C_3. Modifications of C_1.

Record of results.

Scheme for note-book entry.

Properties to be observed in	The original substance Copper	The substance left after the heating of the original copper Burnt copper

(a) Action of heat
    Appearance:
        Colour
        Coherence
    Action of conc. hydrochloric
    acid in the absence of air		

		(i) grams	(ii) grams
(b) Wt. of glass tube + asbestos plug + copper		= 108·06	104·66
,, ,, + burnt + unchanged		= 108·73	105·37
	copper copper		
	Difference	0·67	0·71

Experiment IX. Action of heat on potassium chlorate (or perchlorate). Use the pure recrystallised salt and test with silver nitrate for the absence of chloride in appreciable quantities. If the material available does not fulfil the necessary requirements as to absence of chloride, the salt must be recrystallised until the silver nitrate test shows that the desired result has been attained. If the substance worked with is the perchlorate

Solubility Curves.

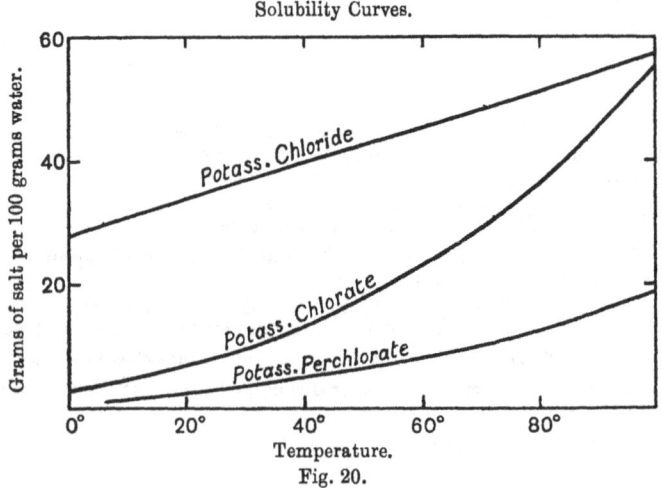

Fig. 20.

the problem is complicated by the necessity of testing also for chlorate, the presence of which is detected by the evolution of chlorine on boiling the solid with hydrochloric acid. The process adopted to remove the chlorate consists in changing it to chloride by evaporating to dryness with hydrochloric acid, and then getting rid of the chloride by first washing with small quantities of cold water and afterwards recrystallising the perchlorate from hot water.

The diagram (fig. 20) of the solubilities of potassium chloride, chlorate and perchlorate indicates the course to be followed in the separation it is desired to effect, viz.: solution in the least possible quantity of boiling water of the mixture of chlorate and chloride, of which latter there is present from the outset a very small amount only; cooling; draining on filter plate, with use of pump (fig. 21), the crystals of chlorate which have separated out; washing the crystals with small amounts of cold water to remove adhering

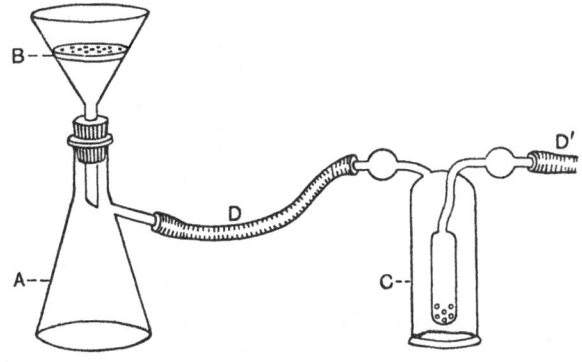

Fig. 21.

A.	Filter bottle.	C.	Empty wash-bottle.
B.	Funnel containing filter plate with hardened filter paper.	D.	Pressure tubing.
		D¹.	,, ,, in connection with pump.

chloride; and repetition of the whole process until a small portion when tested is found free from chloride.

Since the distinguishing properties of the three salts have to be used in the process of purification, and since the product of heating the chlorate or perchlorate is chloride, the table summarising the results of (*a*) is given already filled in (p. 93). For success in the quantitative part of the experiment it is essential that the salts used should be dry, which is ensured by reducing the crystals to fine powder, keeping them for some hours in an oven heated to a little above 100°C., putting them whilst still hot into a desiccator and keeping them there until wanted.

(*a*)   Heat a small amount, about 1 gram, of the salt (which need not have been specially dried) in a hard glass tube, and test the gas evolved. Heat another small amount in an open crucible or small dish (supported on clay triangle) and note the sequence of occurrences :—the liquefaction of the salt;

the evolution of gas, which in the case of the chlorate evidently occurs in two stages, of which the second is the more vigorous; the solidification of the mass after the gas evolution has subsided; and, *provided the temperature is raised very considerably*[1], the final fusion to a clear liquid which seems to remain unchanged on the continued application of heat. After the solid product has been heated strongly for ten minutes or so, it is allowed to cool, removed from the containing vessel (which removal may involve the necessity of breaking the vessel), powdered and again strongly heated. It should be obvious that the procedure here recommended aims at providing material for the qualitative tests in the quickest and simplest possible

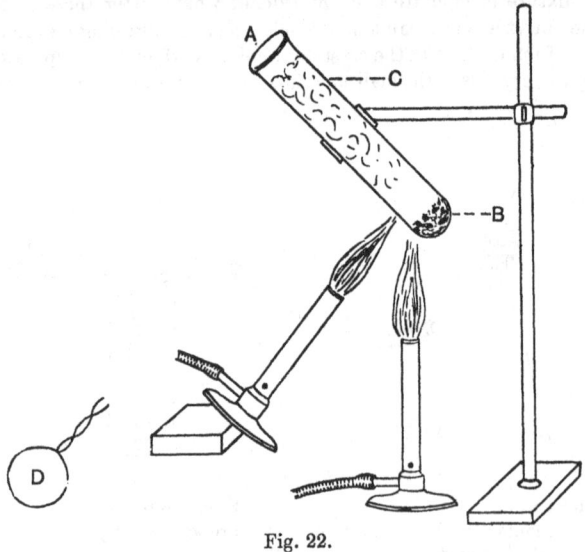

Fig. 22.

A. Hard glass tube of about 50 c.c. capacity.
B. Powdered crystals of pure dry chlorate.
C. Plug of ignited asbestos.
D. Loop of copper wire for suspending the tube from the balance. This loop must be removed whilst the process of heating is being carried out.

manner, discarding precautions against mechanical loss, which will be considerable, since the escaping gas carries with it particles of the solid, showing as white fumes.

Powder some of the solid left behind after the process of heating, and carry out the comparative examination which (together with its results) is summarised in the following table:

---

[1] This can be done by heating with two good Bunsen burners, placed so as to produce their maximum effect on a small amount of the salt contained in a test-tube of thick Jena glass, as in fig. 22.

The Action of Heat on (i) Potassium chlorate or (ii) perchlorate.

	(i) Potassium chlorate	(ii) Potassium perchlorate
	Original substance	White solid left after the heating has been continued until no further change seemed to occur	Original substance
Effect of heat	Fuses easily and then decomposes with evolution of gas which occurs in two stages, then solidifies and fuses again at much higher temperature. Gas evolved rekindles glowing splint (oxygen)	Fuses at much higher temperature	Fuses at much higher temperature	Fuses easily and then decomposes with evolution of gas, then solidifies and fuses again at much higher temperature. Gas evolved rekindles glowing splint (oxygen)
Action of water : cold hot	Slightly soluble	Fairly easily soluble	Fairly easily soluble	Easily soluble. Solution deposits crystals on cooling	Slightly soluble
Heating of solid with conc. hydrochloric acid	No effect	No effect	No effect	Evolution of greenish gas with pungent smell, which bleaches litmus (chlorine)	No effect
Action of cold conc. sulphuric acid on solid. (In case of chlorate use a very small crystal only)	No effect	Evolution of fuming acid gas which turns milky a drop of silver nitrate held at end of rod	Evolution of fuming acid gas which turns milky a drop of silver nitrate held at end of rod	Evolution of orange gas which spontaneously or on slight heating explodes, giving chlorine	No effect
Action of silver nitrate solution on solution of substance	No effect	Curdy white precipitate which turns violet on exposure to light, is insoluble in nitric acid but easily soluble in ammonia	Curdy white precipitate which turns violet on exposure to light, is insoluble in nitric acid but easily soluble in ammonia	No effect	

(*b*) According to the time available and the stage reached by the students, this part of the experiment may be done in one of two ways :

(*a*) Finding out simply whether there has been a decrease or increase in weight, in which case it will not be necessary to complete the reaction, or even to know the amount of salt taken ; moreover, though it is of course desirable to reduce errors due to impurities and to mechanical loss, yet provided the effect of such errors is much smaller than the change measured, time need not be spent in trying to eliminate them completely.

grams
Wt. of glass tube + asbestos + suspending wire + potass. chlorate　　= 28·73
,,　　　　　,,　　　　　,,　 + changed potass. chlorate + unchanged-potass.　,,　} = 27·94
Difference　　= −0·79

(*β*) Determining with the utmost accuracy attainable under the special conditions the ratio of chlorate (or perchlorate) to the substance formed, which is potassium chloride. The modifications in the experimental procedure required for accomplishing this are : (i) an extra weighing to ascertain the actual amount of substance used ; (ii) some *efficient* arrangement for preventing loss of weight due to mechanical dispersion of solid material, *e.g.* one long or two shorter plugs of fine asbestos fibre which have been carefully ignited and then kept until wanted in a desiccator ; (iii) making sure that the whole of the material has been changed by repeating the heating and subsequent weighing when cold (the tubes should be cooled and kept in a desiccator) until two consecutive weighings do not differ by an amount greater than that of the experimental error of the weighings.

Summary of results of class.

Ratio potass. chlorate : potass. chloride $= 100 : x$

Results obtained by a class of 5 students

	(i)	(ii)	(iii)	(iv)	
Student A	60·94	61·37	61·15	60·28	
B	60·81	62·68			
C	60·45	60·86	(83·13)		Mean of 11 experiments **60·77**
D	60·04	59·93	(65·29)		
E	(69·7)	60·0			

Standard Values

Berzelius,	1826	60·851	
Penny,	1839	60·823	General Mean from all series, representing 40 experiments
Pelouze,	1842	60·843	
Merignac,	1842	60·8392	**60·846**
Maumené,	1846	60·791	Stas' Mean **60·846**
Stas, 1st series,	1860	60·8428	Stähler and Meyer, 1911 **60·834**
,,　2nd　,,		60·849	

Record of results.

Specimens of note-book entries.

Determination of the weight of solid residue (potassium chloride) yielded by the heating of 100 parts of potassium chlorate or perchlorate

	Chlorate			Perchlorate	
	(1)	(2)	(3)	(1)	(2)
Wt. of empty hard glass tube + wire.........$= a$	31·8405	35·1893	20·7070	35·165	24·3335
Wt. of hard glass tube + wire + substance $= b$	37·0366	39·6980	21·8303	39·1568	25·4428
∴ Wt. of substance .... $= b - a$	5·1961	4·5087	1·1233	3·9918	1·1093
Wt. of tube + wire + substance + asbestos plug :					
before heating ....$= c$	37·4422	40·3040	22·4680	39·8075	25·8632
after heating .... (i)	35·4088	38·5337	22·0282	37·9686	25·3509
(ii) $\Big\}= d$	35·4095	38·5332	22·0280	37·9670	25·3505
(iii)	35·4055		22·0278	37·9672	25·3505
(iv)	35·4050		22·0278		
∴ Wt. of solid residue (chloride) ....$= b - a - (c - d)$	3·1589	2·7379	·6831	2·1515	·5966
Wt. of chloride produced from 100 of original substance ....$= \dfrac{d - a - (c - b)}{b - a} \times 100$	60·79	60·73	60·81	53·90	53·79
Mean..		**60·78**		**53·84**	

Experiment X. The action on silver of nitric acid, followed by that of hydrochloric acid. Production of silver chloride[1].

For (*a*), the qualitative part of the experiment, the silver may be used in the form of 'leaf,' which dissolves very rapidly and is comparatively inexpensive, but for (*b*), the quantitative part, the material must be the purest foil that can be obtained. In Merck's *Purity of Chemical Reagents* the following directions are given for testing for the more common impurities in metallic silver:

Foreign Metals :—Dissolve two grams of silver in the smallest possible quantity of nitric acid (spec. grav. 1·2). The solution should be colourless and there should be no insoluble residue (antimony and tin). Dilute with about 200 c.c. of water (no turbidity should occur after standing one hour —bismuth), and precipitate the silver by adding hydrochloric acid to the boiling solution. Allow the precipitate to settle in a dark place, filter and evaporate the filtrate. No weighable residue should remain.

(*a*) Treat one leaf or a tiny scrap of foil contained in a largish test-tube with a few c.c. of dilute nitric acid (ordinary strength of bench reagent), and warm. Note the various occurrences accompanying the process of solution; boil until the evolution of red gas ceases; add solution of hydrochloric acid; filter the white curdy precipitate formed; wash with hot water until a few drops of the filtrate give no precipitate with silver nitrate solution; investigate the properties of this solid and embody the results in tabular form.

	Silver	Solid obtained from the silver by the action of nitric acid followed by that of hydrochloric acid
Appearance : Colour Coherence		
Effect of heat		
Action on solid of : Nitric acid Ammonia		
Mixture with sodium carbonate and potass. cyanide heated on charcoal in blowpipe flame		

[1] " Silver chloride occurs as the mineral...horn silver....This ore was known to the older mineralogists. Gesner in 1565 terms [it] *argentum cornu pellucido simile,* and Matthesius in his *Berg-postilla,* published in 1585, terms it 'glass-ore, transparent like horn in a lantern.'...The method of preparing silver chloride artificially was probably known to the old alchemists, but is first distinctly mentioned in the works of 'Basil Valentine,' who says : 'Common salt throws down ☽.'...When it was found that the precipitate was fusible and solidified to a transparent horn-like mass, the name *luna cornea* was given to it.

(b) (a) It is not difficult to carry out fairly quickly an experiment in which a small round-bottomed long-necked hard glass flask is used for dissolving the silver, precipitating the silver chloride, removing the liquid by evaporation and finally fusing the silver chloride, the flask with its contents being weighed before and after the reaction; but the difference in time required for making the experiment quantitative in the extended sense of (β) makes it barely worth while to spend time on showing merely that the chemical change from metal to silver chloride has been accompanied by increase in weight.

A more profitable course would be to do the converse experiment, to produce metallic silver from dry silver chloride contained in a hard glass tube by heating in a current of hydrogen and establishing the decrease in weight. This may be a convenient point for the student to make acquaintance with the various devices for the so-called continuous evolution of gases, which can be produced rapidly and without the application of heat by the action of a liquid on a solid. The catalogues of the various dealers in chemical apparatus may be usefully consulted. Some simple forms of such apparatus are shown in fig. 23, p. 98.

Record of results.

Wt. of tube + asbestos plug + silver chloride = 20·5678 gms.

,,　　　　,,　　+ metallic silver = 20·3020 gms.

Difference　=　·2658 gms.

(β) On the other hand, the determination by the student, with the greatest accuracy of which he is capable, of the ratio silver : silver chloride is an experiment of the utmost importance. It embodies the technique of the gravimetric determination of silver and of chlorides respectively, and it also shows how one of the constants of fundamental importance in combining (atomic) weight determination is ascertained (*Study of Chemical Composition*, p. 209).

It is important that the solution of the metal should be carried out so as to minimise mechanical loss by the carrying away of some of the solution in the form of spray. The acid used should be dilute, so as to prevent violence of action, and not more than is necessary should be used, in order

Libavius stated that the substance obtained by precipitating silver solution with common salt weighed less than the silver itself, but this statement was contradicted by Boyle; and Kunkel in his *Laboratorium Chymicum* says that many substances are difficult to separate from one another: ' Such is seen in ☽ cornea. as 12 oz. ☽ retain out of the common salt 4 ozs. terra and salt.' " (Roscoe and Schorlemmer, *Treatise on Chemistry*, vol. II.) In this determination Kunkel (1630—1702) seems to have attained a very high degree of accuracy, the ratio 12 : 16 being equal to 100 : 133·3, whilst the present-day standard ratio (*post*, p. 103) is 100 : 132·86.

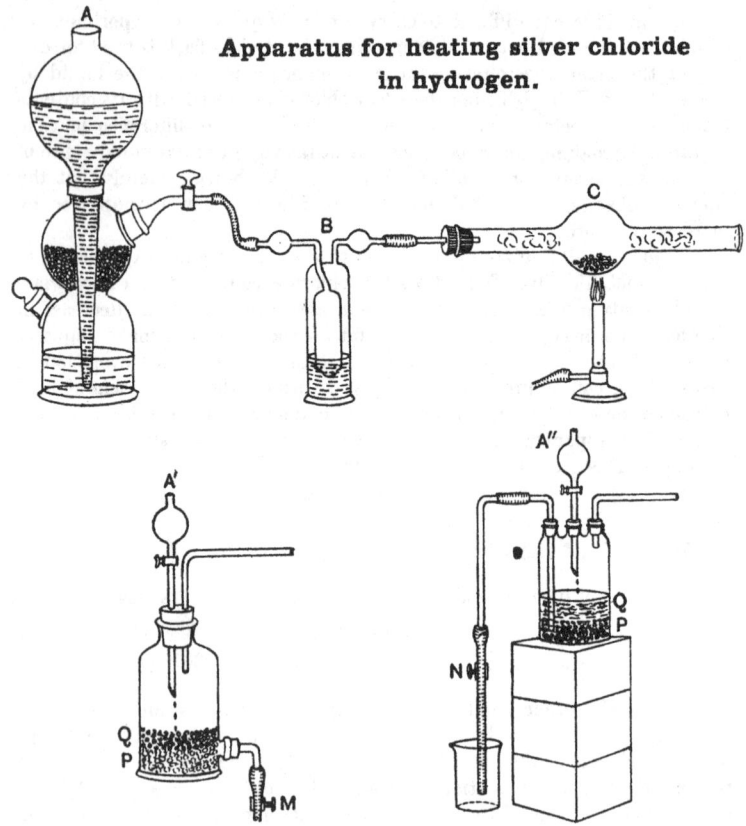

Apparatus for heating silver chloride in hydrogen.

Fig. 23.

A.　Apparatus for the generation of hydrogen by the action of dilute sulphuric acid on zinc. This special form of gas-generating apparatus goes by the name of its inventor Kipp; it may be used in all those cases when a gas is readily evolved by the action of a liquid on a solid without the necessity of applying heat, *e.g.* carbon di-oxide from marble and dilute hydrochloric acid. The advantages of the arrangement are that the evolution, which is described as continuous, can be stopped and started again, and its rate adjusted, at will.

A′, A″.　Simple forms of apparatus that can be used as substitutes for A; by adjusting the rate at which the liquid is dropped on to the solid, the rate of evolution of the gas can be controlled, and the spent liquor can be removed by opening the screw clips M and N. It is advantageous to put a layer P of small glass marbles, or lumps of some other inactive solid, at the bottom of the generating vessel, under the layer of zinc Q.

B.　Gas wash bottle filled with concentrated sulphuric acid, to dry the hydrogen and show the rate at which it is passing.

C.　Bulbed tube of hard glass, containing the silver chloride in the bulb and plugs of ignited asbestos in the sides.

to reduce the time required for the removal of liquid by evaporation. 5 c.c. of acid of strength 3 N for every 1 gram of metal is a suitable quantity.

The arrangement depicted in fig. 24, in which, during the process of solution, the flask is kept in a slanting position and carries a small funnel, is devised with the object of retaining spray, which primarily would impinge on the wall of the flask or the stem of the funnel, whence it would fall or be washed back. Solution should be initiated by gentle warming, after which it should be allowed to proceed unaided; when complete, the solution should be boiled gently for a few minutes to remove nitrous fumes. An approximately equal volume of warm hydrochloric acid solution is then gradually added to the hot silver solution, with constant shaking to prevent the inclusion of any unchanged solution by the curdy precipitate formed; the precipitate is allowed to settle, and a few more drops of hydrochloric acid are added to see whether precipitation has been complete. When this has been accomplished, the isolation of the precipitate for purposes of weighing may be effected

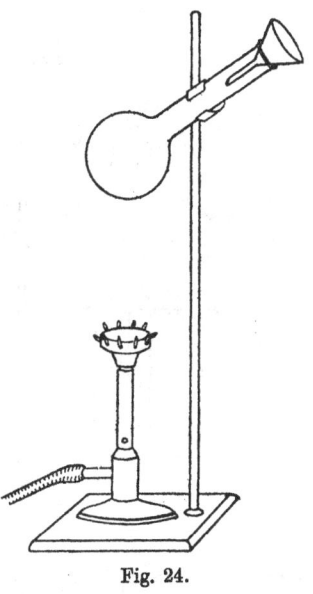

Fig. 24.

in one of two ways: (i) evaporation and subsequent fusion in the flask itself, all the other substances present, viz. nitric acid and hydrochloric acid, being volatile; (ii) the use of a Gooch crucible for the processes of filtering, washing, drying to constant weight and weighing.

(i) The crucial point in this method is to accomplish the evaporation without loss by spurting. The danger of spurting can be reduced and the process accelerated by some arrangement for the continuous sucking out of the water vapour and its replacement by dry air, in which case the liquid need not be heated to the boiling point. The heating is best done, not by the direct localised action of a gas flame, but by some arrangement for surrounding the whole of the lower portion of the flask with heated air (see *A*, fig. 25).

A device used by Sir Edward Thorpe in his determination of the combining weight of radium may with advantage be used in a simple form, to remove by suction as much as possible of the liquid before evaporation is begun (fig. 26).

When evaporation is completed, the solid is heated to incipient fusion by the direct action of a small flame on the flask, the flask is cooled and then weighed, and the process repeated until the weight has become constant.

(ii) The procedure as far as the solution of a known weight of the silver is the same as in (i), after which the precipitate is collected in a

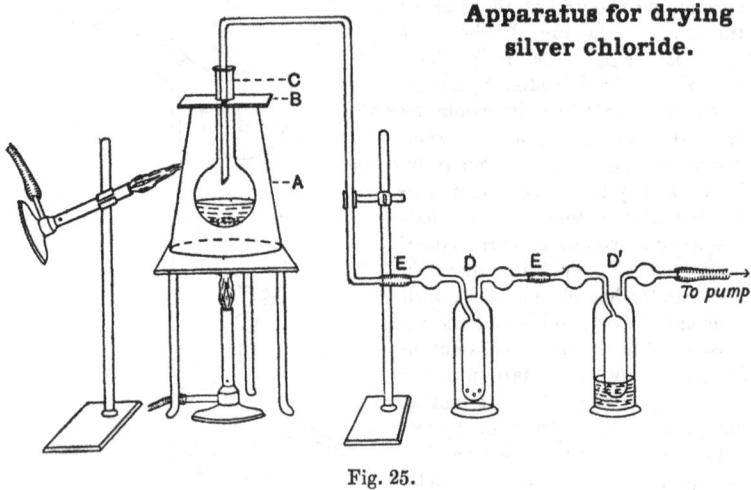

Apparatus for drying silver chloride.

Fig. 25.

A. Metal cone such as is used for the drying of precipitates in funnels, resting on an iron plate.

B. Asbestos sheet with hole which holds the flask C, slipped in through slit in the sheet.

C. Long-necked round-bottomed hard glass flask of about 100 c.c. capacity, holding the precipitate and supernatant liquid.

D, D′. Wash-bottles for the condensation in D, and solution in water in D′, of the acids distilling over from C and prevention of these from attacking the pump.

E. India-rubber joints—it is important that the two pieces of glass joined should touch, so as to minimise the surface of rubber exposed to the action of nitrous fumes.

weighed Gooch crucible, washed, and dried to constant weight. The technique of preparing the Gooch crucible, filtering and drying, is given in most text books of quantitative analysis (*e.g.* Clowes and Coleman).

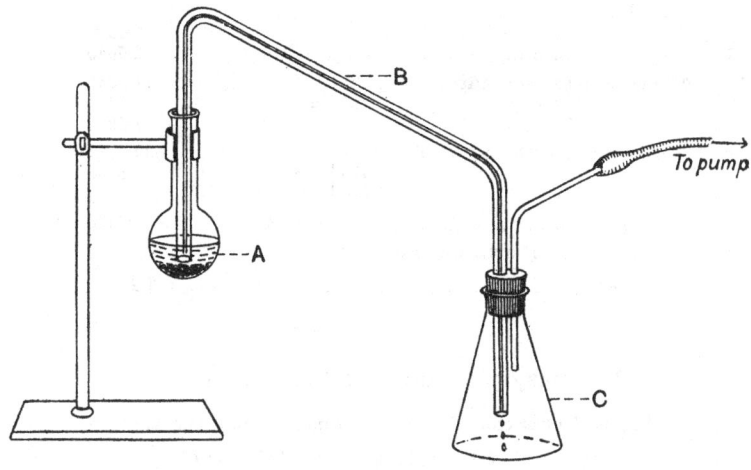

Fig. 26.

A. Flask containing precipitate and clear supernatant liquid, held by clamp which can be easily moved up and down, making it possible to bring the flask into a position such that the capillary tube B is just above the precipitate.

B. Capillary tube of very fine bore, arranged as a siphon through which the liquid is drawn off so slowly that the solid is not stirred up by currents produced in the liquid.

C. Erlenmeyer flask with well-fitting cork, through which passes the capillary and also an ordinary glass tube, suction at the open end of which starts the flow of liquid.

Record of results.

Specimens of note-book entries.

Weight of silver chloride obtained from 100 grams of silver

By method i.		(1)	(2)
		grams	grams
Wt. of empty flask and suspending wire ....$= a$		$35 \cdot 4782$	$38 \cdot 1575$
,, ,, ,, $+ \text{silver} = b$		$37 \cdot 7765$	$41 \cdot 8844$
∴ Wt. of silver.....................................$= b - a$		$2 \cdot 2983$	$3 \cdot 7269$
Wt. of empty flask + silver chloride........(i)		$38 \cdot 5382$	$43 \cdot 1080$
(ii)	$= c$	$38 \cdot 5356$ ⎱ $38 \cdot 5357$	$43 \cdot 1070$
(iii)		$38 \cdot 5358$ ⎰	$43 \cdot 1052$
(iv)			$43 \cdot 1050$
∴ Wt. of silver chloride ....$= c - a$		$3 \cdot 0575$	$4 \cdot 9475$
Wt. of chloride obtained from 100 grams of silver...$= \dfrac{(c - a)\,100}{b - a}$		**133·03**	**132·75**

By method ii.

	grams
Wt. of silver foil [1] .... $= a$	0·5640
Wt. of crucible with asbestos after heating (i) (ii) (iii) $\left.\right\} = b$	11·5937 11·5925 11·5930
Wt. of crucible with asbestos + precipitate (i) (ii) (iii) $\left.\right\} = c$	12·3365 12·3358
∴ Wt. of precipitate .... $= c - b$	0·7428
Wt. of silver chloride obtained from 100 grams of silver .... $= \dfrac{(c-b)\,100}{a}$	**131·70**

Summary of results obtained by class.

Weight of silver chloride obtained from 100 grams of silver

	(1) Weighing of precipitate in flask in which it was produced.	(2) Weighing of precipitate collected in Gooch crucible.
Student A	133·24	131·84
B	132·39	132·56
C	132·87	133·18
D	133·62	132·26
E	131·08	131·15
F	131·94	
G	131·82	133·05
H	[130·5]	
I	132·9	132·3
J		131·6
K		[130·8]
L	131·25	
Mean	**132·26**	**132·24**

Standard values.

Stas, 1860:
  by burning metal in chlorine .... 132·843
  by solution of metal in nitric acid, followed by
    precipitation with gaseous hydrochloric acid .... 132·848
    precipitation with solution of hydrochloric acid.............. 132·848
    precipitation with solution of ammonium chloride .... 132·842
Richards and Wells, 1906:
  by weighing chloride
    (i) in Gooch crucible .... 132·861
    (ii) in quartz vessel in which precipitation had been
      effected without transference of material .... 132·867

---

[1] This can be weighed on the balance pan.

In considering the summary given on the preceding page it is interesting to contrast the close agreement between the two sets of the students' results, viz. 132·26 and 132·24, with the disagreement between their mean value, 132·25 and the standard value 132·864 (Richards and Wells), a disagreement which is certainly due in part to the constant error produced by the impurity of the silver foil used.

III. The results of the study of the quantitative aspect of chemical change used for purposes of classification.

The great interest and importance of the problem dealt with in this chapter lie in its intimate connection with that part of chemical classification which divides all substances into (i) complex and (ii) simple.

(i) Complex substances are characterised by the fact that though they may combine or interact with one or more other substances to form new substances more complex still, and therefore weighing more than the original, they can also by decomposition yield new *substances weighing less than the original.*

(ii) Simple substances, or elements, on the other hand, can undergo the one kind of change only in which the product is more complex—yielding *substances weighing more than the original.*

IV. The conception of 'element.'

'Simple substance' is synonymous with 'element,' the name given from olden times to fundamental constituents or units, from which material all the substances surrounding us and apprehended by our senses are compounded. And there is no more interesting chapter in the history of chemistry than that dealing with the conception of 'element' as held at different times, with the transition from the metaphysical views of the early Greek philosophers to our present position, based entirely on empiricism. The views of the early schools of Greek philosophers are characterised by their simplicity, by the postulation of one single kind of matter capable of transformation into all the other substances which constitute

the world. For Thales (640—546 B.C.) this was water, for
Anaximenes (560—500 B.C.) air, for Heraclitus (536—470 B.C.)
fire; for Pythagoras and his school (end of sixth century B.C.)
number was the permanent thing underlying all change;
Empedocles of Agrigentum (490—430 B.C.) builds up the universe from the four elements *fire, earth, air* and *water* by
the action of the outside activating principles *love* and *hate*;
Anaxagoras assumes an almost infinite number of elements,
considering as specific all substances which by repeated
division yield parts having the properties of the whole; the
atomistic school of philosophy, with which are associated the
names of Leucippus (sixth century B.C.), Democritus (*circ.*
460—360 B.C.) and Epicurus (341—270 B.C.) and the tenets of
which have been handed down to us in the poem *De Rerum
Natura* of Lucretius (98—55 B.C.), made unchangeable and
indivisible 'atoms,' different in shape, size and arrangement,
the building stones in the structure of the different substances
met with in nature. The 4-element hypothesis of Empedocles
was accepted by Plato (427—344 B.C.), who considers fire, air,
earth and water as the material used in the creation of the
body of the Universe; and it passed into the natural philosophy of Aristotle, under whose mighty name it reigned
supreme, dominating science all through the Middle Ages
and right up to the end of the seventeenth century and
even now met with in popular thought and speech[1]. In the
Aristotelian doctrine, matter of one kind is associated with
qualities, the different conjunctions of which give rise to the
different substances; for some reason not very apparent, the
essential properties of which this fundamental matter acts as
carrier are hot and its contrary cold, wet and its contrary dry;
and the possible combinations of these qualities in sets of two
give rise to the four elements, earth and fire, air and water.

---

[1] "Vier Elemente, innig gesellt,
  Bilden das Leben, bauen die Welt." (Schiller.)
  "Elements four, united at core,
  All life enfold, the world uphold."

The Aristotelian Four Elements
and their relation to
The Paracelsian Three Principles

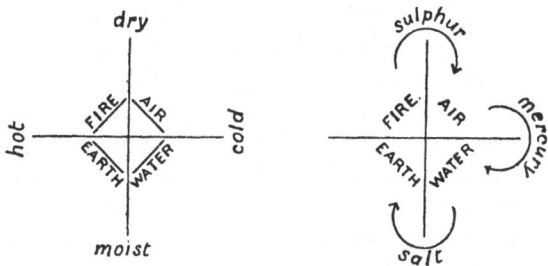

Fig. 27.

Of each element there can be different kinds, due to different degrees of the same properties; moreover, owing to the abstraction of certain qualities and the substitution of others, the elements can change one into another. Great as was the respect for the Aristotelian view of the nature of elements and the ultimate constitution of matter, the recognition of its inadequacy to explain in practice all the observed diversities in matter led to the grafting on to it of the doctrine of the three principles which, initiated by the Arab Geber in the eighth century, found so strong a champion in Paracelsus (1493—1541) that it came to be known by his name. For the determining properties, dry and wet, hot and cold, were substituted others which experience had revealed as characteristic of different classes of substances. Of the three principles postulated, mercury was taken as representative of the specifically metallic properties (ductility, fusibility and lustre); sulphur stood for combustibility, and salt for permanence under the action of heat. What the exact relation between the Aristotelian and the Paracelsian doctrines was, is difficult to decide; but it is certain that a good many writers considered the three principles as the more proximate components of matter which had themselves been evolved from the ultimate constituents, the four elements (see fig. 27).

Revolt against these doctrines, which had become common

towards the end of the sixteenth century, and with which
are associated the great names of Van Helmont (1577—
1644), Bacon (1561—1626), Gassendi (1592—1655), Descartes
(1596—1650), was led to victory by Robert Boyle (1627—
1691). In the *Sceptical Chymist* (1661)[1] both the Aristotelian and the Paracelsian doctrines are refuted by appeal
to experiment, by the demonstration that the results of
analyses and syntheses are not in agreement with them,
that the so-called 'elements' cannot always be produced
from all substances and that they may themselves be
resolved into simpler things still. But Boyle's work is not
only destructive ; its constructive part consists in the introduction into the science of that clear and definite conception
of 'element' which survives in its original form, and in the
course of more than two centuries has had nothing added
to it or taken from it, and which, since its basis is purely
empirical, will no doubt continue to adapt itself to the requirements of a growing science.

I mean by elements certain primitive and simple, or perfectly unmingled
bodies; which not being made of any other bodies or of one another, are the
ingredients of which all those called perfectly mixt bodies are immediately
compounded and into which they are ultimately resolved.

According to this view, the purely empirical nature of
which cannot be too much insisted upon, our so-called elements are not substances which *cannot* be decomposed, but
merely substances which so far *have not* been decomposed.

I should apprehend there are a considerable number of what may
properly be called *elementary* principles, which can never be metamor-
phosed one into another by any power we can control. We ought,
however, to avail ourselves of every means to reduce the number of bodies
or principles of this appearance as much as possible ; and, after all, we may
not know what elements are absolutely undecomposable and what are
refractory, because we do not apply the proper means for their reduction.
                                                (Dalton, 1810.)

In the views that I have ventured to develop, neither oxygen, chlorine
or fluorine are asserted to be elements ; it is only asserted that, as yet, they
have not been decomposed. (Sir Humphry Davy, 1814.)

Moreover at the present day we believe that in the case
of the majority of the substances termed elements we deal
with the *same* stage in the complexity of matter, though this

[1] Reprinted lately in *Everyman's Library* (Dent).

need not be, and probably is not, the simplest agglomeration of matter. The justification for this view is based on the observation of common regularities in these substances, such as the constancy of atomic heat[1] (Newth, *Inorganic Chemistry,* 1907, Part I, VI, p. 46; *Study of Chemical Composition*, ch. XIV), and the variation in properties as a periodic function of the atomic weight (Newth, Part I, XII ; *Study of Chemical Composition*, ch. XVI).

V. Criteria for recognising a substance as complex or elementary.

This then being our conception of 'element,' we must next consider that aspect of the problem which from our point of view is of greatest interest, namely, the experimental criteria for recognising and classing a substance as complex or simple (elementary). The first point calling for attention is that *one* experiment may be sufficient for the recognition of the complex nature of a substance, but that a large number of experiments are required before a substance can be considered as elementary, and that even then from its very nature the classification retains something of the provisional. The quotation given above, in which Sir Humphry Davy summarises the results of his researches on chlorine, brings out this point most admirably. Clearly it will not be possible to comprise in a series of students' experiments anything that would even so much as illustrate the process of recognising the elementary nature of a substance, because to do this would require a very great amount of work, the repetition of many experiments. Take the case of iron: the fact that heating it *in vacuo* or in air, or that the action of dilute sulphuric acid on it, does not

---

[1] "Imagine the simplest bodies, probably as yet unknown to us, the true chemical elements, forming a horizontal spreading layer, and piled above them the simpler, and then the more complicated compounds; the universal validity of Dulong and Petit's law would include the proof that all elements at present assumed lay in the same layer, and that chemistry in recognising hydrogen, oxygen, sulphur, chlorine and the different metals as undecomposable bodies, had penetrated to the same depth in that field of enquiry, and had found at the same depth the limit to its penetration." (Hermann Kopp, 1865.)

produce anything weighing less, anything simpler than the original iron, is of course a quite insufficient basis for a generalisation as to the elementary nature of iron; on the other hand, the *one* experiment on the heating of potassium chlorate proves this substance to be complex.

VI. Classical researches in this province.

A few words remain to be said about classical researches dealing with the problem under consideration. There are the cases of:

1. The isolation of elements from substances the complex nature of which had been suspected or actually accepted from analogies with other substances known to be complex.

2. The proof of the elementary nature of substances either erroneously considered as complex or newly discovered and requiring to be classified.

1. In 1809 the world of science was thrilled by the achievement of Sir Humphry Davy, who in the laboratory of the then recently founded Royal Institution, by the aid of an electric current produced by a huge galvanic pile, had obtained from potash and soda metals named by him *potassium* and *sodium*. This achievement corroborated the view first enunciated by Lavoisier concerning the complex nature of the parent substances potash and soda, a view based on analogy of properties between these substances and other bases known to be metallic oxides. Again from its similarity to hydrochloric acid, there was no doubt whatever that the acid obtained from fluor-spar by the action of sulphuric acid was a compound of hydrogen and an element closely related to chlorine ; the recognition of this went back to the early part of the nineteenth century, but it was not until 1886 that the French chemist Moissan overcame the great experimental difficulties encountered and succeeded in isolating the element fluorine. Similarly when we are told about the discovery in 1898 by Monsieur and Madame Curie of the new element radium, this is strictly speaking not a correct statement of

the facts. What was found was a substance obviously complex as shown by its similarity to other chlorides, and recognised as containing a new element because of its new properties. This element, radium, was however not isolated by Mme Curie and Debierne till 1910.

The general characteristic of the above-mentioned three achievements in the field of analysis is the devising of means for the decomposition of the complex substance under conditions such that the element shall not have the opportunity to form another complex substance. Thus in the case of the isolation of fluorine, the difficulties encountered were due to the fact that this element is chemically the most active of all, and that the isolation had to be effected by the electrolysis of substances free from traces of moisture—liquid hydrogen fluoride made conductive by the addition of some potassium fluoride—in vessels made of fluor-spar and platinum, that is, material not attacked by fluorine. It was a case of finding suitable conditions for the performing of *one* experiment.

2. In marked contrast to this are the cases in which through a comprehensive study of a substance and its derivatives by means of work involving a great number of experiments, proof is adduced of the elementary nature of a substance either newly discovered and requiring to be classified, or of one hitherto erroneously considered as complex.

In 1811 Courtois discovered in the ash of seaweed a new substance named by him iodine. Gay-Lussac undertook the investigation of this new substance, and in 1814 published his results in the form of a monograph which still represents the ideal of what such work should be. A very large number of experiments, qualitative and quantitative, were performed with the iodine and its derivatives such as the hydride (hydriodic acid), the oxide (iodic acid), the compounds with nitrogen, phosphorus and the various metals, moreover the arrangement of the huge amount of material and the lucidity with which it was presented have certainly never been surpassed, probably never equalled. The final inference was:

If we again pass in review the experiments which I have described in this paper, it will be easy to convince ourselves that not one single one amongst them justifies us in considering iodine as complex, and least of all in considering it as containing oxygen. (*Researches on Iodine.*)

Gay-Lussac's name is associated with another great achievement of the same nature. It was the work done by him in conjunction with Thénard (1809) that really established the elementary nature of the substance named oxymuriatic acid, a name changed to chlorine by Sir Humphry Davy, who repeated many of Gay-Lussac's and Thénard's experiments, confirming their results, and made a few original ones of a similar nature.

Like Gay-Lussac and Thénard, he tried all the means then known for the abstraction of the oxygen present in a complex substance, but without success. Even charcoal, the great affinity of which for oxygen is used in the production of metals from their oxides, produced no effect. (*Ante*, p. 78.)

[Davy's] chief service lay in his consistent advocacy of the view that since oxygen could not be obtained from it, [the so-called oxymuriatic acid] ought to be regarded as a simple or undecomposed substance.

Some authors continue to write and speak with scepticism on the subject, and demand stronger evidence of chlorine being undecompounded. These evidences it is impossible to give. It has resisted all attempts at decomposition. In this respect, it agrees with gold, and silver, and hydrogen, and oxygen. Persons may doubt, whether these are elementary bodies; but it is not philosophical to doubt, whether they have not been resolved into other forms of matter.

The chemists in the middle of the last century had an idea, that all inflammable bodies contained phlogiston or hydrogen. It was the glory of Lavoisier to lay the foundations of a sound logic in chemistry, by showing that the existence of this principle, or of other principles, should not be assumed where they could not be detected. (Davy, *The Elementary Nature of Chlorine*, Alembic Club Reprints, No. 9.)

CHAPTER III

CLASSIFICATION OF COMPLEX SUBSTANCES INTO MIXTURES AND COMPOUNDS

I. Recognition of the characteristics of the two types of complex substances.

In the preceding chapter the changes in weight accompanying the transformation of one kind of matter into another were shown to lie at the basis of the division of all substances into the two groups *complex* and *elementary*, a classification which is continued in the further differentiation of the complex into mixtures and compounds. Complex substances may as regards their structure be either homógeneous (*e.g.* calcium nitrate, potassium chlorate, silver chloride, sulphuric acid solution) or heterogeneous (*e.g.* granite, milk in which the cream is separating out, a badly made mayonnaise sauce, turbid river water). It has been stated in an earlier chapter (p. 52) that the chemist selects for separate consideration the homogeneous constituents of heterogeneous agglomerations, but this of course does not do away with the desirability, or rather the necessity, of including heterogeneous substances within the scope of work, the object of which is the experimental study of complex substances leading to that highest type of chemical classification in which properties are correlated with composition. Hence for heterogeneous agglomerations no less than for homogeneous complex substances the course taken by chemical science has been to investigate: (i) the means available for the resolution into homogeneous constituents, (ii) the relation between the properties of the complex structure and those of its constituents. A comparison of the results obtained under each of these counts, for heterogeneous and homogeneous complex substances respectively, supplies the criteria for the recognition of different kinds of complexity. As regards homogeneous complex substances, these are of two kinds, comprising those substances whose properties, qualitative and quantitative, are identically the same for different specimens, and solutions, whose quantitative properties vary but do so as a continuous function of the composition (*ante*, p. 80). It will simplify matters if at this stage we omit solutions from the scope of our consideration, except when to do so would create a wrong impression or produce looseness in definition.

1. **Method of obtaining the data used in the subdivision of complex substances.**

We may either work synthetically, bringing together and mixing intimately substances of known properties, and after having subjected them to conditions favourable to the occurrence of chemical change, investigate the properties of the resulting mass. Or we may work analytically, starting with substances found in nature or prepared artificially, discover by trial the means for resolving them into their proximate constituents, and compare the properties of the original complex substance with those of its constituents. Let us consider each of these methods separately.

(*a*) *Synthesis.*

Experiment XI. The comparative study of

(i)   iron,

(ii) sulphur,

(iii) iron and sulphur intimately mixed,

(iv) iron and sulphur intimately mixed and heated.

This is the stock example inevitably met with by every student of chemistry at a very early stage ; but it is an example so telling, so obviously superior to any other, that it would be foolish to discard its use, to attempt originality at the cost of simplicity.

Use the iron in the form of powdery reduced iron, the sulphur in the form of very finely powdered roll sulphur sifted through muslin. For (iii) use the two solids in the ratio of 3 parts by weight of iron to 2 parts by weight of sulphur and grind together, mixing as intimately as possible.

For (iv) heat about 10 grams of the mixture under conditions which minimise the access of air, using either a crucible covered with a well-fitting lid, or a glass tube suitably drawn out (fig. 28); since the removal of the black spongy mass formed generally necessitates the breaking of the containing vessel, considerations of economy are in favour of a test-tube, all the more as this is an opportunity for disposing of those broken at the top ; and there is the further advantage that a test-tube allows what happens inside to be seen. In any case it is sufficient to start the process by applying heat in one place, at the bottom of the test-tube or crucible, after which the reaction will proceed spontaneously, the whole mass becoming red-hot, owing to the heat liberated during combination. The process is in this respect analogous to the burning of coal in the oxygen of the air, which also must be initiated by the application of outside heat, after which it proceeds spontaneously ; but with this difference, that in the case of the burning coal, since the resulting and one of the reacting substances are gaseous, flame is produced. The table given below summarises the tests which experience has shown to yield results pertinent to the special point investigated.

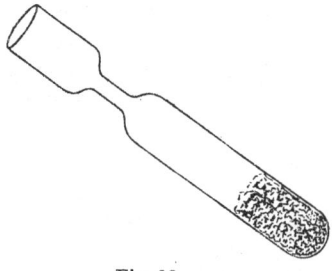

Fig. 28.

The Comparative Study of

	(i) Iron	(ii) Sulphur	(iii) Iron and sulphur mixed	(iv) Iron and sulphur mixed and then heated
Appearance : Colour State of aggregation Structure when examined with lens				
Effect of heat				
Action of magnet				
Effect of suspension in water				
Action of dilute hydrochloric acid, and testing of any gas evolved with regard to: Inflammability Action on lead acetate paper, and smell				
Action of carbon bisulphide				
Inference				

8

Experiment XII. The comparative study of

(i) oxygen,

(ii) sulphurous anhydride,

(iii) a mixture of these two gases,

(iv) the same mixture after it has been passed over heated platinised asbestos and collected in a vessel surrounded by a freezing mixture.

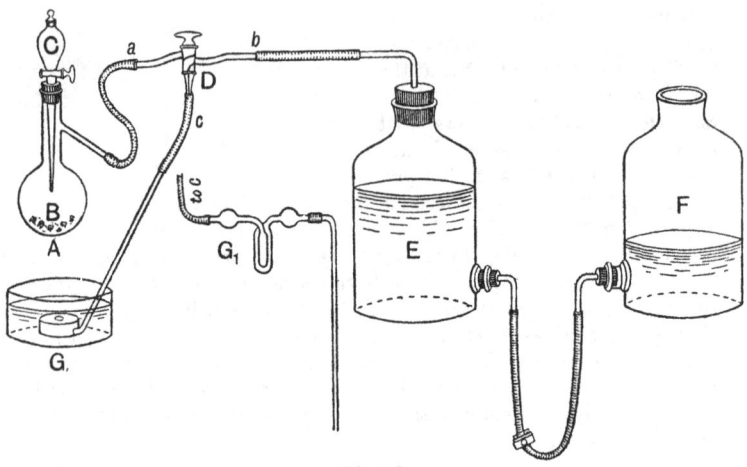

Fig. 29.

A. Round-bottomed distilling flask of about 300 c.c. capacity, holding B, some dry sodium peroxide and fitted with the dropping funnel C, which contains water.

D. Three-way stop-cock connected through a with A, through b with the aspirators E and F, and through c with the delivery arrangement G.

E, F. Aspirators for the storing and delivery of the oxygen; each holding 2—4 litres.

G, G_1. Suitable arrangements for delivering the oxygen: either a pneumatic trough for water or mercury, or a small gas wash-bottle ending in a long tube for downward displacement.

Though by comparison with the preceding one this experiment makes greater demand on apparatus and experimental skill, it should not be difficult to demonstrate the point at issue; and it has the additional advantage and interest of supplying incidentally an illustration of the 'contact method' for the production of sulphuric acid[1], a process the commercial importance of which has grown so rapidly of late as to make it a serious rival of the lead chamber process.

[1] Newth, *Inorganic Chemistry*, 1907, p. 432; Molinari, *General and Industrial Inorganic Chemistry*, pp. 276 *et seq.*

(i) Use the commercial oxygen from a cylinder, or specially prepare some by the action of water on sodium peroxide, and store for use in an aspirator as follows (see fig. 29).

With D placed so as to connect b and c, fill E and all the connecting tubes up to D with water by raising F. Place D so as to connect a and c, and start the evolution of oxygen by letting water from C fall drop by drop on to the peroxide B. The pneumatic trough can be used to indicate the expulsion of the air, the volume of gas coming off being actually measured by filling a cylinder a certain number of times and roughly estimating its volume relatively to that of A. When a volume of gas at least 3 times as great as that of A has escaped, we may assume that the oxygen is free from air.

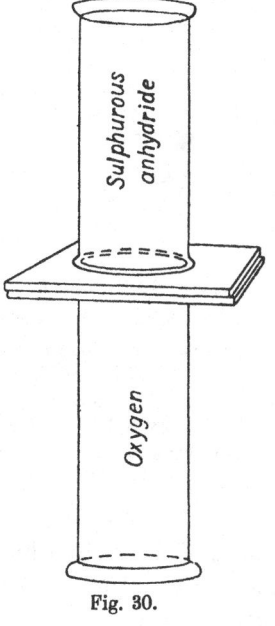

Place D so as to connect a and b; put F, which at the beginning should contain very little water, on the same or a lower level than E, when the oxygen evolved in A will drive over the water from E into F.

(ii) Procure the gas from one of the commercial syphons, or prepare it as required by the action of dilute sulphuric acid on crystals of sodium bisulphite (see fig. 23, p. 98); collect either over mercury or by downward displacement of air.

(iii) Make a mixture of the two gases, either by successive collection over mercury or by bringing together two cylinders covered with plates, withdrawing the plates, and effecting mixture by inverting the cylinders several times (fig. 30). The

Fig. 30.

student should arrange the sequence of the tests enumerated in the table (p. 117) so as to use the same sample for as many different purposes as possible, and thus to minimise the number of fresh lots of mixture wanted.

(iv) The apparatus depicted in fig. 31 will be found suitable for the purpose of subjecting a mixture of oxygen and sulphurous anhydride to the action of heated platinised asbestos. Accepting the result of investigations by which it has been proved that in such reaction as occurs the platinum and asbestos remain unchanged, we can proceed thus. With the condensing vessel detached, and adjusting the rates at which the two gases bubble through the concentrated sulphuric acid so as to make them about equal, no effect is observed as long as the platinised asbestos remains cold ; as soon as heat is applied, white fumes issue at the open end of D, indicating the occurrence of a chemical change ; E is then attached, and the reaction allowed to go on until an amount of white crystals has been deposited presumably sufficient for the purposes of the tests to be done with them. It is desirable to attach to *E* a tube to lead uncondensed fumes into a vessel containing water.

Apparatus for the passing of a mixture of oxygen and sulphurous anhydride over heated platinised asbestos.

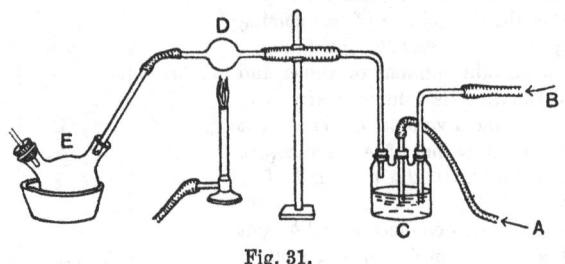

Fig. 31.

A.   Supply tube for oxygen.

B.   Supply tube for sulphurous anhydride.

C.   Concentrated sulphuric acid for drying the gases and showing the rate at which they are delivered.

D.   Bulb holding the platinised asbestos, which can be heated when the flow of the two gases has been started.

E.   Receiver surrounded by freezing mixture for the condensation of the substance produced in the bulb.

The table given on p. 117 summarises the tests and the results of these by which the difference of the relation of (i) and (ii) to (iii), and (i) and (ii) to (iv) respectively is brought out.

(*b*)   *Analysis.*   Examples are supplied by

(1)   A number of technical processes:

(i)   River water is freed from suspended solids by filtration through beds of sand; (ii) magnetic iron ore is picked out by the action of powerful magnets from a previously crushed mass of rock ; (iii) native sulphur is separated from earthy admixture by applying heat and letting the molten sulphur flow away; (iv) by the slow cooling of melted beef or mutton suet and removal of the constituents of higher melting point which solidify first, a product is obtained which is used in the manufacture of margarine; (v) gold dust is separated from

The comparative study of certain properties of (i) oxygen, (ii) sulphurous anhydride, (iii) a mixture of these gases, (iv) the same mixture after it has been passed over heated platinised asbestos and then collected in a vessel surrounded by a freezing mixture.

	(i) Oxygen	(ii) Sulphurous acid anhydride	(iii) A mixture of these two gases	(iv) The mixture after passage over the heated platinised asbestos
State of aggregation and colour	Colourless gas	Colourless gas	Colourless gas	Solid; white silky needles fuming in air
Smell	None	Pungent, like that of burning sulphur	Pungent, like that of burning sulphur	Fumes very irritating
Combustion-supporting power	Glowing splint rekindled, burning with bright flame	Puts out brightly burning splint	Substances burning brightly continue to burn for a time	Fumes extinguish flame
Action of water	Very slightly soluble	Very easily soluble	Partially soluble—reduction in volume equal to volume of (ii) present in the mixture	Dissolves with hissing sound—action very violent
Action of dilute solution of alkali	No effect	Absorption with destruction of smell	Same absorption as above—remaining gas pure oxygen	Same as with water
Action on moist blue litmus paper	No effect	Change to red	Change to red	Change to red if dilute solution used, otherwise charring
Action on solution of potassium chromate or bichromate	No effect	Reduction to green chromic salt	Reduction to green chromic salt	No effect unless white crystals are used in large excess
Inference			Shows additively the properties of (i) and (ii)	Has acquired new properties

river sand by 'washing,' (vi) wheat from chaff by 'winnowing'; the last two are mechanical processes in which the lighter constituents are removed and the heavier retained.

(2) The comparative study of air and nitrous oxide. These are two gaseous substances which by suitable means can be resolved into the same constituents, viz. oxygen and nitrogen, but which exhibit marked typical differences under each of the headings enumerated at the beginning of this chapter.

(i) The resolution of air into its constituents is a process which technologically has become of very great importance because of the industrial utilisation of each of the gases; oxygen being used extensively, together with hydrogen or acetylene in the melting and welding of metals, for the production of ozone on the commercial scale, in the preparation of synthetic nitrates and nitric acid, and for medical purposes; nitrogen being used in large quantities in the production of synthetic nitrates, nitrites, cyanides, ammonium compounds, etc., which are needed in the manufacture of fertilizers, explosives, fine chemicals and drugs, for obtaining gold from its ores, etc. The student is recommended to read the account of the various methods employed, in Thorpe's *Dictionary of Applied Chemistry* or Molinari's *Chemistry*, Vol. I (Inorganic), paying special attention to the division of these methods into two classes: one which makes use of physical processes, *e.g.* differences in volatility or in solubility; and another which is chemical in nature, and in which substances possessed of great affinity for it are made to combine with the oxygen, leaving behind the nitrogen, the oxygen being in some cases subsequently regained. In the evaporation of liquid air the nitrogen boils off first (B.P. $-195 \cdot 5°$ C.), leaving behind a liquid which as it becomes richer and richer in oxygen (B.P. $-182 \cdot 5°$ C.) assumes more and more of a blue tint; cocoanut charcoal cooled by means of a bath of liquid air absorbs the oxygen much more readily than the nitrogen, and on removal from the bath gives up this oxygen nearly pure; copper heated in air completely absorbs the oxygen (*ante*, p. 90), and so does ferrous hydrate in the cold ; Lavoisier, in his complete quantitative synthesis and analysis of mercury calx, regained the oxygen from the calx by heating it to a temperature higher than that at which it had been formed (*ante*, p. 3) ; in the Brin process air is led over heated barium oxide, which combines with the oxygen to form a higher oxide, from which the oxygen is again liberated by raising the temperature or lowering the pressure.

Careful comparative study of air and the two constituents into which it can be resolved by the means above enumerated shows that in every respect, physical and chemical, the properties of the complex substance are additively those of the components: that air retains and exhibits the properties of both the oxygen and the nitrogen in proportion to the amount of each present. Thus the power of air to support combustion is intermediate between that of nitrogen, which extinguishes a flame, and oxygen, which re-kindles a glowing splint. Again, oxygen being nearly twice as soluble in water as is nitrogen, when air is acted upon by water each of these gases is absorbed according to its own coefficient of solubility[1], as shown by the

[1] The advantage of making the investigation of a quantitative property the final criterion in the interpretation of chemical phenomena is well shown by the results of a determination of the solubility in water of the various gases under consideration. The calculation of the solubility of a gaseous mixture of known composition is made on the basis of: (1) Henry's Law (1803), which gives the relation between solubility and pressure, and which asserts that the volume of gas absorbed by a liquid is directly proportional to the pressure of the gas, and (2) Dalton's Law of Partial Pressures (1805), which gives the relation between the volume composition of a gaseous mixture and the pressure, and which asserts that each gas contributes towards the total pressure an amount equal to the pressure which it would exert if it occupied the whole space.

Solubility in water at 20° (i.e. volume of gas reduced to 0° C. and 760 mm. absorbed by 100 volumes of water when the pressure of the gas itself amounts to 760 mm.) of

Nitrogen	Oxygen	Mixture of nitrogen and oxygen in the volume ratio $0 \cdot 8 : 0 \cdot 2$ (calculated)	Air which contains nitrogen and oxygen in the (approx.) volume ratio $0 \cdot 8 : 0 \cdot 2$	Nitrous oxide which contains nitrogen and oxygen in the volume ratio $0 \cdot 66 : 0 \cdot 33$
$1 \cdot 54$	$3 \cdot 17$	$(1 \cdot 54 \times \cdot 8) + (3 \cdot 17 \times \cdot 2)$ $= 1 \cdot 23 + \cdot 63$ $= 1 \cdot 86$	$1 \cdot 87$	$66 \cdot 5$

fact that the gas expelled from water by boiling is richer in oxygen than is air[1].

(ii) The same line of investigation followed out with the gas called nitrous oxide (discovered by Priestley in 1772) brings to light the following facts: Nitrous oxide is an excellent supporter of combustion, in this respect not very inferior to oxygen itself, and when carbon or phosphorus or certain metals (*e.g.* potassium) are burned in it, the result is the production of: (*a*) a substance identical with that obtained by burning in air or in oxygen, and (*β*) a gas apparently identical[2] with the nitrogen left behind when air is used as the supporter of combustion. But though thus shown to be made up of the same constituents as air, viz. oxygen and nitrogen, nitrous oxide is fundamentally different from air: when liquefied (B.P. $-89 \cdot 8°$ C.) or solidified (M.P. $-103 \cdot 7°$ C.) the whole mass changes back to gas or liquid at one definite temperature, and when absorbed or dissolved it is acted upon as a whole (*ante*, p. 119); again, whilst on the one hand it has properties which are not possessed by either of its constituents, *e.g.* produces unconsciousness when inhaled[3], on the other hand certain of the properties of the constituents are absent, *e.g.* the colourless gas nitric oxide, when mixed with oxygen,

[1] The carbon dioxide, which represents by far the greater part of the volume, must of course be removed first. For a description of an experiment on the analysis of the gas expelled from water by boiling, see Clowes and Coleman, *Quantitative Analysis*, 10th ed., 1914, p. 469.

[2] The residual gas obtained after substances have been burnt in air contains argon.

[3] This property of nitrous oxide was discovered by Humphry Davy in 1799. He clearly foresaw the use of the gas as an anaesthetic:

"As nitrous oxide in its extensive operation appears capable of destroying physical pain, it may probably be used with advantage during surgical operations in which no great effusion of blood takes place,"

but his account of his sensations after inhaling it is more concerned with its psychological than its physiological effect:

"By degrees...I lost all connection with external things; trains of vivid visible images rapidly passed through my mind, and were connected with words in such a manner, as to produce perceptions perfectly novel. I existed in a world of newly connected and newly modified ideas: I theorised, I imagined, I made discoveries."

On another occasion, with a view to seeing whether the gas would increase his stock of the divine afflatus, the poet-philosopher deliberately inhaled some on the banks of the Avon by moonlight! The resulting effusion was, however, hardly of a quality to encourage him to repeat the experiment.

either pure or in the diluted form of air, gives a red gas easily soluble in water, but no such effect is produced on mixing it with nitrous oxide.

(3) Production of silver from (i) argentiferous lead, (ii) silver chloride.

(i) Argentiferous lead is produced either in the reduction to metal of argentiferous galena (native lead sulphide) or in the extraction of silver from silver ores by means of smelting with lead, in which process the silver is reduced to the metallic state and then dissolves in the excess of lead forming an alloy. Various methods are employed industrially for the 'desilverisation' of the lead alloy, which in this case means production of the pure silver. Amongst these are some which make use of the fact that in the alloy each of the metals retains its own characteristic properties. Thus lead, which has a lower melting point than silver, crystallises out first from the molten mass, and if removed leaves behind a liquid richer in silver; a repetition of the process produces greater and greater concentration of the silver; and if a physical property of the resulting alloys, such as, say, the density, be measured, it will be found to vary with the composition. In fact the alloy, the complex substance containing lead and silver, exhibits the properties of the constituents : *e.g.* metallic lustre, resistance to the action of hydrochloric acid, solubility in nitric acid, etc., etc. and is in this respect fundamentally different from a compound of silver such as silver chloride.

(ii) Silver chloride. This substance has been the subject of a qualitative and quantitative experiment in the preceding chapter; the results there described show the relation of its properties to those of its constituents, and that they are entirely different from the properties of metallic silver, which can be isolated from it only by a chemical change.

2. Use of the data above obtained to subdivide complex substances into mixtures and compounds.

The results of the two methods of investigation dealt with lead to the same inference concerning the nature of complex substances, namely the existence of two classes, according to: (i) the relation between the properties of the complex and those of its constituents, *i.e.* whether the properties of the complex substance are or are not additively those of the constituents; (ii) the means applicable for the resolution of the complex substance into its constituents, *i.e.* whether each of the differences in the properties of the constituents can or cannot be made a means of separation. Mixtures and compounds are the names distinctive of the two classes of complex substances thus differentiated.

Mixtures have the characteristic properties which differentiate the heterogeneous agglomeration of iron and sulphur from the homogeneous iron sulphide ; which differentiate the homogeneous gas air from the homogeneous gas nitrous oxide. We found that : (i) the properties of the complex substances iron + sulphur, and air are additively those of the components. (ii) The differences in the properties of the components— whether physical or chemical—supply a means of resolution. The colour of the mixture of iron and sulphur is intermediate between the grey of the iron and the yellow of the sulphur; the value found experimentally for the solubility of air in water agrees with that calculated from the solubilities of oxygen and nitrogen and the composition of air. The iron and sulphur can be separated by the action of carbon bisulphide, which dissolves the sulphur, leaving the iron unchanged ; liquefied air can be separated into oxygen (B.P. $-182 \cdot 5°$ C.) and nitrogen (B.P. $-195 \cdot 5°$ C.) by fractional distillation.

Compounds are always homogeneous, and taking iron sulphide and nitrous oxide as types, we found that the properties investigated bear no definite relation to those of the constituents. The value of any quantitative property is not that obtained by calculation from the values for the components and the amounts of these present; a greater or lesser number of the qualitative properties possessed by one or other of the components are not exhibited by the compounds, and *vice versa*. The separation into components cannot be effected by utilising differences in physical properties. The production of iron sulphide (density $= 4 \cdot 84$)

from iron (density $= 7.6$) and sulphur (density $= 2.0$) in the ratio of 56 of iron to 32 of sulphur is, as the above data show, accompanied by an increase of volume[1]; the data and the calculation given on p. 119 show that the solubility of nitrous oxide in water is not only not intermediate between those of its constituents nitrogen and oxygen, but so much greater than either as to be of an altogether different order of magnitude; dilute hydrochloric acid, which has no effect on sulphur, dissolves iron with the evolution of hydrogen, and iron sulphide with the evolution of sulphuretted hydrogen; whilst free oxygen combines with nitric oxide to form a red gas easily soluble in water, the compound nitrous oxide has no such effect. Moreover, iron sulphide cannot be separated into its components by the action of carbon bisulphide, which dissolves free sulphur, nor can oxygen and nitrogen be obtained from nitrous oxide (B.P. $-90°$C.) by fractional distillation.

II. Method followed and practical work required in recognising an unknown substance as a mixture or a compound.

The next point calling for discussion is that of the deductive application of the principle of classification arrived at by the inductive process above outlined, that is, consideration of the procedure followed in practice, when confronted with the problem of finding out whether a substance under investigation is a mixture or a compound, which of course involves antecedent or concurrent classification as complex. Two groups of cases may arise, corresponding in principle to the synthesis and analysis division of the preceding section.

A. The constituents are known, and the problem for solution consists in finding out whether these substances, possessed of known or easily ascertainable properties, when put under definite conditions react, losing their own specific

[1] Calculated density $= \dfrac{(7.6 \times 56) + (2.0 \times 32)}{56 + 32} = 5.5.$

properties and producing a new substance possessed of new properties; or whether they remain in a state of mixture which shows additively the properties of the constituents. Obviously this problem is identical with that dealt with at length in Chapter I concerning the occurrence or non-occurrence of a chemical change, and everything said there is equally applicable here. The question as to whether substances when brought together under certain conditions form a mixture or a compound is merely a verbal variant of "has a chemical change occurred or not?" The following may serve as illustrations of the principle followed in investigating the result of an attempted synthesis.

1. *Recognition of an organic compound as saturated or unsaturated by means of bromine.* In the case of investigations in organic chemistry, a common test consists in the addition of a few drops of bromine water, when the disappearance of the characteristic colour of the bromine—without the evolution of fumes of hydrobromic acid—indicates the occurrence of a chemical change consisting in the absorption of the bromine due to the unsaturated nature of the organic compound. Thus the hydrocarbon methane does not decolourise bromine water, but ethylene or acetylene does.

Experiment XIII. The action of bromine on different types of organic compounds. If on adding a few drops of bromine water and shaking, the colour of the bromine persists, the inference is that the bromine is merely mixed with the original substance, and that no chemical change has occurred.

(1) (i) Fill a medium-sized gas cylinder by upward displacement with coal gas, close with a glass plate and shake with a few drops of bromine water ; or make some acetylene by the action of water on a scrap of calcium carbide (fig. 32): put the carbide into a crucible (or small beaker or tiny crystallising dish), lower by means of a string or wire into a glass cylinder, drop on some water from a pipette, withdraw the crucible and test with bromine. (ii) Perform the same test with some marsh gas prepared by heating in a hard glass test-tube a mixture of fused sodium acetate and soda lime (fig. 33).

(2) Make solutions of (i) some stearin, (ii) some olein or some cinnamic acid. To each add a few drops of bromine water and shake vigorously.

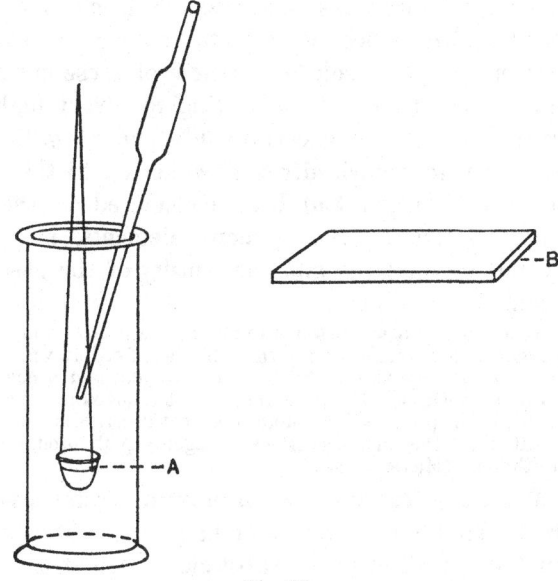

Fig. 32.

A. Crucible containing calcium carbide.
B. Glass plate for closing cylinder.

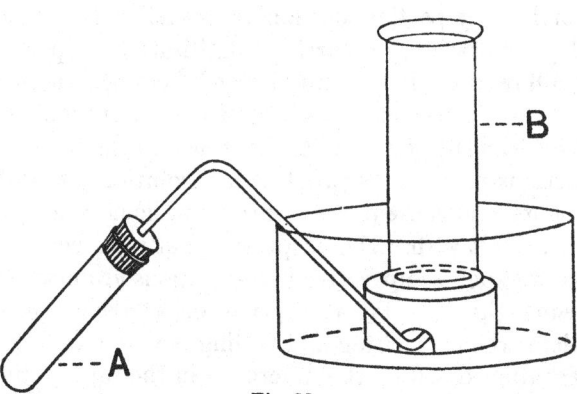

Fig. 33.

A. Hard glass tube containing mixture of fused sodium acetate and soda lime.
B. Cylinder to collect marsh gas.

2. *Attempts to produce compounds of argon and helium.*
Reference has already been made to this subject in Chapter I.
The criterion for the absolute inertness of these substances
was found in the fact that on heating to a very high temperature in the presence of certain substances (*e.g.* titanium,
lithium) known to absorb nitrogen, which up to the time of
the discovery of argon had been considered as the most
chemically inactive of the gaseous elements, no volume
change could be observed and the density of the gases present remained the same.

Titanium as pure as it was possible to obtain it was heated in an atmosphere of argon to the temperature of the softening of ordinary glass; after
30 minutes' heating there was no diminution of volume, and hence no likelihood of any combination. The titanium had not changed in appearance.

Argon heated in presence of lithium does not decrease in volume, and
we know that under the same conditions nitrogen rapidly produces a solid
nitride of lithium. (Moissan, 1895.)

B. The constituents are unknown. Instances of this
class are by far the most frequent in practice; the problem
arises in the 'purification of substances,' a matter of such
great importance in chemical research that the relative and
even the absolute amount of the time given to it exceeds that
devoted to any other process. The work takes the form of
(i) fractionation, *i.e.* division of the substance investigated
into portions by partial solution, or partial solidification, or
partial precipitation, or partial volatilisation, or partial diffusion, followed by (ii) a comparison of the properties of these
portions with each other or with those of the parent substance.
The establishment of a definite difference in the properties of
the fractions is taken as proof of a resolution into different
components, and consequently of the non-homogeneity of the
substance dealt with. The properties that have been used in
practice for such comparative investigations are most varied,
comprising the density, absorption or emission spectrum,
equivalent weight, melting and boiling points, etc.

Separation based on the difference in the rate of diffusion
of the components of a gaseous mixture, followed by a comparison of the densities of the fractions thus obtained, is a
method of great importance in dealing with gases; it was
resorted to by Lord Rayleigh and Sir William Ramsay in the

course of their work[1] which led to the detection of argon, but whilst it lends itself well to the establishment of non-homogeneity, it is not usually suitable for effecting complete separation. In the investigation above referred to, diffusion was effected by means of a number of 'churchwarden' tobacco pipes contained in an outer vessel in which a partial vacuum was maintained. The atmospheric nitrogen was allowed to pass slowly through these porous pipes and was collected in a vessel in which its density could be determined. In two experiments it was found that if the gas were placed under conditions favourable to the removal by diffusion of any lighter constituent, the weight of a certain volume of the residual gas exceeded by 3·7 and 3·3 mg. respectively the weight of an equal volume of the original gas, results which were thus interpreted :

The conclusion seems inevitable that 'atmospheric nitrogen' is a mixture and not a simple body.

Fractional crystallisation is probably the process most widely used in the purification of salts and in the recognition of a substance dealt with as a mixture. The purification of potassium chlorate, *i.e.* the separation of the chlorate from the chloride mixed with it (*ante*, p. 91), is a case in point, and so is the removal of almost pure lead from argentiferous lead, referred to on p. 121. When the quantities of the substances to be separated are not very different from one another, and the solubilities are nearly equal, the process of separation becomes a very laborious one, especially if, as is often the case, it is necessary to obtain each of the constituents in a state of purity. If all that is required is the isolation in the pure state of *one* of the constituents, considerable simplification is possible, since the fractions containing the other constituents may be rejected. Madame Curie's isolation of radium chloride in 1903 is an instance of the second type of case, the fractions rich in barium being rejected, without the necessity of freeing them absolutely from any radium salt. The presence of a substance possessed of the then new property of radio-activity having been demonstrated, its

---

[1] See *Nature*, **51**, 1895, p. 347.

separation from the barium with which it was associated, the
amount of barium being vastly in excess, was effected by some
thousands of fractional crystallisations of the chlorides, first
from water and then from dilute hydrochloric acid.

Beautiful crystals form...and the supernatant saturated solution is easily
decanted. If part of this solution be evaporated to dryness, the chloride
obtained is found to be about 5 times less active than that which has crystallised out. The chloride is thus divided into two portions, A and B—portion
A being more active than portion B. The operation is now repeated with
each of the chlorides A and B, and in each case two new portions are obtained.
When the crystallisation is finished, the less active fraction of chloride A is
added to the more active fraction of chloride B, these two having approximately the same activity. Thus there are now three portions to undergo
afresh the same treatment.

The number of portions is not allowed to increase indefinitely. The
activity of the most soluble portion diminishes as the number increases.
When its activity becomes inconsiderable, it is withdrawn from the fractionation.

An instance of a case of the first type is afforded by the very
extended fractional crystallisation resorted to in the separation from each other of cerium, didymium and lanthanum,
elements belonging to the class termed 'rare earths.' This
is a series comprising elements which resemble each other
very closely, the properties of one merging by extremely small
differences into those of the next ; they are of interest both
on the theoretical and the practical side. The theoretical
interest is associated with their position in the periodic table,
and their practical importance dates from the year 1884, when
a patent was taken out, the first of that long series representing the development of the incandescent gas mantle, the
efficiency of which depends on the presence in varying amounts
of different oxides of the rare earths. The actual separation
of the different members of the rare earth series from each
other is effected by fractional precipitation, or fractional
decomposition of the nitrates, or fractional crystallisation.
For fractional precipitation, solution of a base such as ammonia or potash is added to the solution containing the rare
earths in an amount insufficient to produce complete precipitation. The two portions obtained, the precipitate and the
mother liquor, contain the rare earth compounds in different
amounts. The composition of the precipitate will depend
directly, and hence that of the mother liquor indirectly, on

what is called the *strength* of the oxides, *i.e.* the relative tendency the elements possess to form the hydrates, as measured by the relative amount of the ammonia or potash which each secures in the competition for a share in an amount insufficient to precipitate them all. The precipitate is then dissolved, and spectroscopic examination of each of the two portions is used to decide whether there has or has not been a resolution of the original substance. If this is found to have occurred, and the production of pure substances is required, the process is repeated until after further fractionation the spectrum is unaltered. In the method of separation by fractional crystallisation, the salts which have been used most successfully are the nitrate and oxalate, but others such as the formate, bromate, acetoacetate have also been employed; it is a case of finding by trial the special acid whose salts with the various rare earths show the greatest differences in solubility.

Experiment XIV. The differential decomposition of nitrates by means of heat may be used to show that a silver coin is a mixture of at least two metals.

Dissolve a threepenny-piece in dilute nitric acid, evaporate the blue solution in a small porcelain dish ; when dry place on a deep sand bath and heat carefully until no more fumes appear. Extract a small portion of the black mass obtained with hot water ; if the solution still shows any trace of blue colour, heat the bulk some time longer. When by successive testing there has been demonstrated the complete change into an insoluble black residue of the constituent to which was due the blue colour of the solution, extract the whole mass with water, and evaporate the extract nearly to dryness. A white residue will be left, identical in its properties with silver nitrate, the substance obtained by the solution of pure silver in nitric acid.

Experiment XV. Fractional distillation and fractional crystallisation.

(1) Fractional distillation.

Any of the illustrative experiments suggested in the different text books on organic chemistry will serve the purpose. The stock examples deal with: the separation of alcohol (B.P. 78·3° C.) from admixed water (B.P. 100° C.); the purification of crude benzene; the separation into its constituents of a mixture of benzene (B.P. 80·4° C.) and xylene (B.P. 140° C.). (Price and Twiss, *A course of Practical Organic Chemistry*, 2nd ed., 1914, pp. 86 *et seq.* ; J. B. Cohen, *Practical Organic Chemistry for Advanced Students*, 1904, pp. 120 *et seq.* ; Sudborough and James, *Practical Organic Chemistry*, pp. 8 *et seq.*)

The salient point of the technique is to distil at a regular speed, collecting in separate vessels the distillate coming over at different ranges of temperature, between every 2°, or 3°, or 5°, or 10°, as may appear suitable from the original volume of liquid and the quantities collected. A label is affixed to each fraction indicating the range of temperature within which

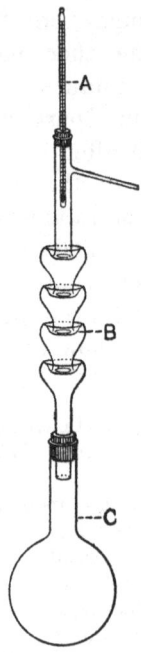

Fig. 34.

Apparatus for Fractional Distillation.

A. Thermometer.
B. 4 pear fractionating column.
C. Flask containing liquid to be fractionally distilled.

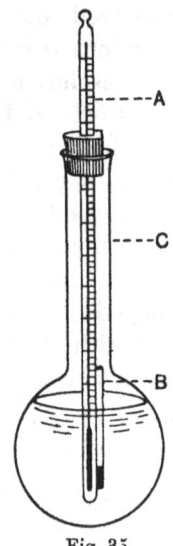

Fig. 35.

Melting point apparatus.

A. Thermometer.
B. Very small thin glass tube containing the substance whose melting point is to be determined.
C. Flask containing a liquid which can be heated without boiling to well above the melting point it is required to determine (conc. sulphuric acid is generally used).

it has been collected. The first of these fractions is then re-distilled until the higher limit of temperature at which it previously came over is reached, when the next is added to the residue in the distilling flask; the distillation is continued till the higher limit of temperature at which the second fraction came over is reached, when the third portion is added, and so on. If by the addition of any portion to a residue a fraction is obtained boiling at a lower temperature than that reached in the process of fractionating the previous portion, it is added to the fraction to which from its boiling point it belongs.

(2) Fractional crystallisation.

Illustrative experiment. The partial separation of the acids of which common soap is a mixture of the sodium salts.

Dissolve about 15 grams of ordinary soap in hot water; add hydrochloric acid, thereby liberating a mixture of so-called 'fatty acids,' which rises to the surface as an oil and solidifies on cooling; the hydrochloric acid used should contain about one part of concentrated acid to one of water, and should be added until present in excess, as shown by the action on the liquid of litmus paper (fatty acids have no action on litmus under these conditions). Collect the solid that has separated out, drain and dry it and find its melting point (fig. 35). Recrystallise from hot alcohol (methylated spirit will do) and repeat the process until the solid separating out has a constant melting point.

By the process of fractional crystallisation outlined above, the stearic acid has been separated from the other two constituents of the mixture in a pure condition, as is indicated by the constancy of the melting point.

Acids liberated from soap by the action of hydrochloric acid (or sulphuric acid) in excess on the aqueous solution of soap.

Summary of results.

		Melting point of		
	Solid thus obtained	Solid after crystallisation from hot alcohol		
		1st crystallisation	2nd crystallisation	3rd crystallisation
Student A	45° – 52°	65°	67°	67°
„ B	59°	65°	68·5°	68·5°
„ C	below 50°	62°	69°	69°

The substance rising to the top when soap is treated with hydrochloric acid is a mixture of three acids, viz.: Palmitic acid, M.P. 62°C.; Stearic acid, M.P. 69°C.; Oleic acid, M.P. 14°C.

When the object of the recrystallisation is a separation as complete as possible, with retention of each of the components, the process must of course be modified so as to take within the compass of the work the solute contained in the mother liquor, which is done by evaporation in stages and collection of the successive fractions of solid separating out. The scheme reproduced from Stewart, *Recent Advances in Physical and Inorganic Chemistry*, makes this clear.

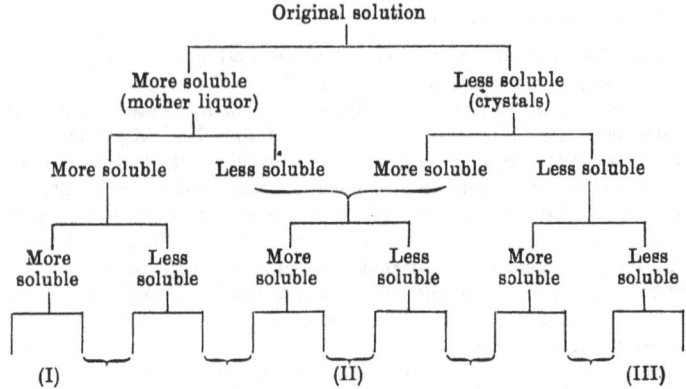

The brackets indicate that two fractions have been mixed together before recrystallisation.

A glance at the scheme will show that we are accumulating the more soluble salts at (I), the salts of intermediate solubility at (II), and the least soluble salts at (III).

III. Classical researches undertaken with the object of recognising whether a special substance is a mixture or a pure substance.

Rayleigh and Ramsay's resolution of atmospheric nitrogen by atmolysis is a case in point that has already been dealt with, pp. 126—127.

1. The resolution of didymium.

Didymium, a substance closely allied to cerium and lanthanum, from which it was first separated in 1841, was in 1885 shown by Auer von Welsbach to be itself a mixture. By long-continued fractional crystallisation of the double salt ammonium-didymium nitrate he obtained two portions which he named respectively praseodymium and neodymium, and which differed not only in the colour of their salts (green and red) but also in their spectra.   Later workers claimed to have found indications of the presence in didymium, not of two but of as many as eight different elements ; these results have however not been confirmed, Demarcay for one having claimed the establishment of the homogeneity of neodymium.

I subjected a considerable quantity of didymium oxide, purified from lanthanum and cerium, to a series of fractionations.

...a double nitrate [was obtained] which...no longer showed a trace of the principal praseodymium band.   This nitrate was the starting-point for a fresh series of fractionations....,I thus obtained more than 20 successive portions which from all points of view exhibited the most complete identity. ...These portions have the same absorption spectrum with the same intensity.   Not the slightest difference can be detected in any of the bands.

Hence I conclude that...neodymium is a simple substance and not a mixture of elements.   (Demarçay, 1898.)

This research is of special interest because it supplies as good an example as any we can get of the relationship between pure science and applied science, the absurdity of considering from the outset 'utility' as the important factor in the direction given to study and research, the impossibility of foretelling the scope of the practical application of any addition to knowledge.

To the laboratory of Bunsen there came Karl Auer von Welsbach in the spirit of an unalloyed philosopher, eager to solve some problems about the group of chemical elements that seemed of all the most remote from any daily human needs.   He noticed the remarkable glow of the mixed oxides when a flame impinged on them, and so begat the mantle....Where would the gas industries of to-day be without this windfall from the tree of scientific knowledge ?   (A. Smithells, 1911.)

2.   The discovery of neon, krypton and xenon.

When in 1895 Lord Rayleigh and Professor Ramsay had shown the presence in air of about one per cent. of a hitherto unknown constituent, to which they gave the name *argon*, the question soon arose as to what the complexity of air really was, and whether the addition of a third constituent to the two that had been known for more than a century completed the tale or not.   It was to be expected from the known values for the densities of air, oxygen and nitrogen, that the amounts present of any unknown gas could be but small, and also that in the light of the newly acquired knowledge concerning argon, it would not be possible to effect a separation by making use of differences in chemical affinities, as is done when the oxygen is first withdrawn by the action of hot copper, and then the nitrogen by the action of hot magnesium.

In 1896 the gas occluded in the mineral cleveite and

liberated from it by boiling with sulphuric acid was investigated by Ramsay, who found its spectrum lines—chief amongst which was one in the yellow, near the sodium line but not coincident with it—to be different from those of argon or of any other known gas, but identical with those which had on several occasions been observed in the chromosphere of the sun during a total eclipse, and which had been assigned to an element terrestrially unknown and named *helium* (ήλιος = the sun). The density of the new gas referred to hydrogen was found to be 2, and it resembled argon closely in that it was chemically quite inactive and monatomic[1]. Here, then, were two elements belonging to a new class, for which a place had to be found in the Periodic Table[2]; a new column was therefore inserted, called Group 0, thus:

Group 0	Group 1
He	Li
4	7
?	Na
	23
A	K
40	39
?	Rb
	85
?	Cs
	133

But this addition involved gaps for an element intermediate between helium and argon, and for others of higher atomic weight. A systematic search was therefore made (*a*) for a chemically inert gas whose relative density $_{H=1}$ should be intermediate between 2 (helium) and 20 (argon), and (*b*) for others, also chemically inactive, of density greater than 20. The search has been thus described:

> There is a proverb about looking for a needle in a haystack; modern science, with the aid of suitable magnetic appliances, would, if the reward were sufficient, make short work of that proverbial needle. But here is a supposed unknown gas, endowed no doubt with negative properties, and the whole world to find it in.

[1] *Study of Chemical Composition*, p. 498; Newth, *Inorganic Chemistry*, 1907, pp. 267 *et seq.*; Ostwald, *Principles of Inorganic Chemistry*, 4th ed., 1914, p. 463.

[2] *Study of Chemical Composition*, pp. 497 *et seq.*; Newth, *Ibid.*, Part I, ch. XII.

The substances first investigated were minerals, mineral waters and meteorites ; helium obtained from various minerals was tested for its homogeneity by diffusion, but without much result.

...the systematic diffusion of argon, however, gave a faint indication of where to seek for the missing element, for the density of the more rapidly diffusing portion was 19·93, while that of the portion which diffused more slowly was 20·01.

But when the proportion of the constituent wanted is low, and when moreover the absolute quantities available are small, separation by diffusion cannot be made complete. The successful issue of Ramsay's search was due to the fact that just then the technical means for the production of liquid air and liquid hydrogen in considerable quantity were perfected, and it was thus made possible to apply the method of fractionation to the liquids obtained by cooling to a very low temperature the gases whose complexity was to be investigated.

...on distilling liquid argon, the first portions of gas to boil off were found to be lighter than argon, and on allowing liquid air to boil slowly away, heavier gases came off at the last.

...It was easy to recognise the gases by help of the spectroscope, for the light gas, to which we gave the name *neon*, or 'the new one,' when electrically excited emits a brilliant flame-coloured light; and one of the heavy gases, which we called *krypton*, or 'the hidden one,' is characterised by two brilliant lines, one in the yellow and one in the green part of the spectrum. The third gas, named *xenon*, or 'the stranger,' gives out a greenish-blue light and is remarkable for a very complex spectrum, in which blue lines are conspicuous.

The neon obtained from the atmosphere showed lines belonging to the spectrum of helium, the presence of which gas in the air was thus demonstrated. On fractionating liquefied air the more volatile portion, after removal of the nitrogen and traces of oxygen present, consisted of helium, neon and argon.

...the purification from argon...a lengthy process...was accomplished by repeated distillation, the lighter portions being always collected separately from the heavier portions and again distilled by themselves.

The final separation of the neon and helium, neither of which could be condensed by liquid air, was effected by fractional distillation at the temperature of liquid hydrogen.

...while neon is practically non-volatile when cooled by liquid hydrogen remaining in the state of solid or liquid,...the gaseous helium could be pumped away from the non-gaseous neon, and the latter was obtained in a pure state.

The other two new gases, krypton and xenon, which, as is shown by their production from the *residue* of liquid air, are the most easily condensable constitutents, were separated by another long and tedious process.

...when most of the argon had been removed, the residue solidified; but while it was possible to remove the krypton by pumping,...very little xenon accompanied it, for at that temperature xenon is hardly at all volatile.

The properties of the gases thus separated were then investigated ; the chief of the results obtained are set out in the following table :

	Density$_{H=1}$	Boiling Point	Melting Point	Amount in c.c. present in 1 litre of air
Helium	1·98	−268·5°	—	0·001
Neon	10·02	−246°	−252°	0·01
Nitrogen	13·90	−195·5°	−213°	769·500
Oxygen	15·88	−182·5°	−223°	206·594
Argon	19·78	−186·1°	−187·9°	9·37
Krypton	41·12	−151·7°	−169°	0·001
Xenon	64·59	−109·1°	−140°	0·00005

Thus within an extremely short time there was demonstrated the presence in air of five new elements. Is there any likelihood of this number falling short of the reality ? The extremely small quantities of krypton and xenon found show that if gases of greater density should be present, this can only be to an almost infinitesimal amount, and a special research has actually revealed the limiting value.

219 c.c. of xenon were obtained from 119 tons of air.

The xenon, after being sparked with oxygen, was fractionated....Both the ordinary and the spark spectra of [the] two last fractions [⅓ c.c. each] were photographed and compared on the same plate with the spectra given by the first fractionation at the opposite end of the series. The lines obtained from the 3 samples of gas were identical....

The results obtained indicate that if a stable element heavier than xenon does exist in the atmosphere, the volume present compared to that of xenon is extremely small. As at least 10 °/₀ of a new gas could probably have been detected in the spectra examined, therefore 0·03 c.c. of such a gas could have been detected. As this existed in 100 tons of air it represents 1 part in 2,560,000,000 by volume.   (Moore, 1908.)

3.   Establishment of the homogeneous nature of helium.

The homogeneous nature of helium was proved by its discoverer by the use of the method of diffusion. Working with the gas obtained from cleveite,

...it was found possible to separate helium into two portions of different rates of diffusion, and consequently of different density, by this means. The limits of separation, however, were not very great. On the one hand, we obtained gas of a density close on 2·0; on the other, a sample of density 2·4 or thereabouts....The density of the lightest portion of these gases was 1·98; and after other 15 diffusions, the density of the lightest portion had not decreased....This substance, forming by far the larger part of the whole amount of the gas, must, in the present state of our knowledge, be regarded as pure helium.

4. Establishment of the homogeneous nature of tellurium.

Between 1869 and 1871 the German Lothar Meyer and the Russian Mendeléeff propounded a principle of the classification of the elements which makes every property of the elements and their compounds a periodic function of their atomic weights, and which under the name of the Periodic System or Periodic Table has become the chemist's most effective tool in the correlation of old and the systematic addition of new knowledge[1]. Considering the enormous scope of the system, and the wonderful harmony revealed by its application to the classification of almost innumerable facts, it is strange that the number of anomalies, which no doubt represent cases as yet imperfectly understood, should be so small. One such anomaly is furnished by the relative position of iodine and tellurium. From its relationships with sulphur and selenium, tellurium should come before iodine ; but its atomic weight is greater than that of iodine.

N	O	F
14·01	16	19
P	S	Cl
31·04	32·07	35·46
As	Se	Br
74·96	79·2	79·92
Sb	Te	I
120·2	**127·5**	**126·92**

Here is an anomaly which calls for an explanation, and the obvious one would seem to be that the substance named tellurium is not simple, but a mixture of at least two elements, one of atomic weight about 128, and another of atomic weight about 125, the true tellurium. During the last 20 years

---

[1] Newth, *Inorganic Chemistry*, 1907, Part I, pp. 112 *et seq.*; *Study of Chemical Composition*, pp. 454 *et seq.*; Ostwald, *Principles of Inorganic Chemistry*, 4th ed., 1914, pp. 794 *et seq.*

numerous attempts have been made to decide the question as to the elementary nature of tellurium. Some investigators approached the subject with a bias so strong that it restricted their outlook, and made them see in the result of some constant error a proof of the resolution they were so eager to effect; others who, whatever their theoretical leanings, allowed themselves to be led by fact, and who had the skill, ingenuity and honesty to detect and eliminate sources of constant error in their own and other people's work, obtained results which lent no support to the hypothesis of the complex nature of tellurium. The last research of great importance on this subject was published in 1907 by Brereton Baker and Bennett[1]. Six different fractionating processes were employed, including: fractional crystallisation of telluric acid; fractional precipitation of the tetrachloride by water (principle the same as that dealt with *ante*, p. 128); fractional electrolysis of the tetrachloride and tetrabromide; fractional distillation of the element; fractional distillation of the tetrachloride; and fractional distillation of the dioxide. The various fractions were then compared by ascertaining the percentage composition of a certain compound prepared from them. One method of comparative measurement consisted in the indirect determination of the percentage of oxygen in tellurium dioxide (corresponding to sulphurous anhydride), and the concordance of the results obtained is shown by the following example:

Material obtained by the fractional crystallisation of telluric acid.

Fraction	Percentage of oxygen in the dioxide
1	20·055
2	20 034
3	20·046
4	20·053
5	20·062
6	20·044

In no single instance was there found any indication of resolution, and the final results were summed up in the statement:

No difference could be distinguished in the atomic weight of tellurium when:

1. Telluric acid, obtained by two distinct methods, was fractionally crystallised.

2. Barium tellurate was dissolved in water.

[1] *J.C.S.* 1907, **91**, 1849.

3. Tellurium was fractionally distilled.

4. Tellurium tetrachloride was fractionally distilled.

5. Tellurium dioxide was fractionally distilled.

6. Tellurium was converted into the hydride, which was fractionally decomposed.

7. Tellurium tetrabromide and tetrachloride were submitted to fractional electrolysis.

8. Tellurium tetrachloride was fractionally precipitated by water.

The atomic weight of tellurium is 127·60.

IV. Recent developments in the technique of the purification of gases.

Till comparatively recently, quantitative work of the highest degree of accuracy could not be carried out with gases, because of the extreme difficulty of producing pure material, *i.e.* removing from the gaseous mixture the last traces of the constituent present in small amount which constitutes the impurity. Diffusion, valuable as the process is for establishing complexity (*ante*, p. 126), does not take us sufficiently far ; differences in solubility have in some cases been used to great advantage; the most important of these is the solubility of hydrogen in palladium, by means of which the hydrogen, which is eventually driven off from the palladium by the action of heat, is completely freed from oxygen and from nitrogen. But whilst this represents an isolated case due to the specific relationship between hydrogen and palladium, a method of general applicability has become available through the facilities for the production of low temperatures (*ante*, p. 17), and the consequent extension of the methods of fractional crystallisation and fractional distillation to the solids and liquids obtained by the condensation of the gases. The purification of the hydrogen chloride used by Gray and Burt in their determination of the density of this gas has been referred to already (*ante*, p. 22). Nitric and nitrous oxides supply another example. In a subsequent chapter (*post*, p. 316) it will be shown how the accurate determination of the equivalent weight of an element, *i.e.* of the amount combining with or replacing the standard amount of the standard element, should whenever possible be made by *direct* reference to the

standard.   With oxygen = 16·000 as standard, this involves
the accurate determination of the composition of a compound
which contains the element under investigation and oxygen ;
but though nitrogen forms several oxides, until recently none
of them had been prepared in a state of sufficient purity,
and hence only indirect methods were available, involving
the use of a number of antecedent data.   By utilisation of
the low temperatures now obtainable by the use of liquid air
it has been found possible to overcome this difficulty; in
quick succession there have been published determinations
of the composition (and density) of nitrous and nitric oxides,
and the results obtained for the atomic weight of nitrogen
by this method rank in degree of accuracy with the best
results obtained for solids.   Gray's work[1] dealt with nitric
oxide[2].   The gas obtained in the usual manner is contaminated
with nitrous oxide (B.P.−89·8°C.,M.P.−103·7°C.; for properties
see *ante*, p. 120) and nitrogen (B.P. −195·5°C., M.P. −213°C.).
The purification was accomplished by systematic fractionation
of the liquid and solid produced by cooling with liquid air.
The various steps in the process, and the results of the control
determinations of the densities of the various fractions are

[1] Gray, 'Atomic Weight of Nitrogen,' *J.C.S.*, **89**, 1905, pp. 1601 *et seq.*

[2] This substance, which is a reduction product of nitric and nitrous acid, is
obtained by a variety of methods, all of them based on the common principle of
the action of a reducing agent present initially or produced in the reaction, *e.g.*:
(i) Action of copper on nitric acid.   No hydrogen is evolved, as is usual in the
action of metals on acids, but some of the nitric acid is reduced to lower oxides
and even nitrogen, the relative amount of these depending on the temperature
and the concentration of the acid.   If the strength of the acid is suitably chosen,
nitric oxide is the chief product, but the red fumes which always accompany
the reaction show the simultaneous formation of higher oxides, which are removed by bubbling the gas through water, or, better still, through dilute potash
(or soda) ; the removal of the nitrous oxide and nitrogen is a much more difficult
matter.   (ii) Action of ferrous sulphate and concentrated sulphuric acid on
potassium nitrate.   (iii) Action of potassium ferrocyanide and acetic acid on
potassium nitrite.   Nitric oxide is a colourless gas, liquefied at −154° C. and
760 mm. and solidified at −167° C., soluble in water, easily absorbed by cold
ferrous sulphate with the formation of a dark compound (principle of dark ring
test for nitrates), from which it is liberated again by the action of heat; it unites
with uncombined oxygen to form red fumes of the higher oxides, which are
soluble in water or alkali (use in air analysis).   The composition, qualitative
and quantitative, can be demonstrated by analysis, the action of certain heated
metals (*e.g.* iron, nickel) producing the oxide of the metal and nitrogen.

contained in the following table, taken from Stewart, *Recent Advances in Physical and Inorganic Chemistry* :

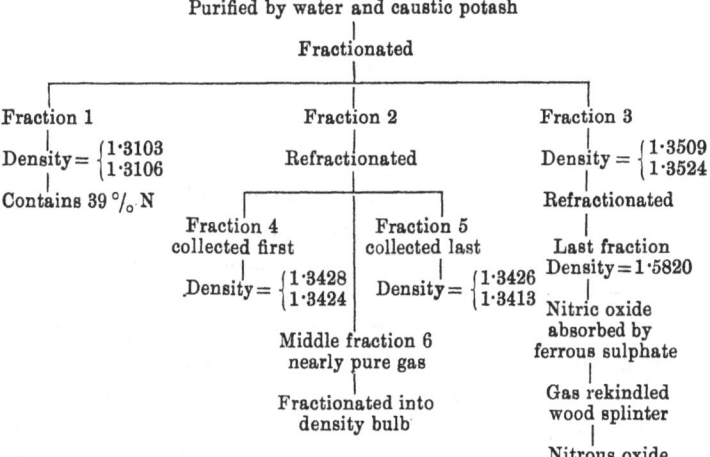

Crude gas evolved from potass. nitrite, potass. ferrocyanide and acetic acid
|
Purified by water and caustic potash
|
Fractionated
|

Fraction 1
Density = $\begin{cases} 1\cdot3103 \\ 1\cdot3106 \end{cases}$
Contains 39 % N

Fraction 2
Refractionated

Fraction 4 collected first
Density = $\begin{cases} 1\cdot3428 \\ 1\cdot3424 \end{cases}$

Fraction 5 collected last
Density = $\begin{cases} 1\cdot3426 \\ 1\cdot3413 \end{cases}$

Middle fraction 6 nearly pure gas

Fractionated into density bulb

Fraction 3
Density = $\begin{cases} 1\cdot3509 \\ 1\cdot3524 \end{cases}$
Refractionated
Last fraction
Density = 1·5820
Nitric oxide absorbed by ferrous sulphate
Gas rekindled wood splinter
Nitrous oxide

Berthelot has given the very suitable and attractive title of *La Révolution Chimique* to the book in which he tells the story of Lavoisier's work on the nature of combustion, work which includes and involves the elucidation of the composition of air. The separation of gaseous mixtures into their constituents by the new means which low temperature research has recently made available may also be described as a revolution, and having in this chapter shown how by this method the composition of air has been ascertained, it will not appear out of place to devote the next chapter to the consideration of how the same problem has been solved by the use of chemical methods.

CHAPTER IV

THE PART WHICH AIR PLAYS IN COMBUSTION

Various considerations point to the advisability of dealing with the phenomena of combustion at this stage. First and foremost, the investigation is complementary to that already made concerning the effect produced on certain substances (*e.g.* magnesium, iron and copper, see *ante*, p. 86) when heated in air. Secondly, it furnishes an admirable example of the method followed in the solution of a difficult chemical problem, viz.: the logical sequence in a sustained argument which correlates and interprets certain definite aspects of a number of apparently very diverse phenomena. And finally, there is the great historical interest due to the fact that, though the phenomena of combustion had for centuries attracted the attention of the philosopher, the alchemist and the chemist, there have been many doublings back on the road of progress, many excursions up blind alleys, and that when success was finally achieved, this was due to an unusual combination of circumstances, viz. the concentrated attention of the genius of one man and a chance discovery made by another.

I. Historical.

The whole story has been told in isolated form[1]; special sections are devoted to it in every history of chemistry[2]; and portions of it are referred to in almost every text book[3].

---

[1] Berthelot, *La Révolution Chimique.*

[2] Thorpe, *Essays in Historical Chemistry* ; E. von Meyer, *History of Chemistry* ; Thorpe, *History of Chemistry*, vol. I ; Lowry, *Historical Introduction to Chemistry*, ch. III.

[3] Newth, *Inorganic Chemistry*, 1907, ch. XI, pp. 321, 322; Perkin and Lean, *Introduction to Chemistry and Physics*, vol. II, chs. II, IV.

Here no more will be attempted than an indication of the chief events and features in the elucidation of one of the most important phenomena of nature, a quest which, beginning in dim antiquity, did not reach its goal until the end of the eighteenth century. It may conduce to clearness in the presentation of the history of combustion prior to Lavoisier if we keep separate the discovery and the establishment of facts from the attempts made to explain these facts.

1. The observation goes back to early times that whilst certain substances such as wood, sulphur, lead, copper are permanently changed when heated, they are in this respect different from others, such as sand and gold. The Arabian alchemist Geber (second half of the eighth century) knew that 'calcination' requires the presence of air; this fact, as well as the correlated fact that the product of heating the metal in air weighs more than the original metal, was known in the Middle Ages, but—probably owing to the absence of a satisfactory explanation—it was disregarded and repeatedly rediscovered. Metals had been obtained from earliest times by heating their calxes with charcoal, itself a very combustible substance (*e.g.* production of tin, Pliny's *cassiteron*, from the cassiterite, tinstone, found in Cornwall). Stahl (*post*, p. 147) showed in 1718 that the supposed analogy between the burning of sulphur and the calcination of a metal like lead was justified, because, just as the metal could be regained directly by heating the calx with charcoal, so could the sulphur be regained from sulphuric acid by a process which, though more indirect, involved as an important part the heating with charcoal. The increase in weight of lead and tin on calcination was made the subject of a special research by Rey in 1630. Boyle showed in 1672 that sulphur could not burn in an evacuated vessel. Mayow in 1674 remarked that a part only of the air supports combustion, because a candle burning under a bell-jar goes out while there is still gas present. Cavendish in 1766, by the action of dilute oil of vitriol on iron, produced a gas which he called 'inflammable air,' our hydrogen. Following on this discovery it was shown that when metals dissolve in acids with evolution of hydrogen, the

substance remaining in solution is identical with that obtained when the calx of the same metal dissolves in the same acid without any production of hydrogen ; moreover, heating the dry calx in a stream of 'inflammable air' regenerates the metal. Rutherford in 1772 showed that when a candle burns in a limited supply of air, combustion is limited, that 'fixed air,' our carbon dioxide, is formed and that after the removal of this 'fixed air' by potash, a gas remains behind which has mephitic properties. Rutherford's discovery was confirmed by Priestley, who named this residual gas 'phlogisticated air,' a name changed later on to 'nitrogen,' because of the gas being contained in nitre.

In 1777 the Swedish chemist Scheele published a memoir entitled *A Chemical Treatise on Air and Fire*[1], in which he gives an account of a long series of experiments on the part played by air in combustion, experiments which confirmed facts previously known and added a number of new ones. It is recognised that air has weight, and that its presence is essential for the production of flame. Experiments are described which indicate that air is composed of two constituents, one acting as the supporter of combustion and occupying between a third and a fourth part of the total volume, and one which will not allow a candle to burn or the smallest spark to glow ; this latter gas is proved to be less dense than the original air. The withdrawal of the part of the air which supports combustion was accomplished by Scheele in so great a variety of ways that his enumeration falls but little short of that contained in a modern treatise : he uses alkaline persulphides (liver of sulphur), nitric oxide, ferrous hydrate, moist iron filings, cuprous chloride ; the burning of phosphorus, of hydrogen, of a candle. By the distillation of 'fuming acid of nitre,' nitric acid, he obtains a gas—he called it an 'air,' as air was then a generic term for all gases—which he finds to be a better supporter of combustion than common air, and which differs from common air in the further fact that it is completely absorbed in all those cases when the

[1] Portions of this are reproduced in the Alembic Club Reprints, No. 8, under the title of *The Discovery of Oxygen*, Part 2.

use of common air produces a residue having between two-thirds and three-quarters the volume of the original air. This same gas he obtains also by the action of concentrated sulphuric acid on pyrolusite, as well as by heating a large number of substances, including the nitrate of magnesium or of mercury, common nitre, gold or silver calx, red precipitate or *mercurius precipitatus per se*. He notes the fact that, whilst *mercurius precipitatus per se* is formed at one temperature by the slow calcination of the metal in air, it is completely decomposed at another higher temperature. He determines the density of the new gas, called by him 'fire air,' and finds that an amount occupying the volume of 20 ounces of water is almost two grains heavier than the same bulk of common air. Mixing one part of 'fire air' with three parts of that kind of air "in which fire would not burn," he obtains an air in every respect like common air. He notices the difference in solubility in water of the two constituents of air—the fire air being withdrawn and the vitiated air left behind. Indeed, Scheele's chemical treatise on air and fire is a monograph on air astounding in its completeness.

Though Scheele's results were not published until 1777, it is generally accepted that the actual discovery of his 'fire air'—our oxygen—was made in 1772. Thus he anticipated Priestley who, on August 1st, 1774, by concentrating the sun's rays with a burning glass on red precipitate obtained the identical gas, which he called 'dephlogisticated air[1].' Concerning this discovery, Priestley tells us how he believed that "more is owing to what we call chance than to any proper design or preconceived theory in this business," and how when he proceeded to investigate the startling properties of the new substance, he was continually surprised by something unexpected happening. Priestley succeeded in preparing the same gas in a variety of other ways and investigated its properties, but the exposition of his results shows these to have been far inferior in completeness to those obtained by Scheele.

---

[1] Alembic Club Reprints, No. 7.

2. Turning now to the hypotheses concerning the nature
of combustion, we find that these also go back to the earliest
times. According to Greek philosophy, substances burned
because of the common presence in them of 'fire matter';
the burning of wood and the effect of heating metals such as
lead and copper in air were recognised as being of the same
nature, *cineres* (ashes) being the common name given to the
residue left behind in either case   Later the burnt metal
and the process of obtaining it came to be designated by
'calx' and 'calcination,' names derived from the supposed
analogy with the preparation by the action of heat of *calx*
(lime) from chalk.   During the Middle Ages combustibility,
i.e. the burning of wood and sulphur or charcoal, and the
power possessed by metals to form calxes, were supposed to
be due to the common presence of 'sulphur,' the principle of
combustibility (*ante*, p. 105).   Rey in 1630 repeats the experiments which prove that lead and tin increase in weight on
calcination.   He accounts for this occurrence, as well as for
the necessity of the presence of air and for the limited
combustion-supporting power of the air, by a hypothesis
according to which the air as a whole becomes fixed, *i.e.*
attached to the minute particles of the *calx*, the actual formation of which latter remains unexplained ; but he makes
no attempt to test this hypothesis deductively.   Boyle in *The
Sceptical Chymist*, 1661, tries to account for the increase in
weight on calcination by endowing heat with weight, and
assuming that the fiery corpuscles, entering by the pores of
the glass, associate themselves with the particles of the burning
substance.   Hooke in 1665 explains combustion by the escape
from the burning substance of a sulphurous principle taken
up by the air, the capacity of which in this respect is however
limited.   Mayow in 1674 promulgates the view that atmospheric air is made up of two kinds of aerial particles, one of
which is also present in nitre, supports combustion and the
respiration of animals, whilst the other kind is corrosive and
extinguishes fire.   This view is supported by experimental
evidence, of which that given above (p. 143) is perhaps the
most important and interesting part.   The speculations of Rey,

Hooke, Mayow and Boyle represent attempts to explain limited special aspects of the phenomena observed to occur in combustion. A hypothesis intended to deal with the phenomena in their entirety, a hypothesis so comprehensive that it developed into a theory, was originated by Becher in 1669 and was developed by Stahl (1660—1734). According to this, all combustible substances and all metals are complex, containing the substance resulting from the burning or calcination (which in the case of metals is the calx) combined with the principle of combustibility named *phlogiston*; burning and calcination consist in the driving out of the phlogiston by heat, whereby simpler substances are produced; heating the metal calx with charcoal, a substance specially rich in phlogiston, regenerates the original complex metal by the restitution to it of the phlogiston taken from it in calcination; similarly the addition of phlogiston in the heating of a sulphate with charcoal allows of the indirect regeneration of the sulphur.

The phlogistic hypothesis reigned supreme and unchallenged for about a century; to its original achievement of classing together and giving a simple explanation of such diverse phenomena as the burning of wood, the calcination of metals and the regeneration of metals from their calxes, it added further triumphs by including within its scope respiration and the explanation of the nature of 'inflammable air.' Since the substance produced by the solution in acid of the metal and a metal calx are the same, the gas produced in the first case must be either phlogiston itself or a substance rich in phlogiston, an inference justified by the fact that metal calxes heated in this gas yield the metal. Against these positive achievements had to be set difficulties and anomalies which grew in number and importance as new facts were discovered and old ones studied more closely, especially when this was done from the quantitative point of view. There were the disconcerting facts of the increase of weight on burning, of the necessity for the presence of air and of the limitation to the power of supporting combustion of a given volume of air. To these phenomena known of old were added newly discovered ones, viz.: the production of metallic mercury by heating mercury calx alone, without carbon or other source of phlogiston; the differences observed when substances were burnt in ordinary air and in Priestley's 'dephlogisticated air' respectively, a residual gas (differing in its properties from air) being left in the one case and not in the other. The original promulgators of the hypothesis had not committed themselves concerning the nature and properties of phlogiston, and had left it an open question whether phlogiston was capable of real existence or not, and what was the part played by the air. This left much latitude to the champions of the theory, who continued to fight for it when in the light of fresh discoveries and more severe demands for strict correspondence between hypothesis and fact, its inadequacies and deficiencies were made to stand out more and more prominently.

> Chemists have turned phlogiston into a vague principle, one not rigorously defined and which consequently adapts itself to all the explanations for which it may be required. Sometimes this principle has weight and sometimes it has not; sometimes it is free fire and sometimes it is fire combined with the earthy element; sometimes it passes through the pores of vessels, sometimes these are impervious to it...it is a veritable Proteus, changing in form at each instant. (Lavoisier, 1783.)

Scheele recognised this; he had found that the gas left behind when air was deprived of the part which supports combustion was not only smaller in volume but also less dense than the original air.

> ...it might be believed that the phlogiston united with this air makes it lighter...but since phlogiston is a substance, which always presupposes some weight, I must doubt whether such hypothesis has any foundation.

But Scheele was so imbued with the preconceived belief in the absolute truth of the phlogistic theory that he was prepared to leave the matter at that; it is strange indeed that with all the facts known to him about the production and decomposition of *mercurius precipitatus per se* he should not have realised how superfluous it was to assume the existence of the *principle* phlogiston. He had considered two possible explanations of the cause of the decrease in volume of air which had acted as a supporter of combustion. The first of these was that the air in absorbing the phlogiston had 'lost some elasticity,' *i.e.* had contracted ; this however was not in accordance with the observed loss of weight. The second hypothesis was that the air lost had united with the substances to the interaction with which was due the decrease in volume and weight observed,

but in that case it would necessarily follow that the lost air could be separated again from the materials employed.

Scheele tried to regain such portion of the air as might have been absorbed by the action of liver of sulphur (alkaline polysulphide), but in this he was not successful. Considering what he knew about the preparation of *mercurius precipitatus per se* from mercury and air and what he himself had found out about its decomposition into metallic mercury and 'fire air' when strongly heated, it is almost amazing that he should not have followed up his reasoning and thereby solved the whole problem ; but till his death he remained a believer in phlogiston. It was reserved for another man to achieve the rout of the phlogistic theory, to show that the recognition of the true nature of the occurrence did away with the necessity for assuming the existence of phlogiston or any other hypothetical principle, that it was a case of combinations and decompositions between substances that could be isolated and investigated like any other kind of matter.

3. Lavoisier's work.

In 1772 Lavoisier added to the already recognised analogies between combustion and calcination the proof that phosphorus and sulphur on being burnt absorb a large volume of air[1] and increase in weight just as metals do on calcination. In work published in 1774 but completed two years earlier, he proves by the calcination of tin and lead in hermetically sealed vessels, the total weight of which before and after the reaction is found to be the same, that the increase in weight of the metal during calcination arises neither from fire matter passing through the pores of the vessel (Boyle's hypothesis, *ante,* p. 146), nor from any other matter extraneous to the vessel, but must be due to absorption of part of the air. The

---

[1] This was also proved by Scheele (*ante,* p. 144).

question next pressing for answer obviously concerned the nature of the portion of air absorbed. Priestley's discovery on August 1st, 1774, of a new gas, a better supporter of combustion than air, supplied the necessary clue. In October 1774 Priestley told Lavoisier of his discovery, and in 1775 Lavoisier published a memoir entitled *On the Principle which combines with Metals during Calcination, and which augments their Weight*. He uses mercury calx in the form of *mercurius calcinatus per se* (a substance obtained by the calcination of mercury in air) and red precipitate (obtained by the action of an alkali such as potash or soda on the solution of a mercury salt); he proves that these substances are identical and are a calx, because when heated with charcoal they yield the metal and fixed air, and from them he obtains by heat metallic mercury and a gas which he considers to be the purest portion of the air surrounding us, and to constitute the principle which combines with metals during calcination and which increases their weight. This work must be considered as preliminary to the classical experiment in which by the complete quantitative synthesis and analysis of mercury calx he supplied unassailable proof that calcination consists in the absorption of a definite portion of the air, leaving behind a smaller volume of a gas different in properties from the original air; that the gas absorbed can be completely regained, when its volume is found equal to the observed decrease of volume of the original air (for the actual numerical values see *post*, p. 160); that it is a much better supporter of combustion than ordinary air, and that when mixed with the air remaining behind after the calcination, ordinary air is reproduced. These results, supplemented by those obtained in the measurement of the maximum volume decrease which it is possible to produce in air by the substitution of the more active phosphorus for mercury, settled the point of the composition of air and of the part played in the phenomena of combustion by that constituent whose volume is one-fifth of the total. In 1778, in a memoir entitled *On Combustion*, Lavoisier finally disposes of the need for a special principle of combustibility and enunciates his *views*, or, since there is really nothing hypothetical in the matter, it would be more correct to say his *discovery*. The burnt substance or calx is more complex than the combustible substance or metal; it is formed by the combination of the combustible substance or metal with the gas named by Priestley 'dephlogisticated air,' by Scheele 'fire air,' by Lavoisier himself at first 'eminently pure air' and later 'oxygen'; the increase in weight incidental to the process of burning or calcination is exactly equal to the amount of gas so absorbed; this gas is present in air mixed with the gas termed by Priestley 'phlogisticated air,' now named nitrogen.

The triumph of the new doctrine, though not immediate, could naturally not be long deferred; it was not a case of choosing between rival hypotheses, each with its own advantages and its own distinct limitations and difficulties; it was the substitution of an explanation based entirely on facts verifiable by experiment for a theory which had been of very great advantage in the coordination of a large number of apparently widely diverse phenomena, but which had proved inadequate to its task with the growth of more accurate knowledge.

II. The various phenomena observed in the study of combustion, illustrated by the special case of: (1) phosphorus, (2) iron.

In the preceding section it has been told how early it had been recognised that the change which occurs when a substance such as lead or tin is heated in air is of essentially the same nature as that which takes place when wood or sulphur burn with a flame. Obviously it will be desirable to supply definite evidence, qualitative and quantitative, for the correctness of this view, and it will also help to a more complete understanding of the various aspects of combustion if each of these is illustrated by *two* examples, one in which a so-called 'combustible' substance and another in which a 'metal' is used. Phosphorus is specially suitable as the representative of the first of these classes, and that for the following reasons: it is very easily inflammable; when the reaction has been started it proceeds spontaneously and is completed in a short time ; and finally, the product of combustion is a solid. Iron is chosen because it can be easily obtained in a state of fine division conducive to the occurrence of the change ; because though fairly active (in this respect superior to copper) it is not changed at ordinary temperature (difference from sodium and potassium) ; and finally because at no stage of the reaction is a volatile product formed. The much greater activity of magnesium has proved so attractive that it is often used for demonstrating the essential features of combustion, including the determination of the volume ratio between the original air and the gas remaining after the reaction, for which latter

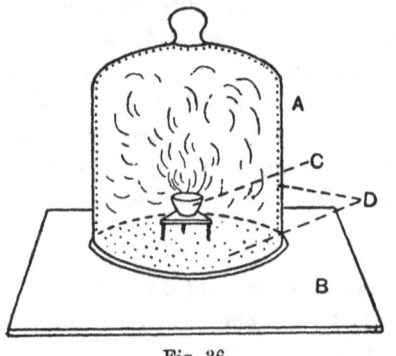

A. Bell-jar of about 6—10 litres capacity, with ground rim.

B. Glass plate, preferably ground.

C. Crucible holding phosphorus, supported on pipe-clay triangle.

D. White deposit of burnt phosphorus forming on the inner surface of the bell-jar and on the glass plate.

Fig. 36.

purpose magnesium is absolutely unsuitable, because owing to its extreme reactivity it forms according to the conditions of the experiment not only the oxide, but also nitride and silicide (see p. 55).

1. **Certain substances when heated in air are permanently changed, giving rise to new kinds of matter.**

The principle and the practice of the work involved have been dealt with fully in Chapter I, where it has been shown what are the criteria for recognising the occurrence of a chemical change, *i.e.* the production of a new kind of matter.

A. Aspirator holding the air which is driven out by water contained in A'.

B₁, B₂, B₃. Wash-bottles containing strong sulphuric acid to dry the air.

C. Reduction tube of hard glass, containing phosphorus in the bulb and asbestos plugs on both sides.

D. Long wide glass tube with cork at bottom on which rests the crucible F, and a doubly perforated cork at the top through which passes the tube E.

E. Long glass tube fitted with cork at top through which passes the reduction tube C. The lower end is nearly flush with the top of the crucible.

F. Crucible for collecting white solid formed in the reaction.

G. Flask for collection of any white fumes not condensed in the crucible.

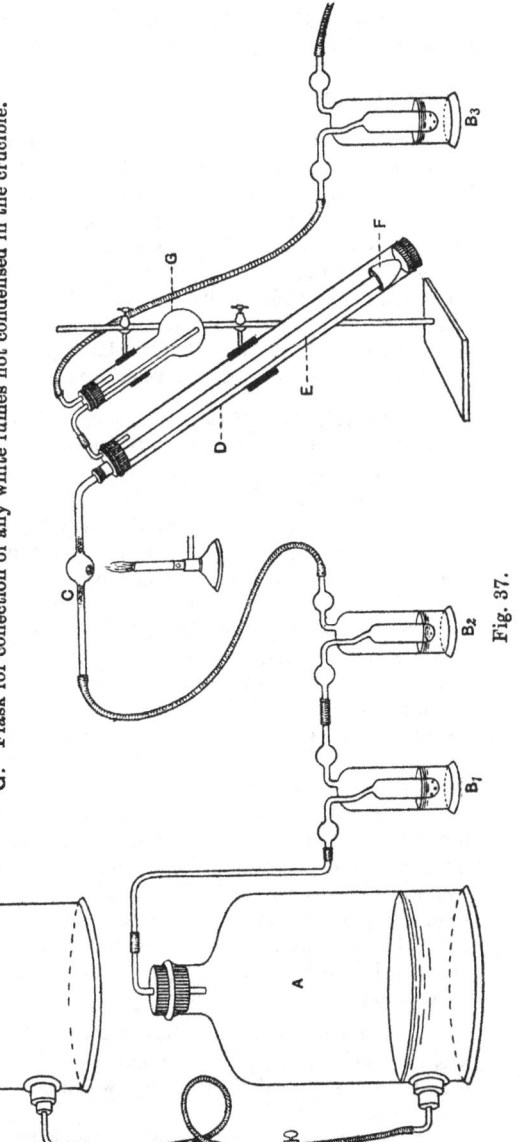

Fig. 37.

(1)   Phosphorus.

Experiment XVI.   Burn some phosphorus in air and demonstrate the occurrence of a chemical change.

Procedure.

(*a*)   Very little heat[1] is needed to start the reaction, which then proceeds spontaneously as long as sufficient air is available.   If the burning is done in a closed vessel such as a bell-jar (see fig. 36), the reaction soon comes to an end, but is started again if by lifting the bell-jar a fresh supply of air is made available.   Note the formation of white fumes which condense on the walls of the vessel, and the change of part of the residual phosphorus into a red mass which can be ignited by heating to a higher temperature and which is the 'allotropic modification' called red phosphorus.   Also note the position of the red phosphorus relatively to the original yellow, which shows that the latter must have volatilised.

(*b*)   The collection of the product of combustion, the white solid produced by condensation of the white fumes, is made somewhat difficult by its very hygroscopic nature.   It is essential that the apparatus should be absolutely dry to begin with, that the air used should be deprived of any moisture, and that the white solid should be kept in a desiccator until wanted.   Figs. 36 and 37 give two arrangements suitable for the purpose. The first is the usual extremely simple contrivance, in which a glass plate and bell-jar are used, together with a crucible for holding the phosphorus. The phosphorus is lighted, the bell-jar put in place and fitted as well as possible on to the plate so as to prevent the escape of fumes.   When combustion has ceased and the whole apparatus has become cold again, the bell-jar is removed and the white solid, which will have deposited mainly on the enclosed space of the glass plate, is scraped off, put on a watch glass and placed in a desiccator.   Fig. 37 depicts a more elaborate contrivance in which the air used can be dried and its rate of supply regulated; moreover the actual amount of air available can be increased at will by successive transferences of the water from one aspirator to the other, detaching both from the rest of the apparatus, reversing the relative position of *A* and *A'*, and when *A* is again full of air beginning anew.   This arrangement is similar in principle to that used in the commercial preparation of phosphoric anhydride, a substance much used by chemists as a dehydrating agent[2], only that in this latter case oxygen instead of air is used as the supporter of combustion.

(*c*)   The following table embodies tests well suited for demonstrating in a simple and quick manner the change in properties which justifies the classifying as chemical of the change which occurs.

---

[1] Generally applied by touching the phosphorus with a heated metal wire.

[2] Newth, *Inorganic Chemistry*, p. 471; Ostwald, *Principles of Inorganic Chemistry*, p. 360.

Record of results.

Specimen of note-book entry.

	Phosphorus	Substance obtained by the burning of phosphorus in dry air
Appearance	Pale yellow waxy solid	White feathery mass
Effect of heat	Out of contact with air melts at low temperature; in contact with air burns with bright flame, producing dense white fumes	Sublimes apparently unchanged
Action of water	None Phosphorus preserved under water to prevent spontaneous ignition in contact with air	Violent action Dissolves with evolution of much heat—solution turns blue litmus red, neutralises potash or soda, produces effervescence with carbonates

It is recommended that the purification of the white solid by sublimation, a process of considerable importance in the technique of the laboratory, should be demonstrated at this stage, even if it should be necessary to supplement the supply of solid actually prepared with some material from the laboratory store room.

(2) Iron. The manner of demonstrating the occurrence of a chemical change on heating in air has been dealt with in Chapter I, Experiment III (ii), p. 55.

2. The so-called 'burnt substance' weighs more than the original substance.

This fact had been observed as far back as the eighth century, and in the seventeenth and eighteenth centuries observations of the occurrence and speculations concerning its cause were numerous (*ante*, pp. 143, 144). All that has been said in Chapter II (p. 86) about the difference in procedure according as to whether the object is merely to establish an increase in weight or to measure the increase per unit weight of substance changed, applies equally here.

(1)　Phosphorus.

Experiment XVII. To show that burnt phosphorus weighs more than the original substance, no attempt being made to burn the whole of the phosphorus nor to ascertain the increase in weight per unit weight of substance burnt.

The apparatus is the same as that used in Chapter II, Experiment VIII, p. 89, but in view of the easy inflammability of phosphorus it is not only desirable but essential to do the weighings with the tube closed by solid corks (see A_1 in the figure, p. 89).

Record of results.

Specimen of note-book entry.

Wt. of bulb + asbestos + corks + phosphorus				56·71	57·02
,,	,,	,,	+ residual phosphorus + burnt phosphorus	57·19	57·28
		Difference		+ 0·48	+ 0·26

(2)　Iron. It has been shown in Chapter II, Experiment VIII (i) (*b*), p. 86, that the 'burnt iron' weighs more than the original iron.

3.　The effects obtained in 1 and 2 require the presence of air.

Knowledge of this dates back to the eighth century, when Geber recommended that calcination should be done in open flat pans. Boyle in 1672 showed that sulphur does not burn in an evacuated flask.

(1)　Phosphorus. Observation of what happens in Experiments XVI and XVII indicates the necessity for the presence of air : the flame of the phosphorus burning under the bell-jar goes out after a time, but reappears if by lifting the bell-jar a fresh supply of air is allowed to enter; the amount of action produced in Experiment XVII depends on the rate at which air is sucked through the apparatus, and the flame appears at the place of ingress of the air; phosphorus in contact with air is liable to ignite spontaneously even at ordinary temperatures, but it can be preserved unaltered and even melted if protected from the air by being kept under water.

(2) Iron.

Experiment XVIII Investigate the change in weight produced in iron when heated (i) *in vacuo*, (ii) in air. Pay special attention to the precautions required to eliminate weight changes due to other causes than the chemical change produced in the iron.

Procedure. Use apparatus shown in fig. 38, p. 158.

Weigh A with the iron and the air contained in it; connect A to the rest of the apparatus as shown in the sketch, set the pump going, watch the manometer, and when the evacuation has reached the highest value possible under the circumstances, heat at first gently and finally strongly, stop heating after about 10 minutes and allow to cool; when quite cold slip out D and cautiously unscrew C, *after* which stop the action of the pump; weigh A and its contents. Repeat the process, with only this difference, that D is not inserted, and C is left open so that by the action of the pump air is drawn over the heated iron; cool with the current of air passing, and weigh again.

Record of results.

Specimen of note-book entry.

Wt. of flask + iron + air	$=a$	79·294[1]	57·281[2]
„ + contents after evacuating, heating, cooling and admitting, air	$=b$	79·285	57·279
Difference	$=b-a$	**−0·009**	**−0·002**
Wt. of flask + contents after heating in air and cooling	$=c$	80·836	57·458
Difference	$=c-a$	**+1·542**	**+0·177**

4. **The increase in the weight of the substance burnt is equal to the decrease in the weight of the air participating in the reaction.**

Some account has already been given of the various attempts that had been made before Lavoisier's time to explain the observed increase in weight accompanying the calcination of metals. Rey, Hooke and Mayow each saw glimpses of the truth when they assumed the participation of the air in the reaction, but what they propounded amounted to no more than hypotheses which were not followed up nor tested deductively. The explanation so long sought was not given until Lavoisier took up the matter and showed the part played by the air, by *proving* that in combustion the increase in one part of the system—the phosphorus, or tin, or lead, or

---

[1] Iron used in the form of fine powder which had been dried, but some of which was seen to be carried away by the inrushing air.

[2] Iron used in the form of fine wire cut into short lengths.

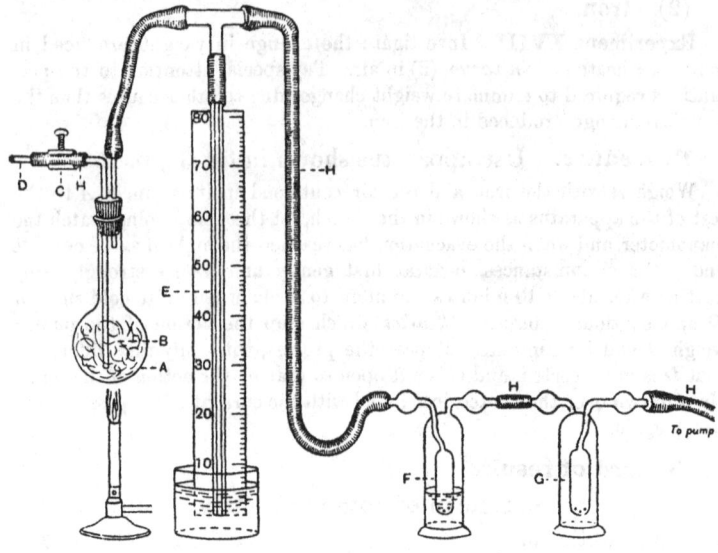

Fig. 38.

A. Round-bottomed flask of thick Jena glass (a thin-walled or flat-bottomed flask would, when evacuated, be shattered by the pressure of the outer air).

B. Iron, in the form of carefully cleaned fine wire (or wool). Iron powder (reduced iron) being more easily acted upon, has the advantage of showing more markedly the increase in weight which occurs when air is admitted (in the second part of the experiment); on the other hand, it always condenses some moisture on its very extended surface, and (unless this is first removed by heating and subsequent cooling in a stream of pure dry hydrogen) heating *in vacuo* shows a loss of weight. There is the further disadvantage that when the clip C is opened, the air rushing in suddenly is apt to blow out some of the finely divided powdery metal.

C. Screw clip.

D. Short piece of glass rod fitting tightly into the pressure tubing. It supplements the action of the clip C in preventing the ingress of air.

E. Manometer, indicating by the height of the mercury column the degree of evacuation attained; the pressure of the air left in A is given by the difference between the height of the barometer and that of the mercury column.

F. Wash-bottle filled with strong sulphuric acid, to prevent access of moisture from pump into A.

G. Empty wash-bottle, to act as a trap for the water which, if the water pump temporarily fails (owing to reduction in the pressure of water supplied from the main), is sucked over into the more highly evacuated part of the apparatus; this bottle is so placed that when the pump again acts up to its maximum capacity the water is sucked out again.

H. Pressure tubing.

iron, or mercury—is the same as the decrease in another part —the air. This proof he supplied in two ways: (*a*) by carrying out the reaction in hermetically sealed vessels, which with their contents were weighed before and after the occurrence of the reaction; (*b*) by means of complete syntheses and analyses.

(*a*) Phosphorus, lead and tin were heated in closed flasks or retorts, and it was shown that the total weight of the system was not altered thereby, no increase being noted until the vessels were opened and air rushed in to take the place of that absorbed (*ante*, p. 149). Working with 8 *onces*[1] of tin, Lavoisier found that the difference in weight of the closed system before and after the calcination was a *loss* of 0·27 grain[1], an amount within the limit of the experimental error. When however the retort was opened and air allowed to rush in, an increase in weight of 3·13 grains was observed, practically identical with 3·12 grains, the increase in weight of the tin as determined separately by weighing the mass composed of the unchanged tin and burnt tin.

(*b*) The same point was proved in the classical experiment on the synthesis and complete analysis of mercury calx (burnt mercury = *mercurius precipitatus per se*, our red oxide of mercury). In this experiment Lavoisier not only established the fact that the increase in weight of the burnt substance is equal to the decrease in weight of the air, but also proved that the reaction consists in the withdrawal of a constituent of the air, which, since it could be regained, made it possible to study the properties of the two portions into which air is separated by the process of combustion.

The synthesis was effected by long-continued gentle heating of mercury in a confined measured volume of air contained in a bell-jar communicating with the retort holding the mercury, the weight of calx present and the decrease in volume of the air being measured.

The red particles were picked off from the surface of the
    unchanged mercury; their weight was found............ $a = 45$ grains
Decrease in the volume of air .... $b = 7$ to 8 cubic inches

---

[1] 1 *once* = 8 *gros*; 1 *gros* = 72 *grains*.

The analysis was effected by heating the red calx, collecting separately the mercury and the gas evolved, and determining the quantity of each.

Weight of calx heated.. $a = 45$ grains
Weight of mercury collected .... $c = 41\frac{1}{2}$ grains
Volume of gas collected .... $d = 7$ to 8 cubic inches
Weight of the above volume of gas, the density of which
    is $\frac{1}{2}$ grain per cubic inch.................................... $e = 7 \times \frac{1}{2} = 3\frac{1}{2}$ grains

Hence since b was found equal to d, viz. 7 to 8 cubic inches, it follows that in the process of calcination this volume of a gaseous constituent of air has been absorbed ; but also, since $c + e = 41\frac{1}{2} + 3\frac{1}{2}$ grains = 45 grains = a, it follows that the weight of the calx is made up of the weight of metal contained in it and the weight of the gaseous constituent which the metal has withdrawn from the air in the process of burning (calcination).

Experiment XIX. The burning of phosphorus in a closed flask.

Procedure.

Burn a small piece (about the size of a pea) of clean dry phosphorus in a clean dry Florence flask of about 300 c.c. capacity, which is closed with a very well-fitting solid rubber cork ; the cork should be greased very slightly and pushed in with a revolving motion as far as it will go. Weigh the flask and its contents accurately (to mgs.) before the reaction and again after it, when sufficient time has been allowed for it to cool. Concerning the desirability of eliminating error due to change in the buoyancy of the air by counterpoising with a closed flask of as nearly as possible the same volume, see *ante*, p. 14. Ignite the phosphorus by heating gently

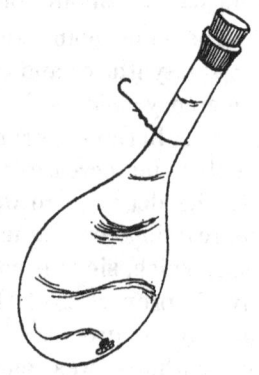

Fig. 39.

with a small flame close to the place where the phosphorus rests, and then twist it round and round, letting some of the molten phosphorus run into the neck[1]. The object of thus extending the active surface of the phosphorus is to make the reaction complete as far as the air available is concerned. Allow to cool and weigh again.

[1] In the description of a similar experiment as given by Newth, instead of heating by a Bunsen burner, the lower portion of the glass vessel is plunged into hot water.

It is desirable not to stop at this point, but to carry the experiment further : either by showing that though the total weight of the system has remained the same, the phosphorus has increased in weight; or by measuring the volume change of the air. For the first purpose (same as 2 (1), Experiment XVII) the flask is opened for a few seconds (the removal of the cork may be a matter of considerable difficulty), air is allowed to rush in, the cork is replaced and a third weighing taken.

For the second purpose (same as 5 (1), Experiment XX) the flask is opened under water ; after some agitation to dissolve the white fumes, it is placed so as to make the level of water inside and outside the same, closed by pressure against the palm of the hand and removed. A large jar or a pail filled with water brought to the temperature of the room by having been kept standing overnight, provides the space necessary for these operations. The volume of the water that had entered is then measured by pouring it into a graduated cylinder, and the total volume of the flask is ascertained in the same manner and taken as equal to that of the original air. The difference between the two gives the volume of the residual air. With the volume measurements made in the manner described, the experiment is at best but rough, and it would serve no purpose to introduce corrections for differences in temperature and pressure. As regards temperature, having taken the precaution to bring the water used in the volume measurements to the temperature of the room, we may assume that there would be little if any difference between that at the beginning of the experiment, when the volume of air (supposed equal to that of the whole flask) was enclosed with the phosphorus, and that at the end of the experiment, when the volume of the residual gas is measured. On the other hand, the assumption that the pressure of the two gases is the same introduces a very appreciable error, the magnitude of which the following considerations allow us to evaluate approximately. Supposing that—as is most likely to be the case—the barometric pressure has not changed appreciably, there remains the fact that whilst the original air was comparatively dry, the residual gas when measured was moist, *i.e.* completely saturated with aqueous vapour; and hence the real pressure P exerted by it was only equal to that of the barometer, B, less τ, the tension of aqueous vapour at the temperature of the experiment[1], *i.e.* $P = B - \tau$. At 18° C. the tension of aqueous vapour is 15 mm., which for a barometric pressure of about 760 mm. introduces an error of about $\dfrac{15}{760}$, which is very nearly 2 °/₀. Since volume is inversely proportional to pressure (Boyle's Law), it follows that the value *taken* for the volume of the residual gas is greater than the true one in about that ratio. But as it happens the final result is not likely to be incorrect to that amount, since in the ratio *residual gas : original air* the volume of the air is affected by errors which though smaller are in the same direction, thus in taking the volume of the original air equal to that

---

[1] Glazebrook, *Hydrostatics*, § 84, p. 165; *Heat*, § 117, p. 131; Clowes and Coleman, *Quantitative Analysis*, 10th ed., 1914, pp. 442 *et seq.*; Newth, *Manual of Chemical Analysis*, 1903, pp. 387, 388, 461.

of the whole interior of the flask, we neglect for one thing the volume of
the solid phosphorus; and for another, in pushing in the solid cork and
compressing a volume supposed to be equal to that of the whole flask to
one slightly less, it is almost certain that there would be some loss, some of
the air escaping before the cork is completely in contact with the neck of
the flask.

Record of results.

A.   Specimen of note-book entry.

	(i)	(ii)	(iii)	(iv)
Wt. of flask+cork+air+phosphorus $=a$	66·852	67·178	100·412	88·184
„ „ +residual gas+residual phosphorus + burnt phosphorus $=b$	66·851	67·180	100·412	88·184
Difference $=b-a$	−0·001	+0·002	+0·000	+0·000
Wt. after opening in air $=c$	66·946	67·270		
Difference $=c-a$	+0·095	+0·090		
Vol. of water that rushes into the flask when the flask is opened under water $=d$			79	60
Vol. of water filling whole flask $=e$ ∴ Vol. of original air $=e$			384	297
∴ Vol. of gas remaining after combustion of phosphorus $=e-d$			305	237
∴ Vol. of residual gas expressed as % of the original air $=\dfrac{(e-d)\,100}{e}$			79·4	79·8

B.   Summary of results obtained by a class of 11 students.

	Difference between the weights of the closed system before and after the reaction	Volume of the residual gas expressed as percentage of the original air
Student A	·000	79·4
„ B	·000	79·8
„ C	·000	78·5
„ D	+(·070)*	
„ E	− ·0005	(82)**
„ F	·000	79·7
„ G	·000	(83·6)
„ H	·000	78·6
„ I	+(·003)	78·5
„ J	− ·001	79·7
„ K	+ ·0017	78·6
Mean	·000	**79·1**

* Continuous increase during second weighing—cork used had not been
greased, and air was obviously leaking in slowly.
** Flask was opened before it had become cool.

Sections 1 and 2 dealt with combustion considered exclusively from the point of view of the changes produced in the burnt substance, viz. the phosphorus or iron ; in 3 and 4 we considered the interdependence of the substance burnt and the air ; the next three sections deal with combustion exclusively from the point of view of the air, viz. with the investigation of the volume and the properties of the residual gas.

Lavoisier, in his investigations on the increase of weight of iron on calcination, determined this increase indirectly by measuring the volume of air absorbed and calculating the weight of it, using as antecedent datum the known density of air. His results were as follows :

Weight of iron before calcination	145·6 grains.
,, ,, after ,, ...	192·0 ,,
Increase in weight due to ,, ...	46·4
Air absorbed	97 cub. in.
1 cubic inch of air weighs	0·47317 grains.
∴ 97 cubic inches weigh	45·9 grains.
Difference between increase in wt. observed and wt. of air absorbed.........................	0·05 grain.

5. **The air decreases in volume, and does so to the same extent for (1) phosphorus and (2) iron.**

Reference has been made in the historical section to the work done on this subject by Scheele, who measured the volume change produced in air by the action of a large number of substances (*e.g.* liver of sulphur, nitric oxide, moist iron filings, phosphorus, hydrogen, a candle, etc. etc. ; *ante*, p. 144); Lavoisier's observation and measurement of the volume changes produced in air by the calcination of mercury and of iron respectively have been dealt with in detail in the preceding section.

Experiment XX. Measure the volume change produced in air by (1) the smouldering of phosphorus, (2) the rusting of iron[1].

[1] The action of the iron in this experiment is not strictly the same as that studied under 1, 2, 3, since at the ordinary temperature the presence not only of air but also of moisture is essential. Hence, as far as the work here mapped out goes, there is omission of the necessary proof that iron 'rusting' in *moist air* at the *ordinary temperature* and iron 'burning' in *dry air* at a *higher temperature* produce the same effect on the air. Continuation of Exp. XX, in which the change of volume in the air would be measured—just as was done for

Let the action occur slowly at the ordinary temperature in measured volumes of air confined over water; measure the volume change. In order to make the comparison of gaseous volumes measured on different days independent of changes due to temperature and pressure, apply Boyle's and Charles' laws to calculating the corresponding volumes at normal temperature and pressure, viz. 0° C. and 760 mm.

Use either graduated tubes (200 c.c. in 1/1) or suitable-sized glass tubes closed at one end (50—80 cm. long, 2—3 cm. wide), in which the space occupied by the gas may be marked off by india-rubber bands and subsequently measured by means of water delivered from a burette or a graduated cylinder; introduce the phosphorus in the form of a thin clean stick 3—4 cm. long, attached to a stout copper wire of suitable length, and the iron as filings contained in a piece of fine-meshed iron gauze folded over at the two ends and pierced by a stout copper wire of suitable length (fig. 40 *A*). The process of inserting and removing the long wires carrying the phosphorus or iron, as well as that of equalising the level of the water inside and outside the tube prior to each volume measurement, will be facilitated by the use of large pails filled with tap water brought to the temperature of the room by having been allowed to stand overnight; for any operation in the course of the experiment which involves the use of water, draw upon this supply. A tall wide gas jar will serve well to hold the tubes during the reaction. The transference from pail to jar, and *vice versa*, can be accomplished with ease and without risk of admitting air, by using a small crucible as shown in fig. 40 *E*. Holding the upper end of the tubes for some time in the warm hand may introduce a quite appreciable error, and hence the readings of the volume measurements should be taken with the tubes held in a suitable clip (fig. 40 *H*). It will depend on the length of the gas measuring tube relatively to the height of the gas jar whether the adjustment of the pressure of the moist gas to atmospheric pressure, accomplished by equalising the inside and outside level of the water, can be done in the gas jar or not. If there is not sufficient depth of water in the jar to do so, the tube must for this purpose be taken to the pail, which should be full of water (an auxiliary supply of water at the right temperature being kept in another pail), so that no error may be introduced through the observer's eye not being at the same level as the meniscus of the water in the tube. But even when by this precaution the level has been correctly adjusted, it may prove difficult to take a reading under these conditions, and it may be preferable only to mark the position of the meniscus by slipping on and placing accurately in position a very thin rubber band of suitable diameter[1]. Several bands can be placed in readiness both at the top and bottom of the tube and pushed into position as required, the readings being done after the transference of the tubes to the gas jar, when the graduations can be seen more clearly.

phosphorus in Exp. XIX—and comparison of the result with that found in this experiment, would supply the proof required.

[1] Such bands can be made by cutting off with sharp scissors thin strips from rubber tubing.

The measurement of the volume change produced in air by the rusting of iron and the smouldering of phosphorus respectively.

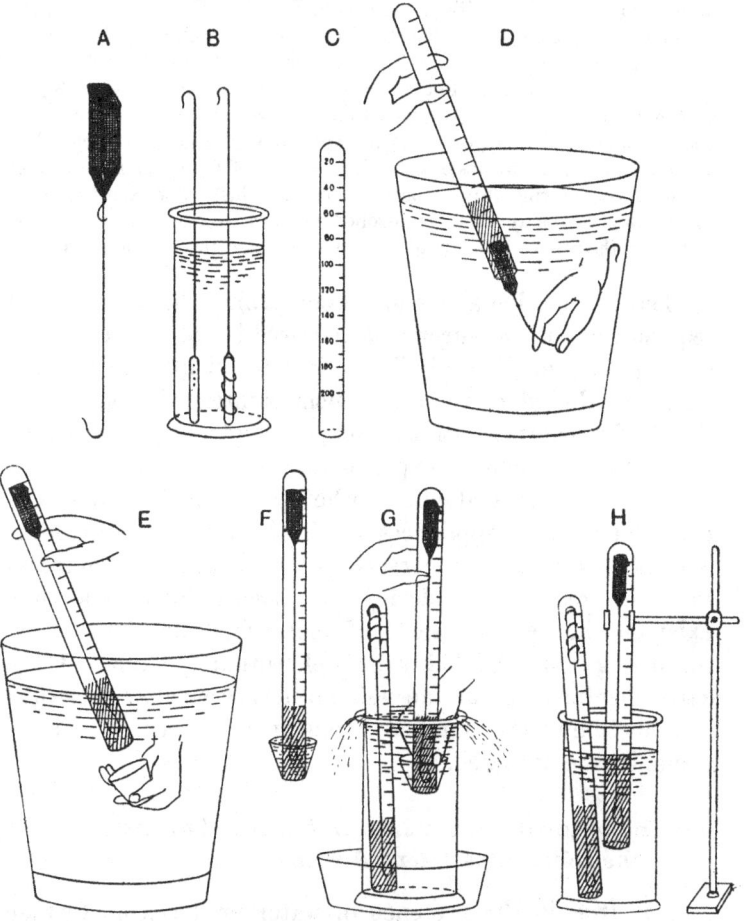

Fig. 40. **Apparatus, and mode of manipulation.**

A. Iron filings wrapped in iron gauze tied up with copper wire, for slipping into graduated tube.

B. Sticks of phosphorus wrapped round with or melted on to copper wire for slipping into graduated tube, and kept under water until wanted.

C. Graduated tube, two of which are required.

D. Method of introducing the solids into the measured volume of gas confined over water.

E, F, G. Method of transferring graduated tube, which contains a measured volume of gas confined over water, from one vessel to another.

H. Volume measurement when the pressure of the gas has been made equal to atmospheric by the equalisation of the level of water inside and outside.

G, H. Arrangement for keeping the tubes whilst the reaction is going on.

Procedure.

Isolate a suitable volume of air by pouring a little water into the graduated tube and then inverting it in the pail. Measure the volume of *moist* air either with or without transference to the jar, with or without the use of a rubber band. Introduce the iron or phosphorus, and take another measurement, which gives the volume of moist air + that of the solid above the level of the water[1]. Read the barometer and the thermometer. Repeat all these operations with the second tube. Leave the tubes to stand in the gas jar for at least a day, or as much longer as may be needed to complete the action, as shown by observation of the level of the water in the tubes (an india-rubber ring slipped on will facilitate this). For each tube read the volume of the residual gas + solid[1]; remove the solid and read the volume of the residual gas; read the barometer and the thermometer.

The information above given concerning the number and sequence of the measurements required is summarised in the table, p. 167, which also indicates how to calculate from these data the value which the experiment is intended to ascertain. This table, together with that on p. 168, gives the records for this as well as the next experiment.

It is more than doubtful whether, when doing this experiment with the appliances and in the manner described, it is legitimate to apply a correction for variations in temperature and pressure, considering that such differences as may exist on the two occasions of measuring the air and the residual gas will probably be well within the limit of the experimental error of the volume measurements ; a critical examination of the numbers found in the specimen experiments is recommended.

6.   **The residual gas obtained in 5 is inactive towards both the solids experimented with.**

(1)   Iron in the presence of water produces no further volume change in the gas remaining after the action of phosphorus on air.

(2)   Phosphorus has no effect on the gas left after the rusting of iron in air.

---

[1] Concerning the advantage of taking these extra readings which do not enter into the calculation, see the remarks in the table on p. 167.

Record of results. Experiments XX and XXI.

A. Specimens of note-book entries.

The volume change produced by :

	(1) The action of Iron on				(2) The action of Phosphorus on			
	(a) air		(b) gas left after slow combustion of phosphorus in air		(a) air		(b) gas left after rusting of iron in air	
Before action of solid.								
Date	22.10.08	23.10.09	24.10.08	25.10.09	22.10.08	23.10.09	24.10.08	25.10.09
Vol. of gas (measured before the introduction of the solid) $=V_1$	162	155	155	140	191	179	130	121
Temperature................ $=T_1$	18°	19°	18°	13·5°	18°	19°	18°	13·5°
Barometric pressure................ $=P_1$	776	758	768	759	776	758	768	759
Vol. of gas+vol. of solid (not required for calculation, but taken in order to get an indication of the progress of the change occurring .... $=S_1*$	171	163	164	149	193	182	132	125
After action of solid.								
Date	24.10.08	25.10.09	27.10.08	28.10.09	24.10.08	25.10.09	27.10.08	28.10.09
Vol. of gas+vol. of solid................ $=S_2*$	139	130	164	150	157	144	132	127
Vol. of residual gas (after withdrawal of solid) .... $=V_2$	130	121	154	141	155	140	131	123
Temperature................ $=T_2$	18°	13·5°	16°	13°	18°	13·5°	16°	13°
Barometric pressure................ $=P_2$	768	759	761	754	768	759	761	754
Calculation.								
Ratio of vol. of residual gas to vol. of original gas (compared under same conditions of temperature and pressure) .... $=\dfrac{V_2 T_1 (P_2-\tau_2)}{V_1 T_2 (P_1-\tau_1)}$	·79	·80	·99	1·00	·80	·80	1·01	1·01

(where τ_1, τ_2 = tension of aqueous vapour at temperatures T_1 and T_2 respectively)

* A useful check for the correctness of the volume readings taken can be applied by remembering that for both phosphorus and iron (entered in columns 1, 2, 5 and 6) $S_1 - V_1$ should be equal to $S_2 - V_2$, and that the volume of each of the solids thus twice determined is given twice more in the corresponding entries in columns 3, 4, 7 and 8.

Experiment XXI. (1) Measure the volume change produced by iron on the gas remaining after the burning of phosphorus in air. (2) Do the same, using phosphorus and the gas left after the rusting of iron in air.

After the last volume measurements made in the preceding experiment, repeat the experiment in exactly the same way, putting the iron in the tube in which the phosphorus had been, and *vice versa*.

B.   Summary of results obtained by classes of 18 and 12 respectively, working in pairs.

Ratio of volume of residual gas to volume of original gas (air) in the action of

| | Iron | | Phosphorus | |
| | on | | on | |
	Air	Gas remaining after action of phosphorus on air	Air	Gas remaining after action of iron on air
Pair A	·81	1·01	·80	1·00
,, B	·78	1·00	·81	1·00
,, C	·81	·99	·79	(·97)
,, D	(·75)	·98	·78	·99
,, E	·77	1·00	(·76)	accident
,, F	·79	·97	·81	·98
,, G	·79	·98	·77	1·00
,, H	(·82)	1·00	(·82)	1·00
,, J	·78	(1·05)	·81	1·00
Mean	**·790**	**·995**	**·795**	**·995**
Pair A	·774	(1·019)	·798	1·008
,, B	·783	1·007	·786	·978
,, C	·784	·988	·813	·989
,, D	·809	·989	·796	·989
,, E	·788	1·000	·798	1·000
,, F	·796	accident	·778	(·841)
Mean	**·789**	**·996**	**·795**	**·993**
Mean obtained by rapid combustion of phosphorus in Exp. XIX			**·791**	

7.   The properties of the gases left after the action on moist air of (1) phosphorus, (2) iron are the same, but differ from those of the original air.

Both Rutherford and Priestley had investigated this gas, which the latter named 'phlogisticated air'; Scheele established the points which differentiate this 'air' from common air (see *ante*, p. 144), viz. its inability to support combustion or respiration, its lower density and smaller solubility in water.

Experiment XXII. The examination of the properties of (*a*) air, (*b*) the gases residual after the action on air of (1) phosphorus, (2) iron, involving incidentally the measurement of gaseous densities.

Fig. 41 shows an apparatus suitable for preparing and storing larger quantities than those produced in the preceding experiments of the gases it is intended to investigate, and for the transference of these gases at an easily regulated rate into other smaller vessels.

The qualitative tests by which these gases are easily and conclusively differentiated from ordinary air and from carbon dioxide respectively are :

(i) The action on a lighted taper.

(ii) The action on lime water. In this test it is of course essential that any water used in the preparation of the gas should itself be free from carbon dioxide (see *post*, p. 172).

(iii) The density. The most conclusive demonstration of the identity of the two residual gases and of their difference from the original air is afforded by a determination, or even a mere comparison, of the densities of the three gases.

The subject of gaseous densities is one of such importance to the chemist that its consideration, both from the theoretical and the experimental point of view, must loom large in any course dealing with the general principles of the science. In the introductory chapter the experimental determination of the densities of hydrogen chloride and chlorine has been dealt with incidentally during the consideration of the large number of assumptions made in a students' experiment of this kind ; in these measurements the density of air was required as an antecedent datum. In order not to interfere with continuity in what is our real object at this point, viz. the investigation of the essential features of combustion in air, it seems best to relegate to an appendix at the end of this chapter the more comprehensive consideration of the subject of gaseous density as a whole. For our immediate purpose the description of apparatus and directions for procedure appended to fig. 41 should provide sufficient information concerning the conduct of the simple experiment by

Apparatus for the preparation, the storage and the delivery of the gas residual after the action on AIR of

IRON

PHOSPHORUS

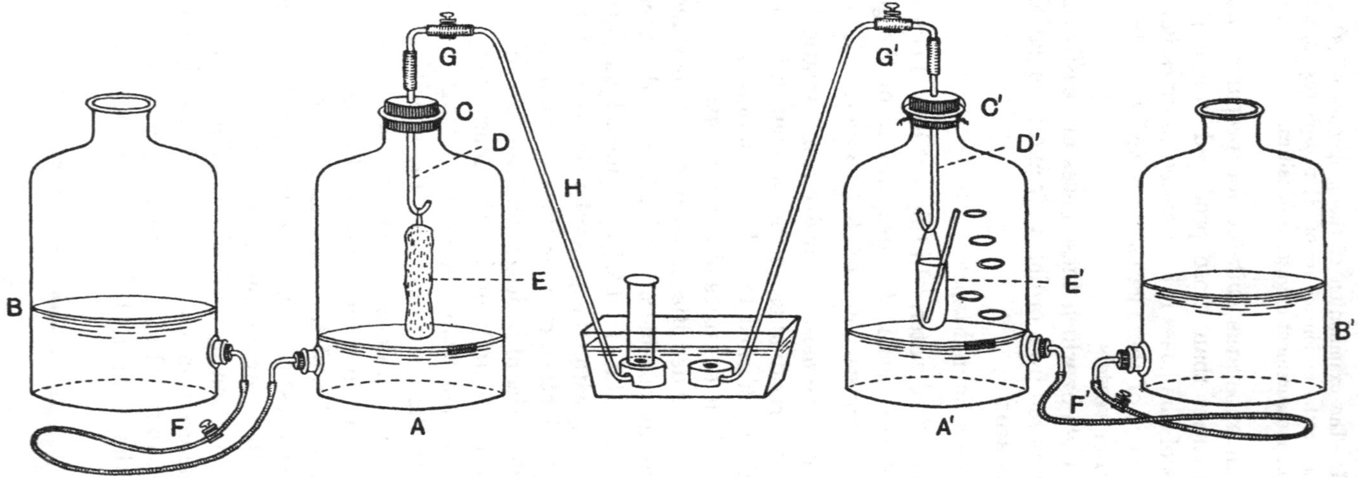

Fig. 41 a.

A. Aspirator holding the air and the residual gas obtained from it. The strips of paper which mark the level of the water and hence the volume of gas make it possible to watch the course of the action and to ascertain when it has been completed and the residual gas is ready for use.   B. Second aspirator acting as reservoir for the water by means of which the pressure of the gas in A is governed.

C. Slightly greased, very well-fitting cork, which is pushed in as far as it will go, and through which passes D.

D. Glass tube bent at one end into a hook, which carries E.   E. Long narrow muslin bag filled with clean fine iron filings.

F, G. Screw clips, by the careful manipulation of which, when B is raised, gas from A can be delivered to a desired amount at a desired rate.

H. Delivery tube used when it is desired to collect some gas over water.

A', B', C', etc. etc., the corresponding arrangement for phosphorus, identical in all respects except that: the cork C' must be wired in securely so that it may not be driven out when, owing to the spontaneous ignition of the phosphorus which is apt to occur, the action becomes violent; E' is a small tube holding one or two thin sticks of clean phosphorus long enough to protrude considerably, so as to allow of the free circulation of the air acted upon.

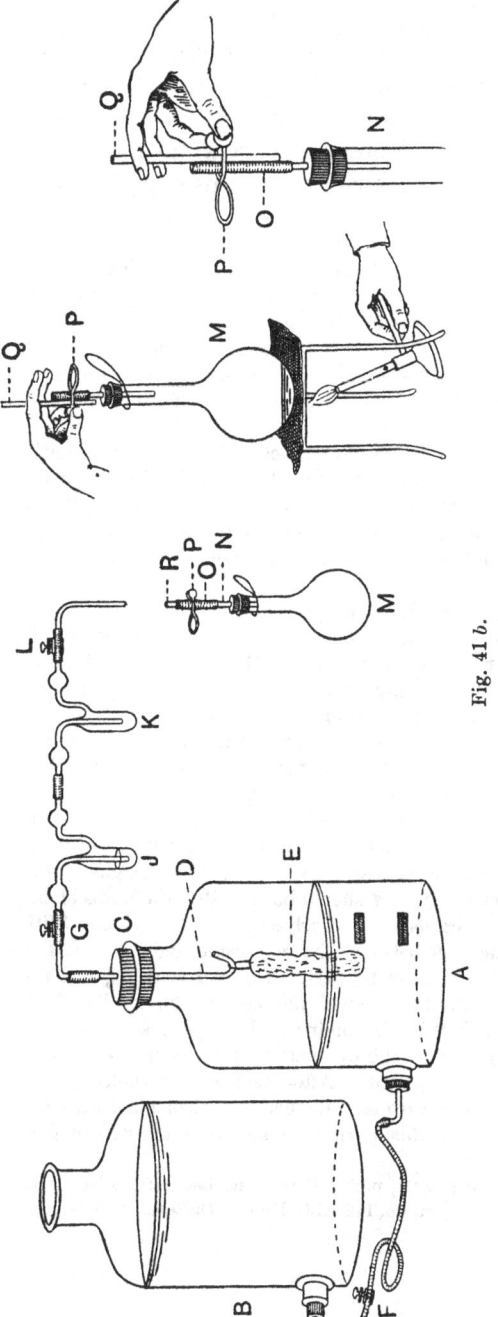

Fig. 41 b.

Determination of the density of the residual gas

J, K, L. Arrangement used when it is desired to transfer some of the gas into the evacuated flask M.

J. Small wash-bottle containing a few drops of sulphuric acid, to show the rate at which the gas is delivered from A.

K. Small empty wash-bottle, to catch any of the contents of J which might be splashed up when on connecting it with the evacuated flask M there is a sudden rush of gas from A.

M. Round-bottomed flask of Jena glass, of capacity about 300 c.c., containing about 50 c.c. of water and closed with a very well-fitting rubber cork. This cork, as shown in the detailed drawing, carries:

N. Short glass tube.   P. Good pinch clip.

O. Piece of fairly thick-walled flawless rubber tubing.

Q. Glass rod which serves to keep the pinch clip open whilst by the boiling of the water all the air is driven out from M, and by the slipping out of which at the instant of removal of the gas flame the rubber tube O is closed.

R. Short piece of glass rod which after the closing of the clip is slipped into the upper end of O to supplement the action of the clip, and to serve as a further safeguard in preventing the leaking of outside air into the evacuated flask.

which it is usual for students to determine the density of
gases such as air and its components, or nitric oxide, or
carbon monoxide, all of which are only slightly soluble in water
and which may therefore be weighed in flasks made vacuous
by the boiling of water. Moreover, a great many of the
current text books contain descriptions of this experiment[1].

Procedure.

If, as is the case at Cambridge, the tap water contains much chalk and
carbon dioxide, complications likely to arise—such as the production of
milkiness in testing the residual gas by shaking up with lime water—must
be prevented by using distilled water, or if this is not available tap water
which has been boiled in an open vessel for at least a quarter of an hour,
then allowed to cool, and finally siphoned (or poured) off from the solid
sediment.

Take out the cork C from A, unscrew the clip F, and lower B until,
when the levels of the water in the two vessels have automatically become
the same, the water in A stands only a little above the tubulure; mark this
level. Close the clip F, and push in the cork C with all its attachments,
the clip G being open, and the tube H dipping under the water in the
trough ; close the clip G.

In the case of phosphorus, whilst the procedure so far is identical, it
is further necessary to wire in the cork carefully and to keep the clip F'
closed until the violent part of the reaction, due to the bursting into flame of
the phosphorus, is over. Two to three days are needed for the completion
of the action of the iron. Mark the level to which the water has risen in A,
and roughly evaluate the volume decrease suffered by the air. In proportion
as gas is withdrawn from A, boiled-out water should be added to B to
replace that which has been used to drive out the gas.

In spite of the precautions above indicated, there is the danger that
outside air may leak in or gas leak out at the cork C. To minimise this, it
is desirable that, except whilst we are actually drawing upon the gas supply
in A, the position of B relatively to A should be such that the levels of the
water contained in the two vessels are as nearly as possible the same. With
this precaution, communication between the gas inside and outside A is
reduced to diffusion through the small apertures in and round C, and the
effect is negligibly small, owing to the small difference in the density of the
two gases, a difference which at the beginning of the experiment is *nil*.

The manipulation required for the evacuation of M should be obvious
from the description of the apparatus. After having been cooled to the
temperature of the room, M is carefully weighed and then filled with the
gas under investigation. For this purpose some gas is allowed to pass

---

[1] Ramsay, *Chemical Theory*, § 24, p. 26 ; Perkin and Lean, *Introduction to
Chemistry and Physics*, 1909, vol. II, pp. 112, 113 ; Fenton, *Outlines of Chemistry*,
p. 16.

through J and K, after which L is closed. R is then removed, the india-rubber tube O slipped over the gasometer delivery tube, the pinch clip P is opened; after this L is opened very cautiously. With the level of water in B higher than that in A, it follows that when equilibrium has been established the gas which has passed into M will be at a pressure slightly greater than that of the atmosphere. Close P and L, detach the flask M with its fittings from the gasometer, open the clip P for an instant to make the pressure of the gas in B equal to that of the atmosphere, insert R and weigh M again. Read the thermometer and barometer. Find the volume of the water contained in M and also its total volume.

Record of results.

A. Specimen of note-book entry.

Properties of:

| | Air | Gas residual after the action on air of | |
		Iron	Phosphorus
(i) Action of lighted taper	continues to burn	goes out (same as in carbon dioxide)	goes out
(ii) Action of lime water	produces no turbidity	produces no turbidity (difference from carbon dioxide)	produces no turbidity
(iii) Density, *i.e.* weight in grams of 1 litre of the gas at 0° C. and 760 mm.	1·293	1·25	1·25

The table on p. 174 indicates the various measurements and the calculations that require to be made in determining the density of the gas.

No inference concerning identity or difference in the three kinds of gas investigated will be valid unless we have some idea of the experimental error, *i.e.* the difference between the results obtained in successive measurements for the same gas, and compare these differences with those between the mean values obtained for each kind of gas. Though with a carefully thought-out plan for the dovetailing of the various operations involved, and the table ready prepared for entering the different measurements, the task should not really prove very formidable, yet the time at the disposal of the average student will generally not be sufficient to allow him to make several such density determinations, and consequently the whole class, or sections of it, will have to work as a syndicate.

Determination of the density, i.e. the weight of 1000 grams at 0° C. and 760 mm. of:

	Air	Gas residual after the action on air of	
		Iron	Phosphorus
Wt. of flask and fittings, and water from which by boiling all the air has been expelled... $= W_1$	100·1125	107·999	121·5115
Wt. of the same flask after the gas investigated has been admitted .... $= W_2$	100·6650	108·419	121·9275
Temperature .... $= t$	12·8° C.	19·1° C.	16·2° C.
Barometer .... $= B$	758 mm.	771 mm.	768 mm.
Tension of aqueous vapour at temperature t .... $= \tau$	11·0	16·4 mm.	13·7 mm.
Vol. of residual water in the flask.............................. $= V_1$	15·6 c.c.	16 c.c.	13 c.c.
Total vol. of the flask .... $= V_2$	470·0 c.c.	375 c.c.	366 c.c.
Vol. of gas admitted at temperature t, and which together with the aqueous vapour in the flask exerts pressure B ...$= V_2 - V_1 = V$	454·4 c.c.	359·0 c.c.	353 c.c.
Vol. of gas calculated to what it would be at 0° and 760 mm. $= V \times \dfrac{273}{273+t} \times \dfrac{(B-\tau)}{760}$	426·8	333·6	330·9
Wt. of the above vol. of gas $= W_2 - W_1 = W$	0·5525	0·420	0·4160
$\therefore$ density of gas $= \dfrac{(W_2 - W_1)\,1000}{(V_2 - V_1)\dfrac{273}{273+t} \cdot \dfrac{B-\tau}{760}}$	1·294	1·259	1·257

The illustrative experiments, which for the sake of clearness have been arranged in a somewhat arbitrarily chosen system of classification under seven headings, cover all that had been established, all that was known on the subject of combustion prior to the discovery of oxygen. In the historical section (*ante*, p. 142) emphasis was laid on the fact that these data were insufficient; that before the problem of the nature of combustion was solved, before it could be solved, the gas absorbed during combustion had to be regained and its properties investigated; thus, and thus only, could support be obtained for the *hypothesis*[1] that the substance burnt had combined with a **portion** of the air.

8. The increase in the weight of the burnt substance and the decrease in the weight and volume of the air are due to the combination of the burning substance with a constituent of the air; air is made up of two constituents, one of which is Priestley's dephlogisticated air (Scheele's 'fire air,' our oxygen), and the other the gas whose properties have been investigated under 7.

It is owing to the difficulty of regaining from the burnt substance the substance with which it has combined and to which it owes its increase in weight that the elucidation of the nature of combustion presented such formidable difficulties, and took so many centuries to accomplish ; and for this same reason this crucial point cannot be satisfactorily illustrated by a students' experiment. A full account has already been given of Lavoisier's classical experiment on the *quantitative* synthesis and the *complete quantitative* analysis of mercury calx (*ante,* p. 160), in which from the substance he had made by heating mercury in air he regained the original mercury and a gas which proved a better supporter of combustion than air, and which when mixed with the gas residual after combustion reproduced a substance identical with the original air. Some years earlier Scheele, working independently with the same substance (*mercurius precipitatus per se*), had demonstrated the same facts and drawn the same inference from them. What constitutes the great superiority of Lavoisier's handling of the same subject is that by working quantitatively he could prove, not only that a certain gas contained in air is absorbed, but also that this is *all* that happens ; by accounting for *all the loss* in the volume and the weight of the air, for *all the gain* in the weight of the mercury, by the volume and the weight of the gas regained

---

[1] Concerning an alternative and equally possible hypothesis, see *post,* p. 179.

from the burnt mercury, the possibility of the reaction being of a more complex nature was eliminated.

Lead when heated in air yields a yellow, easily fusible calx named *litharge* ; litharge when carefully heated for a considerable time in air at a temperature between 300 and 400° C. is changed to the brilliantly coloured *red lead,* which when heated to a still higher temperature gives the original litharge together with a gas which is a better supporter of combustion than ordinary air.   This cycle of changes is capable of being fairly easily demonstrated, and often forms the subject of a students' *qualitative* experiment; it is an experiment well worth doing in this form, provided that we realise clearly its limitations and its insufficiency.   For this experiment to supply, as is done by Lavoisier's experiment on mercury calx, all the links required in the completion of the chain of

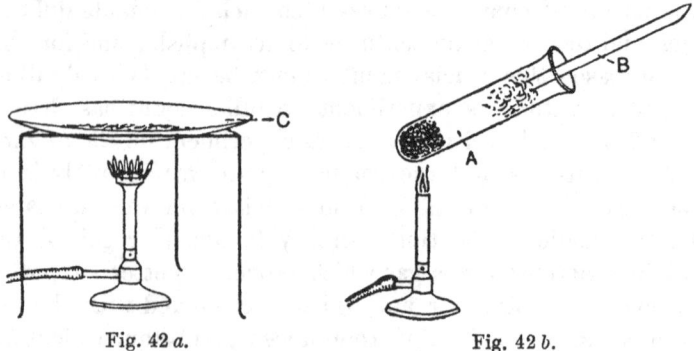

Fig. 42 *a.*   Fig. 42 *b.*

C. Iron tray containing litharge which when heated gently for some hours is converted to red lead.

A. Test-tube containing red lead, heated strongly by a single burner.
B. Glowing splint to test for oxygen.

evidence, it would have to include : (i) the synthesis of the litharge in a confined volume of air, or other indirect proof (probably more complicated) of the relation between lead and air in the formation of the litharge, and (ii) the study of the quantitative relations occurring in the decomposition of the red lead.   Such an experiment involves no inherent insurmountable difficulties, but at best it would be a very laborious undertaking, requiring more time and skill than the average student is likely to possess at the stage when the subject of combustion is presented to him.

Experiment XXIII. The preparation of oxygen from air by the use of litharge.

Some finely ground litharge[1] (about 10 grams) is spread on a thin iron sand tray and heated by a flame from a rose burner so arranged that the tip of the flame is about a millimetre from the tray (fig. 42 *a*). At intervals of about 20 minutes stir so as to present fresh surfaces to the action of the air. When the powder has become deep red, and does not appear to change further, stop the heating, put some of the finely ground substance into a small hard glass tube, heat very strongly, note the colour change and test for oxygen in the usual manner with a glowing splint (fig. 42 *b*).

III. Interpretation of the various phenomena observed in the study of combustion and illustrated by the experiments dealt with in the preceding section.

The summation of the facts derived from the experimental study of combustion as set forth in the preceding section satisfactorily and completely elucidates the nature of the process ; practically nothing remains to be said about the manner in which this is done. It has been shown incidentally how, from purely empirical considerations, there has emerged the recognition that combustion consists in the combination of the burning substance with a portion of the air, that another portion which does not support combustion is left behind, and that the distribution of matter before and after the reaction can be represented by :

Combustible substance + Air = Burnt substance + B
composed of the two composed of the combustible
portions *A* and *B* substance + *A*

But a very great deal could be said, and something must be said, about the manner in which 'interpretation' is carried out in the ordinary kind of elementary course, in which the learners have become acquainted by means of illustrative experiments with a more or less limited number of the facts pertinent to combustion—some would say, have 'discovered' them. Combustion, including as it does the composition of

---

[1] The reaction will occur more readily if in place of the commercial litharge we use freshly precipitated hydroxide or basic carbonate, which in the first stage of heating would be changed to litharge. These substances can be easily obtained by the addition to a solution of lead nitrate (or lead acetate) of either ammonia or ammonium carbonate, followed by the filtering, washing and drying of the precipitate obtained.

air, is a very difficult subject to deal with by a purely inductive method ; the problem presented is very complex, as is
proved by the length and the circuitous nature of the path
which led to its eventual elucidation. Hence if we wish to
follow in its presentation the historical method, great demands
must be made on the logical powers of the learners, who when
the moment comes for final co-ordination and inference, must
keep in mind for simultaneous survey a considerable number
of facts, the establishment of which has involved experimental
work great in amount and most varied in kind.

Moreover even to supply the necessary data for this
difficult mental process, a very great deal *must be told.* It is
very unusual to hear of a class repeating Lavoisier's experiment on the quantitative synthesis and analysis of mercury
calx[1], and yet—as has been pointed out already—without the
data supplied by this experiment the evidence is incomplete,
and the drawing of any conclusion an impossibility. Admitting,
then, that the matter dealt with in paragraphs 1 to 7 of the
preceding section represents the maximum of what *can be*
comprised within the scope of students' experiments—probably
this is a very great deal more than what *is* actually done—let
us try to put ourselves in the position of a person who knows
thus much and no more. As a matter of fact we all do
know about oxygen and nitrogen and air ; hence it is not
easy to follow strictly the heuristic method in this matter,
and experience has shown that without a considerable amount
of 'telling,' the pupil will not be able to draw the correct
conclusion. Moreover it has been pointed out[2] that at this
stage two hypotheses are possible, that there are two ways of
accounting satisfactorily for all the facts enumerated : (i) air
is a complex substance (compound or mixture), part of which

---

[1] Such work as is done with this substance is of a partial and inadequate
nature only, and is often looked upon merely as supplying additional proof,
whilst in reality it is the pivot on which the whole matter turns. Thus, "the
formation of oxygen by the action of heat on red oxide of mercury" (here it
would seem that the composition of this substance is supposed as known—how
has this knowledge been obtained?) "gives further evidence as to the composition of the air."

[2] R. Winderlich, *Logik in der Chemie.*

is absorbed by the burning substance, which thereby is
rendered more complex and heavier. This is the explanation
with which we are familiar, and which somehow we have
come to look upon as the only possible one ; but there is
another equally justifiable on logical grounds, viz.: (ii) whilst
air is a simple substance, the combustible substances are all
compounds containing as their common constituent the gas
which is left behind after combustion, and which in the
process of burning is exchanged for the air. Air is absorbed
as a whole, and its weight being greater than that of the gas
liberated, the burnt substance, though not more complex, is
heavier than the combustible substance. The reaction would
in this respect be absolutely analogous to a great many well-
known and simple ones, such as the expulsion of silver by
zinc from any silver salt, of hydrogen by magnesium from
any acid, of iodine by chlorine from any chloride, of chlorine
by iodine from any chlorate. And it would explain satisfactorily every one of the facts observed : the increase in
weight on calcination, the air absorbed being heavier than
the nitrogen liberated (just as magnesium sulphate is heavier
than hydrogen sulphate); the corresponding decrease in the
weight of the gas present ; the limited and definite power which
air has of supporting combustion; the identity in amount and
kind of the residual gas. The regaining of the gas absorbed
supplies the crucial test required for deciding between these
two possible explanations, and for substituting for two equally
adequate hypotheses actual definite unambiguous knowledge.
But it would seem that in the interpretation of what are
currently called "the learner's own results"—in reality expected results carefully arranged for[1]—there must be a serious
flaw in the reasoning, the impression being conveyed that
some definite conclusion has been arrived at.

How is it that so few teachers realise this ; that for the
last quarter of a century so many of them have blindly
followed fashion and have assigned to a subject so difficult

---

[1] Even to the extent of finding that burning magnesium in air decreases the
volume by 1/5, which if it does happen can only be due to a chance cancelling
of errors.

and complex as combustion a place quite at the beginning, or at least very near the beginning, of the chemistry course ; that they are devoting a disproportionate amount of time to it, considering the inevitable inadequacy of the final result? What is the nature, what the origin of a delusion which culminates in such statements as the following : "The constituents of air are thus discovered, and on further investigation of the metallic oxides the traditional methods of preparing oxygen are discovered and the work of Priestley and Lavoisier discussed"? "Here the history of Priestley's and Lavoisier's experiments is given in confirmation (!!) of the learner's own." Could anything be more unscientific, more subversive of intellectual honesty than all this make-believe about discovery, this pretence about young people emulating the achievements of the heroes of chemistry? Surely one may be allowed to wonder why, if at such an early stage they discover and elucidate such obscure phenomena, they do not go on later to discover things quite simple but new. Moreover the pernicious effect of pretending to discover things *de novo*, when all the time the process is really dominated by the teacher's knowledge of what the facts are, clearly appears in the conventional treatment[1] of that aspect of combustion which is dealt with under 6, viz. the volume change produced in air by the burning of phosphorus, or magnesium, or copper, or iron. The course generally followed uses this experiment as a link in the chain of evidence required to elucidate the nature of combustion and the composition of air ; the result is almost

---

[1] It makes one wonder how this mistake has taken such firm root and become as prevalent as is the case. Thus in an elementary text book deservedly very popular, the usual experiments on the determination of the volumetric composition of air and of nitric oxide are given. In each case the oxygen is removed—in the case of the air by the action of phosphorus, in the case of the nitric oxide by heating in the gas a metal such as iron—and the volume of the residual gas is measured. The experiment with air *precedes* the recognition that air is a mixture ; how then are we to account for the discrepancy in the interpretation of the results?

"1/5 of the air has been taken out by the phosphorus, therefore 1/5 of the air consists of oxygen, which is one of the chief constituents."

"The volume of the nitrogen left is exactly 1/2 the volume of the original gas ; this however does not tell us anything about the volume of oxygen in the compound."

invariably stated in the form: "A gas measuring 1/5 the volume of the original air is withdrawn"—an inference perfectly unjustifiable in the circumstances. Here the assumption is made that volume, like mass, is an additive property, which however is true only for mixtures, and hence the above inference concerning the volume of gas removed can only be justified on the ground of previous knowledge concerning the nature of air—a beautiful instance of argument in a circle. The same reasoning applied with equal legitimacy to the case of other gases would lead to some strange results. That the gas nitrous oxide (laughing gas) supports combustion has been mentioned in a previous chapter (*ante*, p. 120). If the oxygen to which it owes this property is withdrawn, and the volume of the residual nitrogen is measured, this is found equal to the volume of the original gas. What is the volume of oxygen withdrawn? Unless a case is made out to show an essential difference from that of air, the answer would be '*nil.*' Again, supposing we devised an experiment in which ammonia gas is decomposed by the action of heated magnesium, which *withdraws* the nitrogen, we should find the volume of the residual gas $1\frac{1}{2}$ times that of the original ammonia, and the inference would be that the magnesium had *given up* some gas. Examples need not be further multiplied. It must be due to the prevalence of this error that students at a later stage so often have difficulty in grasping the fact that in order to find the volumetric composition of a gas c which is made up of the gaseous constituents a and b, two equations are required for giving the values of

$$a : b \text{ and of } (a + b) : c.$$

However much one generation of students may differ from another in attainments or ability, one thing may be safely predicted: namely that when determining a, the volume of hydrogen obtained from a known volume of hydrogen chloride c, the majority will in their written records of this experiment almost go out of their way to add that the volume of chlorine withdrawn is $(c - a)$.

Is it permissible to stop at this point, or is it the writer's duty to supplement destructive criticism by constructive suggestions as to the 'when' and the 'how' of the treatment of combustion in elementary teaching? Another very long chapter would be required to do this at all adequately, and to devote so much space to what after all in the scheme of this book is a digression, would be unjustifiable. Epitomising what has been expressed or implied in the preceding pages, and adding to it what is required to fill in a mere outline, the following list of warnings and recommendations emerges:

The introduction of the subject at a very early stage is a mistake; the final results will be much better if the learner's reasoning power is first developed by exercise on simpler tasks. Each of the various aspects of combustion can be dealt with by itself as an experimentally verifiable fact, and as such affords excellent opportunity for observation and reasoning within a circumscribed area of manageable size. How many of these 'sections' are dealt with at any one stage must depend on the special conditions, such as the age of the learners and their previous training. Comprehensive treatment of the whole subject at one stage is only justifiable under exceptional circumstances, such as when the learners are fairly mature in mind and can give a considerable amount of time to the study of chemistry. Otherwise the matter drags, and by the end of the term, when it comes to summarising the whole work, the parts done at the beginning will have become vague and shadowy.

It must moreover be clearly recognised that, when we begin to explain the nature of combustion, to make the argument complete the experiments done by the learners and those shown them *must* be supplemented by things merely told, by an account of the results obtained by the 'experts.' Whether the 'telling' is restricted to the least possible, or whether things are told which by the expenditure of more time might be illustrated by experiments, must again depend on the special conditions. The writer here ventures to express the personal opinion that vivid recognition and appreciation of the reasoning involved can be attained, and that considering all that is at stake, it may be well worth while to do a little more than the minimum amount of 'telling.' Of course it is assumed that the 'telling' is done well, not always an easy matter.

The historical method of presentation, in which the attempt (not always successful) is made to suppress artificially all knowledge of the existence and the properties of oxygen until the end[1], is in many respects inferior to the alternative one, in which oxygen and air are from the outset treated side by side, and knowledge concerning the nature of combustion and the composition of air is arrived at through a comparative study of the behaviour and the effects produced by oxygen and air respectively. This path to the goal, which is a much shorter one, is as a matter of fact followed by quite a number of writers of text books and by many teachers. The crux of the method consists in the devising of a really suitable and yet simple experiment for showing that, as regards the burnt substance, the result—qualitative and quantitative—is the same whether oxygen has been used or air.

To those who would urge that the historical method is preferable on the *a priori* ground that students of science should be made to retrace the paths by which discoveries have been made, the writer would reply that even if such a plan could be consistently adhered to, yet when, as is so often the case, these paths are long and devious, it is better to take the shortest way to the goal, this being after all a way by which the discovery might have been made[2].

This chapter has grown to an inordinate length: may it not be urged in extenuation that the subject of combustion, of fire, of flame, is one which from earliest times has loomed large in almost every department of human life, in every man's daily practical activities, in the mythology of all races, in the poetic imagery of all nations, and last not least, in the building up of the edifice of science?

---

[1] Some go even further than this, and try to convey the impression that proceeding thus the learners are led to discover oxygen.

[2] Cf. Winderlich, *Logik in der Chemie.*

APPENDIX TO CHAPTER IV

Gaseous density: what it is and how it is measured.

Density being defined as the mass of unit volume, the numerical value will of course depend on the units used for weight and for volume, and a statement of what these units are should always be given, thus: Density of air $= 1\cdot293$ grams per litre, the gram and the litre (1 litre $= 1000$ c.c.) being the units usually employed in the case of gases. It must be a matter of regret that for some reason unknown, in the case of gases 'density' and 'specific gravity' ($=$ relative density) are not rigorously and consistently differentiated. Specific gravity (or relative density) being defined as the ratio between the weight of any volume of the substance considered and an equal volume of an arbitrarily chosen standard, is a pure number, the value of which does not depend on the units of weight and of volume used, but does depend on the standard selected; hence just as the statement of the value of a density should include that of the units used, so the statement of a specific gravity should include that of the standard to which it is referred. Whilst for solids and liquids the standard is always water (either at $0°$ or at $4°$, which latter is the temperature of its maximum density or at some other definitely specified temperature, such as $18°$ or $20°$), in the cases of gases there is no such uniformity: at one time the standard gas was air $= 1\cdot000$; then hydrogen $= 1\cdot000$ became the standard; now it is customary to use oxygen $= 16\cdot000$, which is equivalent to hydrogen $= 1\cdot0076$, thus:

$$\text{Specific gravity of nitrogen}_{\text{Oxygen}=16\cdot000} = 14\cdot008.$$

But though specific gravity (or relative density) is the correct designation for the value $14\cdot008$—the corresponding value for the density being $1\cdot251$ grams per litre—it has become customary to substitute the term 'density' for 'specific gravity'; this is practically always done in the case of the so-called 'vapour densities,' that is, the relative gaseous densities of substances which at ordinary temperatures are solids or liquids, and for which the determination of this constant involves special arrangements for volatilisation at a suitably high temperature. A critical inspection of any special numerical value given will generally show clearly what is meant, whether the real density (weight in grams of 1 litre or 1 c.c.) or the specific gravity (density relative to that of one or other of the current standards); but it is a pity that there should be any such uncertainty.

Since as a matter of fact all the determinations of the specific gravity of gases or vapours consist in the actual measurement of the (real) density, the transformation into the specific gravity value being a mere matter of additional arithmetic, the scope of our considerations is correspondingly narrowed.

In the expression density $D = \dfrac{\text{weight}}{\text{volume}} = \dfrac{W}{V}$, of the two terms involved W is a quantity which experiment has shown to be practically unaffected by physical conditions: the weight of a definite amount of matter, say a piece of glass or a certain volume of hydrogen chloride confined in a flask, remains the same whatever may be the alteration in temperature or pressure (provided of course that the effects due to changes in the buoyancy of the surrounding air have been counteracted or allowed for[1]). This is not the case with the other term V, the volume, which alters with variations in temperature and pressure, the amount of change produced depending on the state of aggregation and on the specific nature of the substance. Thus for gases the effect is very much greater than for solids and liquids, is in fact of a different order of magnitude, thus:

	Steam	Liquid Water	Ice
Coefficient of expansion	·00336	·00043	·0001125
		(mean value between 0° and 100°)	

---

[1] Glazebrook, *Hydrostatics*, p. 105; see also *ante*, Introductory, pp. 17, 21.

Moreover though accurate measurement reveals distinct specific differences, yet to a first approximation all gases behave alike under variations in pressure (Boyle's law) and variations in temperature (Charles' law), the relation being expressed by the equation

$$V_1 = V_0 \cdot \frac{273 + t}{273} \cdot \frac{760}{p},$$

where V_0 is the volume at 0° C. and 760 mm. pressure. No such simple relation exists for solids and liquids, each substance having its own coefficient of expansion, the extreme values showing considerable differences; the coefficients of compressibility are specific also, but so much smaller in amount that in any but the most accurate work the effect of pressure on the volume of a solid or a liquid is neglected.

Mean coefficient of cubical expansion between 0° and 100° C.

Solids

Carbon (diamond)	·0000360
Platinum	·0000267
Gold	·0000417
Copper	·0000501
Brass	·0000567
Lead	·0000828
Glass (flint)	·0000234
Silica (fused) 0° to 30°	·0000013

Liquids

Water	·00043
Ethyl Alcohol	·001244
Chloroform	·000610
Bromine	·001348
Mercury	·000180

Gases

Air	·003670
Hydrogen	·003661
Nitrogen	·003670
Oxygen	·003674
Carbon dioxide	·003710
Sulphur dioxide	·003903

The bearing of these relations on the subject of density determinations is that in the case of solids and liquids the statement of the value for any special substance must be accompanied by a statement of the temperature to which it refers, thus :

$$\text{Density water}_{18°\text{C.}} = 0\text{·}99862 \text{ gram per c.c.,}$$

and that in the case of gases either the temperature t *and* the pressure p must be stated, or the value found for temperature t and pressure p must be calculated to what it would be at 0° and 760 mm. on the assumption of the applicability of Boyle's and Charles' laws. Thus, if it had been found that at 15° C. and 754 mm. pressure 0·243 gram of oxygen occupies 177·5 c.c.

$$D_{15°\text{C. and }754\text{ mm.}} = \frac{W}{V_{15°\text{C., }754\text{ mm.}}} = \frac{0\text{·}243 \times 1000}{177\text{·}5} = 1\text{·}37 \text{ grs. per litre,}$$

and

$$D_0 \quad = \frac{W}{V_0} = \frac{0\text{·}243 \times 1000}{177\text{·}5 \times \dfrac{273}{273+15} \times \dfrac{754}{760}} = 1\text{·}45 \quad \text{,,} \quad \text{,,}$$

Hence the determination of a gaseous density D involves the making of four measurements, viz.: W, a weight; V, a volume; p, a pressure; t, a temperature. These four quantities are so related that if the gas is brought into conditions in which the value of any three of them is arbitrarily fixed, the fourth will have a certain definite fixed value; thus a certain weight of gas, confined under a pressure and at a temperature arbitrarily fixed, assumes a definite volume. (Findlay, *The Phase Rule*, 3rd ed., 1911, pp. 14 *et seq.*; Ostwald, *Principles of Inorganic Chemistry*, p. 72.)

It appears then theoretically that there should be four methods differing in principle according as to which of these four quantities is selected to be the one having a fixed value dependent on the values assigned to the three arbitrarily variable quantities. In reality, owing to the difficulty, we might say the impossibility, of maintaining the temperature within a closed system for any length of time unaffected by that outside—the conductivity for heat of the containing vessels leading to equalisation—the method the principle of which would be to ascertain the temperature produced in a known weight of gas made to occupy a known volume at known pressure does not enter into practical consideration.

Another method, the principle of which consists in ascertaining the pressure exerted by a known weight of gas made to occupy a known volume at a known temperature, has been applied in some classical experiments[1], but is not used to anything like the same extent as the two remaining ones, viz.:

(i) The determination of the weight of gas or vapour filling a known volume at a known temperature and pressure.

(ii) The determination of the volume occupied by a known weight of gas at a known temperature and pressure.

(i) This is the method perfected by Regnault for gases (*ante*, p. 16), and made applicable to vapours by Dumas. A suitable glass vessel is first weighed either full of air or empty (made vacuous), and again after filling with the gas or vapour. In the first case the density of air requires to be known as an antecedent datum. Various plans are resorted to for effecting evacuation, *e.g.* the action of a water pump or a Töpler[2] mercury pump, or absorption by charcoal cooled with liquid air (*ante*, p. 17), or by steam (*ante*, p. 171). The density determinations dealt with in detail in the preceding chapters illustrate cases of various types of gaseous density determinations based on this principle, and give for each of them the measurements that have to be made and the manner of calculating from these data the value required. The ground covered by these experiments comprises: (*a*) Standard determinations: the density of hydrogen chloride (evacuation effected by cooled cocoanut charcoal). (*b*) Students' experiments: the density of air and of nitrogen (evacuation effected by steam); the density of hydrogen chloride and of chlorine

---

[1] Morley's determination of the density of hydrogen, in which a known weight of hydrogen, obtained by heating palladium hydride, was received in evacuated flasks of known volume, kept at the temperature of melting ice, the pressure produced being read by a suitable manometer, is a case in point.

[2] Glazebrook, *Hydrostatics*, § 105, p. 196.

(density of air used as antecedent datum). As regards the application of this method to vapours, the special technique of vapour density determinations of such substances as chloroform or alcohol by Dumas' method is described in almost every text book[1].

(ii) This method is used for vapours in the two forms given to it by Hofmann[2] and by Victor Meyer[3] respectively.

In Hofmann's method we actually measure the volume of vapour produced from a known weight of solid or liquid by effecting the volatilisation in a confined space over mercury, the special advantage of the method being the possibility of obtaining the vapour at a lower temperature by working under reduced pressure. In Victor Meyer's method, which, because of the quickness with which the work can be done and the comparative simplicity of the apparatus required, is the one most commonly used, we measure the volume of vapour produced by the volatilisation of a known weight of a solid or liquid, not directly but indirectly through the equal volume of air which this vapour displaces ; the case is analogous to that in which we measure the volume of a gas evolved in a certain reaction by the equivalent volume of water it displaces, *e.g.* the volume of carbon dioxide evolved by the action of dilute hydrochloric acid on marble. As regards gases, the method based on the principle of finding the volume occupied by a known weight is of comparatively limited applicability, involving as it mostly does the possibility of starting with the gas in a condensed form, either with a compound which is easily decomposed by heat, yielding the gas required (*e.g.* potassium chlorate or perchlorate, which on heating give oxygen, the weight of which is found by weighing the salt before and after heating),

[1] Clowes and Coleman, *Quantitative Analysis*, 10th ed., 1914, pp. 484 *et seq.*; Lewis, *Inorganic Chemistry*, pp. 139, 140; Walker, *Introduction to Physical Chemistry*, 6th ed., 1910, pp. 195, 196.

[2] Fenton, *Outlines of Chemistry*, Part I, p. 18, 1st ed.; Walker, *ibid.*, pp. 196, 197; Watson, *Practical Physics*, ch. xv, pp. 255 *et seq.*

[3] Clowes and Coleman, *ibid.*, pp. 481 *et seq.*; Walker, *ibid.*, pp. 198 *et seq.*; Watson, *ibid.*, pp. 253 *et seq.*; Newth, *Manual of Chemical Analysis*, 1903, pp. 439 *et seq.*

or with a solid which at one temperature absorbs a considerable amount of the gas, which it gives up readily on heating (*e.g.* hydrogen absorbed in palladium, nitric or nitrous oxide absorbed in cocoanut charcoal). The determination of the density of oxygen given in many elementary courses in which a weighed amount of potassium chlorate is the source of the oxygen, which is collected and measured in a graduated tube, is a case in point, and shows the measurements made and the calculation performed. The same principle underlies a

Determination (rough) of the density of air.

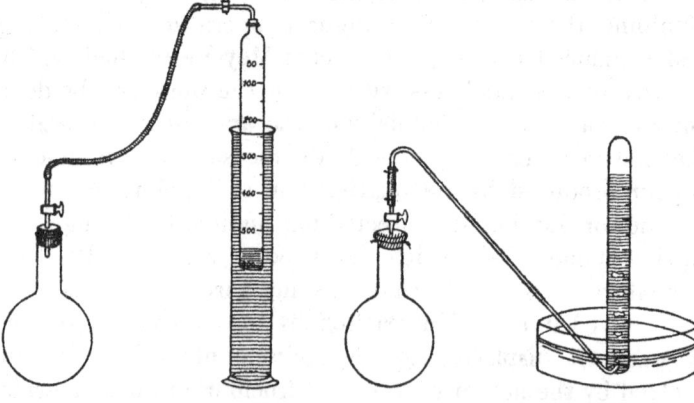

| Measurement of volume of air required to bring back to atmospheric pressure the air in the partially evacuated flask. | Measurement of volume of air escaping when the compressed air in the flask returns to atmospheric pressure. |

Fig. 43.

method of determining the density of air which, though rough, is effective because of its simplicity. It consists in either reducing (by means of a water pump) or increasing (by means of a cycle pump) the pressure of air in a flask of about 1 litre capacity, fitted with a good glass stop-cock (fig. 43); the flask, thus made to contain air under increased or reduced pressure, is weighed, the next measurement being that of the volume of air which enters or which escapes when on opening the stop-cock the pressure is brought to that of the atmosphere; the flask, now full of air at atmospheric pressure, is weighed again.

For the sake of completeness another method, not coming under the above classification, must be referred to The determination is an indirect one : the gas under investigation is evolved in a specific reaction in which a solid obtainable in a pure state is one of the reagents, *e.g.* the production of carbon dioxide from pure dry sodium carbonate and dilute acid (hydrochloric or sulphuric), of hydrogen from aluminium and potash (or magnesium and acid). By determining (*a*) the *volume* of gas evolved from a known weight of the solid, (*b*) the loss of *weight* due to the escape of the gas evolved from a known weight of the solid, we obtain the data required for finding the weight *and* the volume of the gas produced from the same weight of solid, and hence can calculate the density of the gas.

CHAPTER V

THE CONSERVATION OF MASS

Bacon, pre-eminent not only for the power to recognise and classify the fundamental principles of the 'inductive method,' but also for the clearness and beauty of the language in which he urged the cause of the investigation of nature, and for the imagery by which he made his appeals so vivid, had written *circa* 1623 :

> Men should frequently call upon Nature to render her account; that is, when they perceive that a body which was before manifest to the sense, has escaped and disappeared, they should not admit or liquidate the account before it has been shown to them where the body has gone to, and into what it has been received.

It was by action based on this principle that, one and a half centuries later, supreme success was achieved in the elucidation of the problem of combustion, which had so long puzzled and baffled generations of enquirers. In all his experimental work, in all his reasoning, Lavoisier was guided by the clear recognition, by the implicit belief, that in chemical change there is neither creation nor destruction of matter ; and the merit and glory of his achievement are none the less because the principle of conservation of mass, which owes to him the fundamental and dominant position it holds in chemistry, was not a new discovery but an old and generally accepted philosophical tenet.

I. Historical.

1. **The indestructibility of matter assumed axiomatically from earliest times.**

All the different schools of Greek natural philosophy propounded as one of the tenets of their systems that as far as human agency goes, nothing is created and nothing is destroyed.

Wrongly do the Greeks suppose that aught begins or ceases to be ; for nothing comes into being or is destroyed ; but all is an aggregation or secretion of pre-existing things ; so that all becoming might more correctly be called becoming mixed, and all corruption, becoming separate.

(Anaxagoras, c. 450 B.C.)

Lucretius (98—54 B.C.), acting as the exponent of the atomistic philosophy of Democritus (*c.* 460—360 B.C.) and Epicurus (341—270 B.C.), writes :

...the law of nature whose first principle we shall begin by thus stating, nothing is ever gotten out of nothing...if things came from nothing, any kind might be born of anything...nor would time be required for the growth of things.

According to Plato (427—347 B.C.), the whole of the matter available for creation and consisting of the four elements was by God fashioned into the shape of a perfect sphere :

Nothing went forth of it nor entered in anywhere; for there was nothing. For by design was it created to supply its own sustenance by its own wasting.

During the many centuries when alchemy reigned supreme, and when owing to the exigencies of the case all pronouncements concerning the fundamental principles on which was based the ' Art ' were marked by studied vagueness and ambiguity, there seems no doubt that the ideas propounded were not contradictory to the principle of conservation of mass. The alchemists did not pretend that gold could be created, but merely that a pre-existing amount of matter could by the abstraction of some qualities and the addition of others be changed into the perfect metal gold[1]. Bacon, Boyle, Gassendi, the fathers of inductive science, clearly expressed their acceptance of conservation of mass as an *a priori* fact :

There is nothing more true in nature than the twin propositions, that *nothing is produced from nothing* and *nothing is reduced to nothing*, but that the absolute quantum or sum total of matter remains unchanged, without increase or diminution.　　　　　(Bacon, *Novum Organum.*)

It far exceeds the power of meerly naturall agents, and consequently of the fire, to produce anew so much as one atom of matter, which they can but modify and alter, not create ; which is so obvious a truth, that almost

---

[1] " The manufacture of gold, or the transmutation of metals into gold, is to be much doubted of. For of all bodies gold is the heaviest and densest, and therefore to turn anything else into gold there must needs be condensation....But the conversion of quicksilver or lead into silver (which is rarer than either of them) is a thing to be hoped for." (Bacon, *History of Dense and Rare.*)

all sects of philosophers have denied the power of producing matter to second causes, and the Epicureans and some others have done the like, in reference to their gods themselves.   (Boyle, 1661.)

2. The indestructibility of matter becomes a scientific principle.

...it was left to Lavoisier[1] to transform a philosophical tenet into a fruitful scientific principle, and to apply it to the interpretation of chemical phenomena.   (Berthelot, *La Révolution Chimique*.)

The experimental establishment of the law of conservation of mass has followed curious lines: Lavoisier, with whose name the law is justly and fitly associated, did not arrive at it strictly inductively, by generalisation from a large number of cases in which the weight of the substances participating in a chemical reaction was compared with the weight of those resulting from the reaction. The belief then growing among physicists of the imponderable nature of heat[2], together with the old view of the indestructibility of matter in general, supplied him with the basis for an assumption which he used with signal success in the interpretation of the phenomena of combustion. The special researches, which rank as classical, in the history of his establishment of the law of conservation of mass fall into two classes: (i) those in which he proved the constancy in weight of a closed system before and after the occurrence of a chemical change within the system (the calcination of lead and tin in a closed retort, *ante*, pp. 149, 159); (ii) those in which the weight of a compound was compared with the sum of the weights of its constituents (the complete analysis of mercury calx, *ante*, pp. 150, 159, and the complete synthesis of iron calx, *ante*, p. 163). From these experiments he drew deductions concerning the nature of combustion which were verified by the formation of air from the two gases oxygen (dephlogisticated air) and nitrogen (phlogisticated air). Having thus deductively proved the validity of his fundamental assumption, he enunciated it formally in 1785, that is, *after* the completion of his researches on combustion:

---

[1] Black's claim to be considered a forerunner of Lavoisier in this respect has been referred to already (Chapter II, p. 82).

[2] Boyle had held that 'fiery corpuscles' possess weight (*ante*, p. 146).

Nothing is created, either in the operations of art or in those of nature, and it may be considered as a general principle that in every operation there exists an equal quantity of matter before and after the operation; that the quality and quantity of the constituents are the same, and that what happens is only changes, modifications. It is on this principle that is founded all the art of performing chemical experiments; in all such must be assumed a true equality or equation between the constituents of the substances examined, and those resulting from their analysis.

Here, then, we have not only a precise statement of the principle of the conservation of mass, but also the first suggestion for its application in the representation by chemical equations of the quantitative aspect of chemical change. The manner and the extent of the use of chemical equations supply full justification' for the assertion that the whole of the modern science of chemistry is based on the principle of the conservation of mass. With every equation we write we affirm our belief in its truth; in every analysis or synthesis we assume that the weight of a compound C is equal to the sum of the weights of the constituents A and B,

$$\therefore\ B = C - A,$$

e.g. in the gravimetric determination of the composition of water by Dumas' method, we weigh A, the oxygen used (abstracted from copper oxide), and C, the amount of water formed, and infer that

B, the weight of hydrogen combined with A of
$$\text{oxygen} = C - A.$$

II.   The experimental basis of the law, and its accuracy.

What then is the experimental evidence for this relationship in chemical transformations, according to which the total amount of matter participating in a chemical change is equal to the total amount resulting from it? This evidence may be grouped under the three following heads:

1.   The correct results of analyses making use of equations based on this principle.

It is usual when analysing a compound not to isolate the constituents and weigh them as such, but to estimate them

in the form of some compound of known composition; every
one of the hundreds of thousands of such analyses, the results
of which—within the limit of experimental error—add up to
100 °/$_o$, constitutes a verification of the law.   Thus Berzelius,
in his classical research on artificial and native sulphide of
iron (*Study of Chemical Composition*, p. 166), determined
the iron by the amount of iron oxide, and the sulphur either
by difference or by actual measurement of the barium sulphate which could be obtained from a given weight of the
sulphide, the antecedent data required being the composition
of ferric oxide and barium sulphate respectively.   In the
analysis of the artificial sulphide he found:

$$\text{Iron} \dots\dots\dots\dots\dots = 63{\cdot}00 \ °/_o$$
$$\text{Sulphur (by difference)}\dots\dots = 37{\cdot}0 \ \ °/_o$$
$$\text{,,} \quad \text{(actually determined)} = 37{\cdot}1 \ \ °/_o$$

Again, in the analyses of organic substances the oxygen is
never determined directly, but is calculated from the difference
between the total weight of substance taken and the weight of
hydrogen, carbon, nitrogen, etc. found present in this amount;
the results thus obtained have always been justified by the
correctness of the deductions drawn from them.   Moreover,
in the estimation of hydrogen in an organic compound, the
hydrogen is not isolated and weighed as such, but is estimated
from the amount of water the substance yields on complete
combustion, the antecedent datum required being the composition of water.   But until the gravimetric composition of
water had been established by complete synthesis, the percentage of hydrogen present was inferred from the relation

Hydrogen = weight of water formed − weight of oxygen
abstracted from copper oxide.

The experience that in the analyses of hydrocarbons the
amounts of carbon and hydrogen found approximated to
100 per cent. in accordance with the skill of the experimenter, proved the validity of the assumption made in the
determination of the composition of water.

2. The results of complete syntheses and analyses, in which C, the actually determined weight of the compound, is compared with (A + B), the sum of the actually determined weights of the constituents.

A good many such classical investigations are on record; they all have this in common, that they are experiments not specially devised or intended to demonstrate the validity or to test the accuracy of the relation $C = A + B$. Their immediate object is to determine with the greatest accuracy attainable the value of the ratio $A : B$, which, on the assumption of the absolute validity of the principle of conservation of mass, may be done by the measurement of any two of the three quantities A, B and C. But the measurement of all three supplies a test for the degree of success achieved in the prevention of mechanical loss, and in the elimination of impurities. The effect of mechanical loss and of the presence of impurities would be to make the values C', A', B' actually found different from the true values C, A, B and so to destroy the equality between C and $(A + B)$, except in the extremely unlikely contingency that the total loss is exactly balanced by the total gain, or that losses and gains together affect each side of the equation in exactly the same ratio

$$\left(\frac{C}{C'} = \frac{A + B}{A' + B'} \right).$$

Complete analyses and syntheses are becoming more and more the order of the day in research of this kind, and the importance of understanding their object and nature is so great that some further remarks on this subject may be justified. The fundamental importance in chemistry of the ratio in which hydrogen and oxygen combine together to form water, justifies a detailed consideration of the methods by which this ratio has been determined. The method chiefly employed consists in a gravimetric synthesis, in which water in the form of steam is produced by the combination of gaseous hydrogen either with gaseous oxygen (Morley's direct synthesis, see *post*, p. 199), or with the oxygen provided by an easily reducible oxide, such as black oxide of copper (Dumas' method, recently used again in a modified and improved form by Noyes, *post*, p. 200). In order to produce a weight of water sufficiently great for the effect of the probable error in the weighing not to tell disproportionately, the volume of hydrogen used, of which 1 litre weighs only ·090 gram, must be considerable ; hence contamination with even a very small amount of air, of which 1 litre weighs 1·293 grams (which makes the relative density of air $_{H\,=\,I}$ equal to 14·4 approx.), will produce a comparatively large error in the final result. Again, the prevention of all loss in the absorption of the easily volatilised water is a matter of considerable difficulty, since the excess of hydrogen which escapes tends to carry off water vapour with it. These are only samples from a very long list of probable errors which have to be guarded against in this special case. It becomes therefore obvious that it is desirable to find some criteria whereby to judge of the amount of success attained in the attempts made to eliminate the various errors. If conservation of mass holds, then C, the true weight of water, must be equal to $(A + B)$, the sum of the true weights of the hydrogen and the oxygen, and a comparison of the experimentally found value for C with its theoretical value $(A + B)$ will supply the required check, the closeness of the agreement being a measure for the correctness of all three values, provided we neglect the extremely remote possibility discussed above that the errors had affected both sides of the equation in the same ratio.

So much for the object of complete analyses and syntheses. But it is perfectly legitimate to reverse the position and to utilise the data obtained in such complete analyses and syntheses for the purposes of proof of the principle of the conservation of mass; the *assumption* made in this case being that complete success in the elimination of mechanical loss and of impurities has been attained.

(1)   Stas was the first to apply this searching test of the

accuracy of stoichiometrical[1] work in his complete syntheses of silver iodide and silver bromide[2] and the complete analyses of silver iodate[3], the results of which were published in 1865. (For details see *Study of Chemical Composition,* pp. 65 *et seq.*) For more than a quarter of a century Stas' work represented to chemists achievement so high that it could barely be emulated, certainly not surpassed. But towards the end of the nineteenth century attention began to be once more concentrated on the accurate measurement of composition. Equipped with new means for purification and improved instruments for measurement, chemists succeeded in bringing within the compass of most accurate determination gaseous substances which hitherto had presented insurmountable difficulties in the matter of purification, collection and accurate weighing. In the most important and most reliable of such determinations, the example set by Stas has been followed, and the analyses and syntheses have been made *complete,* whereby quite a large number of data has been provided which, by an interchange of what is assumed and what it is intended to test, can be used as evidence for the law of conservation of mass.

(2) In 1895[4] Morley published the results of complete syntheses of water, in which weighed amounts A of hydrogen and B of oxygen were made to combine in a suitable apparatus, and C, the water formed, was collected and

---

[1] A term introduced by the German chemist Richter about 1790, to designate the quantitative relations between chemically interacting substances, from στοιχεῖα, the fundamental constituents, and μέτρον, a measure.

[2] Made by the interaction between known weights of iodine (or bromine) changed by treatment with ammonium sulphite into ammonium iodide (*ante,* p. 56), and a known weight of silver changed by treatment with nitric acid into silver nitrate (*ante,* p. 96). The precipitated silver iodide or bromide is collected, washed, dried and weighed, the treatment being in every point analogous to that followed in the preparation and collection for weighing of silver chloride (*ante,* p. 100).

[3] The decomposition of silver iodate by heat into silver iodide and oxygen is analogous to the decomposition of potassium chlorate (*ante,* p. 90), but in Stas' complete analyses, besides weighing the original iodate and the residual iodide, the oxygen also was weighed by absorption in a tube containing heated copper.

[4] *Nature* (abstract), **53**, 1896, p. 428; *Study of Chemical Composition,* pp. 72 *et seq.*

weighed. In 9 experiments a total weight 33·2435 grams of hydrogen combined with 263·9387 grams of oxygen, making $A + B = 297$·1822 grams; C, the weight of water formed, was 297·1766 grams, which gives for $C - (A + B)$ the value $-$·0056 gram.

Gray's Apparatus for the complete analysis of nitric oxide.

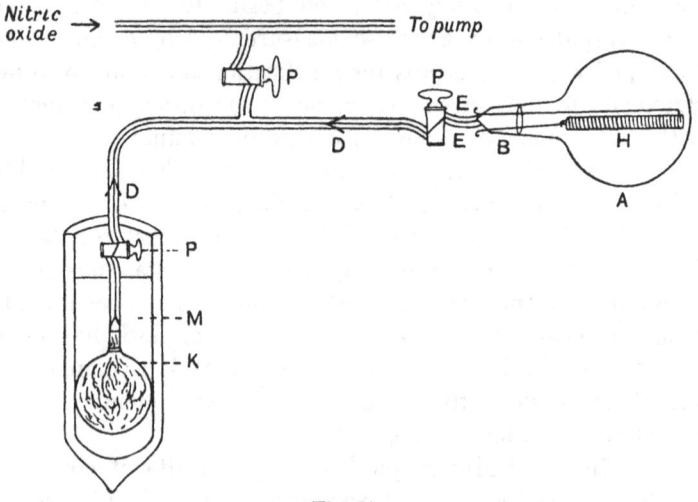

Fig. 44.

A.　Combustion bulb.

B.　Glass stopper carrying two thick platinum electrodes E E.

D, D.　Capillary ground glass joints.

E, E.　Electrodes connected with and supplying heat to wire wound round boat H.

H.　Porcelain boat containing finely divided nickel for decomposition of the nitric oxide, surrounded by coil of platinum wire.

K.　Nitrogen absorption bulb filled with powdered charcoal.

M.　Vessel holding liquid air for cooling charcoal.

P.　Glass tap.

(3) In 1908 Noyes supplied another set of data for the complete gravimetric synthesis of water. A known weight of pure hydrogen, obtained by the heating of palladium hydride (*ante*, p. 139), was burned to water by heated copper oxide, the loss of weight of which represented the oxygen; the water formed in the copper oxide tube was driven over into a receptacle suitable for its complete retention and

subsequent weighing (fig. 45, p. 204). The total amount of hydrogen, oxygen and water dealt with in 6 experiments was:

Hydrogen	Oxygen	Hydrogen + Oxygen	Water formed	
A	B	$(A+B)$	C	$C-(A+B)$
21·32179 gms.	169·24165 gms.	190·56344 gms.	190·56240 gms.	− ·00104 gm.

(4) In 1908[1] Noyes, with the object of ascertaining the ratio hydrogen : chlorine, effected the complete synthesis of hydrogen chloride. The hydrogen was weighed absorbed in palladium, the chlorine in the form of potassium chloroplatinate[2];

the hydrogen is passed over the heated potassium chloroplatinate to form hydrogen chloride. The hydrogen chloride formed is condensed in a third section of the apparatus and weighed.

	Hydrogen	Chlorine	Hydrogen + chlorine
	A	B	$(A+B)$
Total of 11 exps.	6·41925 gms.	225·86022 gms.	232·27947 gms.

	Hydrogen chloride	
	C	$C-(A+B)$
,,	232·27288 gms.	− 0·00659 gm.

(5) Reference has been made in a previous chapter to the widening of the scope of accurate stoichiometrical investigations, due to the possibility of purifying gases by fractional distillation and fractional crystallisation at temperatures low enough for their liquefaction and solidification; the case of nitric oxide was specially instanced (*ante*, Chapter III, p. 140). In 1905 Gray published[3] the results of complete analyses of this substance. Weighed amounts of the most carefully purified oxide were decomposed by finely divided heated nickel, the increase in weight of which supplied a measure for the oxygen abstracted, whilst the nitrogen liberated was weighed by absorbing it in cocoanut charcoal (fig. 44).

The results obtained were:

Nitric oxide decomposed	Oxygen formed	Nitrogen formed	Nitrogen + Oxygen	
C	A	B	$A+B$	$C-(A+B)$
2·93057 gms.	1·56229 gms.	1·36819 gms.	2·93048 gms.	+ 0·00009 gm.

[1] *J. Amer. Chem. Soc.* 1908, **30**, 13.

[2] This is the substance obtained in the ordinary test for potassium with platinic chloride as a yellow crystalline precipitate, very sparingly soluble in alcohol and ether. Heating decomposes it into chlorine, potassium chloride and platinum.

[3] *J.C.S.* 1905, **87**, p. 1601.

(6) A complete analysis of nitrosyl chloride by Guye and Fluss (1908)[1] is interesting because of the variety of the substances used in the absorption of the three constituent gases into which the gaseous nitrosyl chloride is decomposed:

	Nitrosyl chloride	Chlorine absorbed by finely divided silver heated to about 500°	Oxygen absorbed by heated copper	Nitrogen absorbed by heated calcium	
	A	B	C	D	$A-(B+C+D)$
Total of 5 exps.	2·8380 gms.	1·5364 gms.	·6931 gm.	·6067 gm.	+0·0018 gm.

3. **Experiments specially devised and intended to test the accuracy of the law of conservation of mass.**

So far the proofs adduced for the validity of the principle of conservation of mass in chemical reactions have been altogether deductive or indirect, derived from experiments not specially devised for the testing of this relation. As a matter of fact, the experiments in which the results of complete analyses and complete syntheses are tested for their approximation to the requirements of conservation of mass are not well suited for this purpose. It is impossible to avoid some loss of, or some access to, the matter dealt with, and the effect due to this mechanical loss and introduc-

---

[1] The object of this investigation was the determination of the combining weights of nitrogen and chlorine by *direct* reference to the standard oxygen (*post*, Chap. IX, p. 316). Nitrosyl chloride is obtained by the interaction between nitrosulphuric acid (chamber crystals) and sodium chloride; it is an orange-coloured gas condensing at − 5·6° to a blood-red liquid and solidifying at − 61° to a yellow solid. With water it acts in the manner characteristic of the halides of non-metals.

NOCl   (nitrosyl chloride) $+ H_2O = HNO_2$ (nitrous acid) $+$ HCl
                                    (water)   (hydrochloric acid)
POCl   (phosphorous oxychloride) $+ 2H_2O = H_3PO_3$ (phosphorous acid) $+$ HCl
PCl$_3$ (phosphorous chloride) $+ 3H_2O = H_3PO_3$ $+3$HCl
POCl$_3$ (phosphoric oxychloride $+ 3H_2O = H_3PO_4$ (phosphoric acid) $+3$HCl
PCl$_5$ (phosphoric chloride) $+ 4H_2O = H_3PO_4$ $+5$HCl
CrO$_2$Cl$_2$ (chromyl oxychloride) $+ 2H_2O = H_2CrO_4$ (chromic acid) $+2$HCl

Nitrosyl chloride is quantitatively decomposed by finely divided heated silver, which completely absorbs the chlorine, liberating all the nitrogen and oxygen in the form of nitric oxide. *J. Chim. Phys.* 6, 1908, p. 732.

tion of impurities cannot be distinguished from any change of weight that might actually accompany chemical transformation. From the close approximation to 100 °/₀ of the results of analyses, which involves the validity of the principle of conservation of mass, we know that if weight changes occur they must be very small; and hence arises the necessity for very special arrangements if such small deviations are to be detected. Within the last two decades a considerable amount of work has been done on the subject, each new publication being a record of success in tracing and eliminating sources of constant error and in the attainment of increased accuracy in the measurements made. The common feature of all these experiments is: that a reaction is made to occur in a closed vessel, the weight of which is ascertained before and after; that the accuracy of the weighings is pushed to the furthest possible limits; and that the possibility of changes due to changes in physical conditions (chiefly differences in volume of air displaced) is most carefully guarded against. There is no transference from vessel to vessel, with the inevitable attendant loss or gain of matter.

(1) An experiment of Noyes, done incidentally to the determination of the gravimetric composition of water already dealt with (p. 200 and fig. 45) may be quoted here, though it does not strictly conform to the type above specified.

At the beginning of each determination all the connecting tubes are exhausted; when enough hydrogen has been transferred, the copper oxide tube is closed and the palladium tube cooled, when practically all the hydrogen left in the connecting tube E is re-absorbed by the palladium.

It seemed possible so to carry out the work as to secure some evidence with regard to the question of change of weight in a chemical reaction in which a large amount of energy is dissipated...in 25 experiments the same hydrogen was weighed, first as absorbed in palladium, and second after conversion into water by means of the copper oxide. The results were as follows:

Total of 25 exps., in 14 of which the hydrogen as weighed in the palladium appeared heavier, while in 11 the hydrogen after conversion into water appeared heavier	Gain of copper oxide tube	Loss of palladium tube	Difference
	36·72299 gms.	36·72562 gms.	+0·00263 gm.

(2) Landolt's investigations on the occurrence of change

Noyes' Apparatus for the complete gravimetric synthesis of Water.

Part of the apparatus used for collecting and weighing the **water** formed in the copper oxide tube and driven over by immersing part of that tube in hot water.

Part of the apparatus used for the combination of the **hydrogen** (weighed by loss of palladium tube as well as by gain of copper oxide tube) and the **oxygen** (weighed by loss of copper oxide tube).

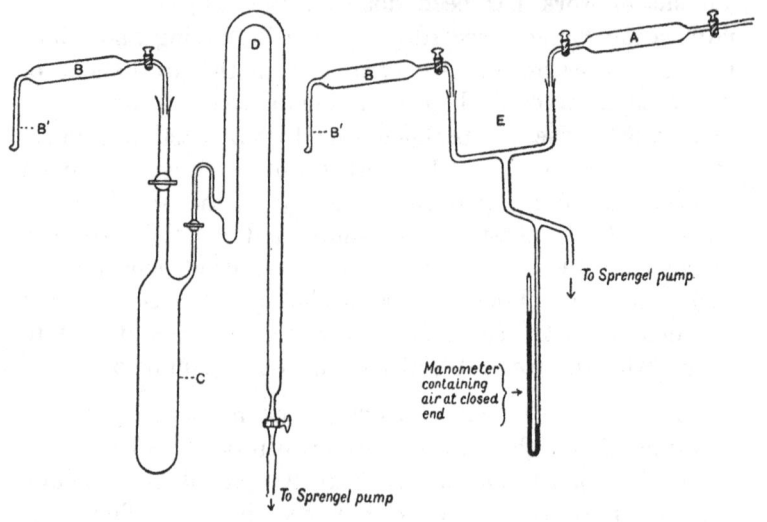

Fig. 45.

A. Palladium charged with hydrogen; the tube containing it can be heated in the same way as the copper oxide tube **B**.

B. Tube containing copper oxide. It is surrounded by an electrically heated air bath (not shown in fig.) by means of which it can be raised to any desired temperature.

B′. Part of copper oxide tube which projects from the air bath and is surrounded by a coil conveying either cold or warm water, according as to whether the object is to condense the water formed by the copper oxide or to drive the water over into another part of the apparatus.

C. Tube surrounded by freezing mixture of ice and sulphuric acid (not shown in fig.), for the collection of the bulk of the water in the form of ice.

D. Phosphorus pentoxide tube for retention of traces of water not condensed in **C**.

E. Arrangement employed for the transfer of the hydrogen to the copper oxide tube.

[1] *J. Amer. Chem. Soc.* 1907, **29**, pp. 1718 *et seq.*

in the total weight of substances undergoing chemical transformation, 1890—1907.

The special features of this research have been already dealt with in some detail in the Introductory chapter (pp. 32 *et seq.*), the object there having been to convey an adequate idea of the manner in which constant errors are searched for, detected and eliminated, and of the mode of interpreting the final results. In 48 experiments, 25 showed a decrease, 23 an increase in weight; in the preponderating number of cases the experimental error of ± 0.03 mg.[1] was not exceeded, whilst the average magnitude of the changes observed was ± 0.012 mg. The final inference (already quoted, p. 34) was:

> The final result of the whole investigation is that in all the 15 decompositions involved it has not been possible to establish a change of weight. The observed deviations from absolute equality before and after the reaction are due to external physical causes and are not the result of chemical reactions.

The average amount of substance undergoing chemical change having been 300—400 grams, and the maximum experimental error ± 0.03 mg., Landolt's result proved that if any real change does occur in the total mass of reacting substances it must be less than ± 0.03 mg. in 350 grams, that is, less than ± 1 in 10,000,000.

(3) Manley, who in 1912 (*Phil. Trans.* 1912 A, 212, 227) published the result of a research the same in principle as Landolt's monumental work, found it possible to effect a further reduction of the experimental error, bringing it down to -0.006 mg. (one-fifth of Landolt's value). The results obtained were such as to prove that in the special case investigated, which was the reaction between barium chloride and sodium sulphate, if any change does occur it must be less than ± 1 part in 100,000,000 parts. Reviewing, then, the whole available evidence,

> we are led to conclude that this present research has but tended to confirm the truth of an almost universally accepted belief that a given total mass is an unchanging and unchangeable quantity[2].

[1] This was the maximum change observed in blank experiments, *i.e.* experiments in which the manipulation of the apparatus was the same as in the reaction experiments, but the substances used were such as did not react.

[2] The bearing of these results on that most important problem of chemistry

III. Students' Illustrative Experiments.

It may conduce to clearness in the recognition of the principles underlying the different experiments, and to a correct appreciation of the scope and the limitation of each special one, if we classify them on a plan similar to that which has been followed in the presentation of the experimental evidence on which the law is based.

1. A reaction is made to occur in a closed system, which is weighed before and after.

(1) Combustion of phosphorus in a confined volume of air.

In the preceding chapter, under II. 4. (p. 160), the burning of phosphorus in a closed flask, done as a students'

which deals with the accurate determination of combining (atomic) weights has been thus expressed by T. W. Richards, the greatest living authority on this subject:

"The subject of atomic weights has acquired new interest recently, because of the striking demonstration by Landolt that the law of the conservation of mass holds true to a great degree of precision in common chemical reactions. The fact that the sum of the reacting weights remains perfectly constant, within the limit of error of the most exact experimentation, strengthens the conviction that each of these reacting weights possesses fundamental significance. Evidently no error is committed by calculating one atomic weight by subtracting another from the molecular weight of a substance containing two elements, and the whole structure of the table of atomic weights is seen to rest on a satisfactory basis.

"These assurances are timely in view of the extraordinary discoveries concerning radio-activity in recent years. Not a few radical thinkers have supposed that these discoveries lessen the importance of exact atomic weight determinations because of the doubt cast on the permanence of the supposed atom, but Landolt's admirable work assures us that under ordinary circumstances the chemical combining proportions are wonderfully permanent, and therefore as full of meaning as they have been supposed to be. The new discoveries concerning radio-activity extend the bounds of knowledge, but in no wise lessen the significance of that which went before."

The justification for these doubts on the part of the 'radical thinkers' is that recent researches have proved beyond doubt that the effects, photographic, electric, chemical, physiological, summed up in the term 'radio-activity' are due to the shooting off from the radio-active substance of material particles of two orders of magnitude, the β rays (mass about $\frac{1}{1750}$ that of hydrogen atom) and the α rays, actually identified as helium atoms (mass about twice that of hydrogen atom).

illustrative experiment, has been dealt with in detail from
the practical point of view; in the Introductory chapter
(p. 29) this experiment has formed the subject of a detailed
discussion concerning (i) the nature and the magnitude of
the probable experimental error, (ii) the impossibility of
evaluating the degree of accuracy to which this special
illustrative experiment can carry us, due to the fact that we
know nothing about the amount of phosphorus changed[1].
The tables on p. 162 give records of results actually obtained.

(2) Interaction between solutions in which an easily
visible effect is produced, *e.g.* a colour change, or precipitation of a solid with or without colour change.

Experiment XXIV. The principle underlying these experiments is
the same as that in Landolt's and Manley's research work, and hence the
reactions chosen must likewise be restricted to such as occur without the
evolution of a gas or the production of much heat.

The apparatus used is the extremely simple and effective arrangement
suggested in Ostwald, *Principles of Inorganic Chemistry*, 4th ed., 1914,
p. 25, and which is depicted in fig. 46. The procedure is so simple and
the sources of error so small in number and so insignificant in their effect[2],
that the results obtained by a class are apt to become dull through the
almost unbroken recurrence of 0·000 for the value of the difference in the
weight of the closed system before and after the reaction; owing to this
there will be no opening for the stimulating process of a critical discussion
of the various likely sources of error, the probable value of each of these,
and their combined effect on the final result, a feature of the work which
in the writer's opinion is the most valuable part of it. The experiment may
however be useful for any specially backward members of the class who
require *purposeful* practice in weighing and in very simple manipulation,
as well as the stimulus and encouragement derived from the getting of
so-called 'good results.' Otherwise the experiment will not be worth
doing unless we arrange matters so as to get from it some incidental
advantage.

(*a*) The class may be led to increase its knowledge of the external
characteristics of different chemical reactions by introducing, according to
a definite system, a great variety in the reactions chosen. Along one bench
there might be a display of precipitates varying in colour from the pale
yellow of lead chromate through the deeper shades of lead iodide and

[1] If however we assume as antecedent data the density of oxygen and the
volume of it contained in air, and measure the volume of the flask, we can
evaluate the transformation of matter which has occurred.

[2] Of course only when judged relatively to the accuracy generally aimed at
and attained in the weighings.

cadmium sulphide to the reddish brown of silver chromate and silver arsenate; whilst along another bench preference might be shown for stronger colour effects produced without the simultaneous formation of precipitates, such as the change from the orange of the bichromate to the green of the chromic salt, of the deep purple of the permanganate to the green of the manganate, etc.

Investigation of the effect on the total weight of a system within which occurs a reaction between silver nitrate and ferrous sulphate, resulting in the precipitation of metallic silver.

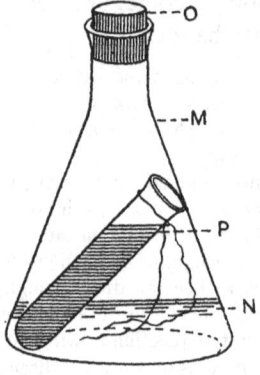 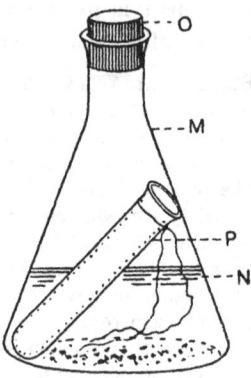

Fig. 46.

A   B
Before the occurrence of the reaction.   After the occurrence of the reaction.

Two vessels A and B, as nearly as possible the same in external volume, are used.

M. Erlenmeyer flask of about 150 c.c. capacity, which contains
N. One of the two solutions for reaction.
O. Well-fitting solid rubber cork.
P. Small test-tube which holds the second of the two solutions for reaction; its size is such that it has not room to fall down; it has a piece of cotton attached for lowering it into position without spilling the contents.

(b) A rough estimation may be made of the amount of change that has occurred, thereby supplying the additional datum required for evaluating the degree of accuracy to which the law has been shown to hold; of course when doing the experiment in this form it becomes essential to increase the accuracy of the weighings to a degree such that errors in weighing begin to tell. The following is a record of such an experiment, in which the reacting substances were ferrous sulphate and silver nitrate.

Procedure.

Quantities of silver nitrate (about 2 grams) weighed to centigrams are put into the small test-tubes; approximately weighed quantities of ferrous sulphate (supposed to be present in excess) are put into the flasks, and quantities of water added which will ensure the complete solution of the solids; A and B, each with its cork by its side, are then put on the two pans of a rough balance and water is added to the one of lesser weight until the weights are very nearly the same; the corks are pushed in firmly. Using the best balance available, ascertain with the utmost accuracy attainable the difference in weight between A and B: (i) before cautiously tilting the flask till the solutions mix; (ii) after the occurrence of the reaction in A; (iii) after the occurrence of the reaction in B.

Record of results.

A. Specimen of note-book entry.

	A	B
	gms.	gms.
Wt. of empty flask ....$= a$	19·77	24·88
Wt. of weighing tube with silver nitrate ......$= b$	16·16	14·02
Wt. of silver nitrate tube after shaking some of its contents into the small test-tube ... $\Big\} = c$	14·02	11·36
∴ Wt. of silver nitrate taken....................$= b - c$	2·14	2·66
Wt. of finely powdered ferrous sulphate placed in the Erlenmeyer flask (weighed on piece of paper on rough balance) .... $\Big\}$	10 (about)	10 (about)
Wt. of the whole apparatus:		
Before the occurrence of the reaction	$A =$	$B + ·6832$
After the occurrence of the reaction in A ...	$A =$	$B + ·6828$
„ „ „ „ B ...	$A =$	$B + ·6834$
Change in wt. accompanying the reaction in A	·0004	—
„ „ „ „ B	—	·0006
Wt. of flask + precipitated silver ....$= d$	21·10	26·52
∴ Wt. of silver precipitated in the reaction...$= d - a$	1·33	1·64
Change of weight accompanying the reaction per 100 of silver precipitated .... $\Big\{$	$\dfrac{·0004 \times 100}{1·33}$ $= ·030$	$\dfrac{·0006 \times 100}{1·64}$ $= ·036$
	$= 1$ in 3000 approx.	
Silver precipitated expressed as percentage of the silver nitrate taken ....$= \dfrac{(d-a)100}{b-c}$	62·1[1]	61·7[1]

[1] The reaction was not complete, as was shown by the formation of a thick white precipitate on the addition of hydrochloric acid to the liquid decanted from the precipitate; hence the deviation from 63·5, the standard value for the percentage of silver in silver nitrate.

The next and last step in the experiment consists in the estimation of the metallic silver precipitated by the ferrous iron, which in this process is oxidised to ferric iron. The precipitate is washed in the flask by decantation (the siphoning device described *ante*, Chapter II, p. 101, could be used with advantage), the washings being poured through a small filter paper ; the washing is continued until the thiocyanate test(Clowes, *Practical Chemistry*, 8th ed., 1908, p. 136 ; Newth, *Manual of Chemical Analysis*, 1907, p 165) shows the absence of all ferric iron. The flask with its contents is then dried in an air bath ; the filter paper, containing any silver removed with the washings, is burned in the usual manner (Clowes and Coleman, *Quantitative Analysis*, 10th ed., 1914, pp. 59 *et seq.*; Newth, *ibid.* pp. 206 *et seq.*), and the ash and silver added to the contents of the flask (the weight of the ash can be neglected), after which the flask and its contents are heated fairly strongly, cooled and weighed.

2. Measurement of the loss of weight in one part and the gain in weight in another part of a system within which a reaction occurs.

Experiment XXV. The action of hydrochloric acid on marble, and the absorption of the gas formed. Comparison of the loss in weight of the marble and acid with the gain in weight of the potash used to absorb the gas evolved in the reaction.

(*a*) The qualitative aspect of the reaction occurring. Show that by the action of hydrochloric acid on marble a gas[1] is produced which :

   (i)   is a non-supporter of combustion ;

   (ii) is heavier than air ;

   (iii) turns lime water milky ;

   (iv) is quickly and completely absorbed by potash.

Variety should be aimed at in the demonstration of the absorbing action of the potash. A very simple device for filling a large test-tube with the gas by means of downward displacement is shown in fig. 47, where the subsequent operations are also indicated.

(*b*) The quantitative part of the experiment.

Fig. 48 represents a suitable arrangement for the evolution and the absorption of the carbon dioxide. As regards the latter process, the apparatus intended for this purpose must be of such a kind that the surface of contact between the gas and the absorbing solution is made as great as possible. The problem is the same as that arising in organic analysis, where the complete absorption of the carbon dioxide formed in the burning of the substance and made the measure of the amount of carbon present is essential to success. Here again it may be useful (and perhaps even a little entertaining) to look up in catalogues various forms of

[1] For various types of apparatus suitable for the continuous production of this gas, see *ante*, Chapter II, p. 98.

potash bulbs, from the original simple kind known by the name of its inventor Liebig (1803—1873) to the much smaller, internally often very elaborate arrangements which are beautiful specimens of the glass-blower's

The absorption of carbon dioxide by potash.

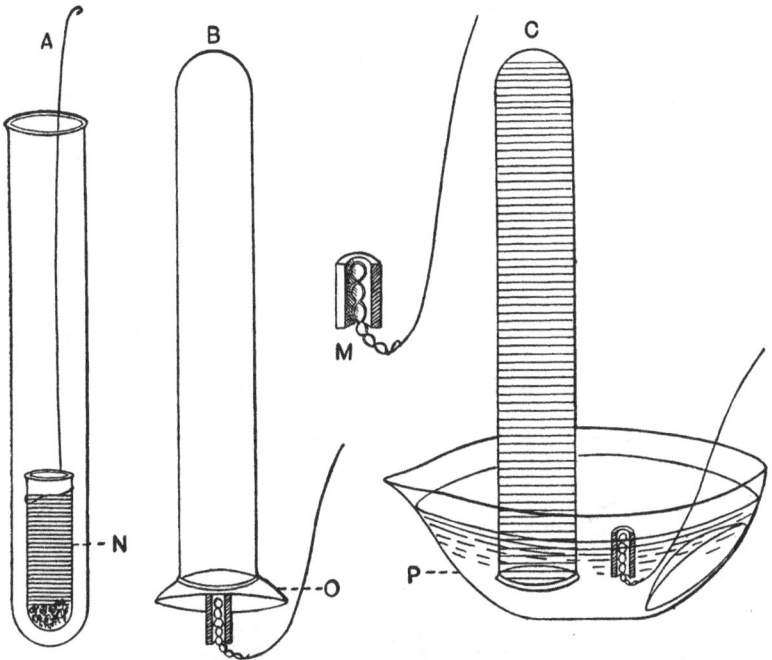

Fig. 47.

N. Small test-tube containing the marble and hydrochloric acid required for the generation of carbon dioxide; the tube is carried by a loop made in a long stout copper or iron wire.

M. Device for holding a plate or watch glass O in position under the opening of the tube or cylinder. It consists in a piece of stout iron or copper wire twisted at one end, over which is slipped a short piece of very stout-walled rubber pressure tubing.

O. Small watch-glass for closing the mouth of the large test-tube. The interposition of a piece of moist filter paper increases the efficiency.

P. Porcelain dish (or small glass crystallising dish) containing a solution of potash which has been coloured by a drop of phenolphthalein.

A. Arrangement for filling the tube with gas.

B.　　,,　　,,　conveying the tube into the dish holding the potash.

C.　　,,　　,,　effecting the absorption of the gas by potash.

art. But however excellent may be the provision made for maximum efficiency in the absorbing power of the potash, success can only be attained by letting the reaction occur slowly, so that the bubbles pass at a rate of

Comparison of the loss of weight in *A* to *E*, the part of the closed system in which carbon dioxide is evolved by the action of hydrochloric acid on marble, with the gain in weight of *F*—*G*, the part of the closed system in which the carbon dioxide is evolved by potash.

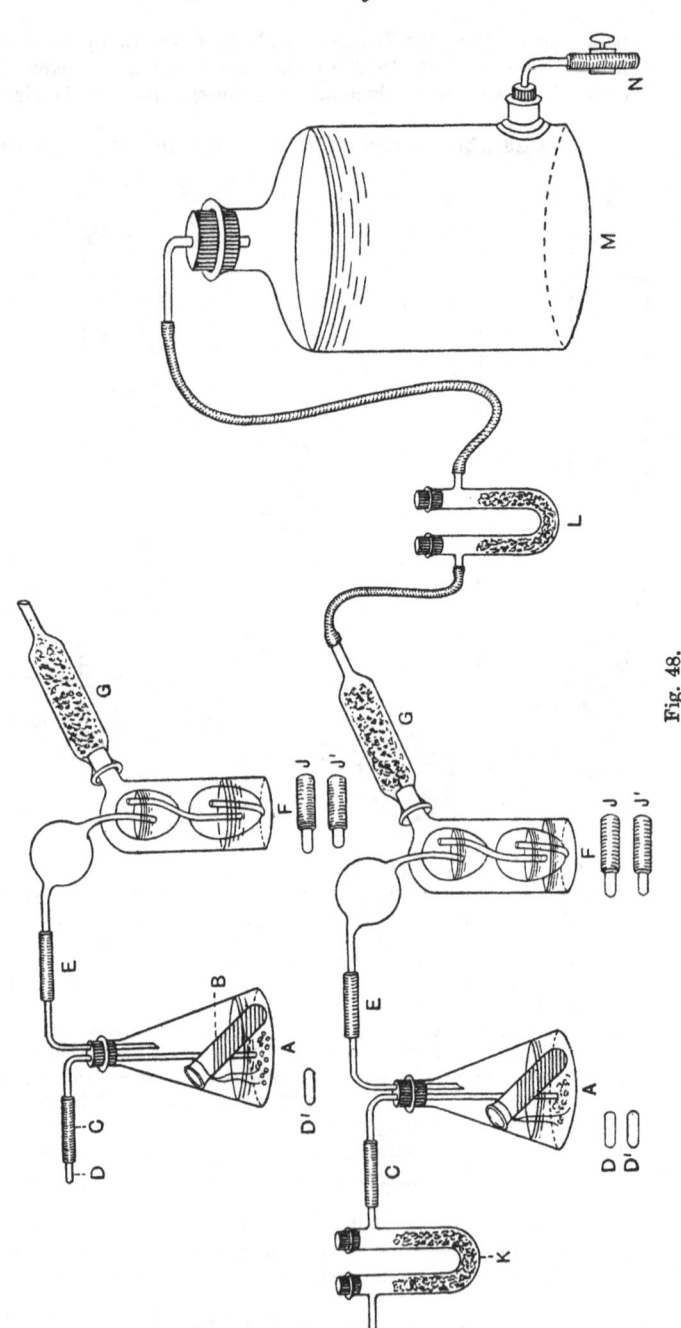

Fig. 48.

A. Erlenmeyer flask of about 150 c.c. capacity, fitted as shown with *good* two-holed rubber cork; of the two glass tubes which pass through the cork, the longer must dip under the surface of the water which together with some small lumps of marble covers the bottom of the flask.

B. Small test-tube of length such that it cannot fall down inside the flask; it is provided with a piece of fine string (or stout cotton) by which it can be lowered carefully, so as to avoid any risk of spilling the concentrated hydrochloric acid with which it is filled. The combination of water in the flask and *concentrated* acid in the tube supplies an adequate amount of acid in the diluted form in which it must be used for the reaction to proceed at a rate sufficiently slow to ensure the complete absorption in F of the carbon dioxide gas.

C and E. Rubber tubes, fairly thick-walled, flawless and well-fitting. Owing to the fact that rubber tubing is not absolutely impermeable to carbon dioxide, it is desirable to make the glass tube from A almost join that from F within E; and since the escaping gas is certain to carry spray from A, of which some will be deposited in the connecting tube, E must constitute part of the closed system and must be weighed with A (or with F) both before and after the reaction.

D and D′. Stoppers made of short pieces of glass rod, for closing one or both of the rubber tubes C and E.

F, Potash bulb filled with strong solution (1 part by weight of solid potash to 2 of water). The volume of solution introduced must not be greater than can be easily held by the big bulb (next to E) into which it gets sucked at the end of the experiment when the evolution of gas in A has ceased.

G. Chloride of calcium tube in which is retained the moisture carried by the air which escapes at the beginning of the experiment. The chloride of calcium used must be in small grains to increase the amount of active surface available, and free from powder which might clog the tube; it is kept in place at each end by plugs of glass wool (or cotton wool).

J and J′. Stoppers made from rubber tubing and glass rod, for closing the ends of F, G.

M. Aspirator filled with water, for drawing dry air through the apparatus at the end of the experiment, whereby to re-establish the original conditions when A was filled with air; this air was expelled at the beginning of the experiment by the carbon dioxide gas evolved, and escaped at the open end of G after giving up all its moisture. The current of dry air entering at C sweeps the carbon dioxide left in A into F, where it is absorbed, and thus is replaced the air that escaped at the beginning. The excess of air that must be used will make no difference, provided it escapes at G in the same state as it entered at C, *i.e.* dry and free from carbon dioxide.

N. Screw clip.

K. U-tube filled with granulated soda lime to deprive the in-going air of moisture and carbon dioxide. The soda lime is kept in place by plugs of glass wool.

L. U-tube filled with chloride of calcium (or soda lime) to prevent access of moisture from M into G.

not more than one in a second, preferably more slowly; to accomplish this the acid used must be dilute, and it must be delivered very gradually. Moreover, since it is essential that except for the *transference* of carbon dioxide (which will carry with it water vapour) everything at the end of the experiment should be the same as at the beginning, any *escape* of water vapour must be carefully guarded against, and special arrangements must be made for replacing at the end of the experiment the air which at the beginning was expelled by the carbon dioxide evolved; whilst on the one hand the escaping air must be quite dry, the moisture carried by it being retained in the apparatus, the air led in at the end of the experiment to take its place must be dry and free from carbon dioxide. It is of course quite immaterial whether all the marble is dissolved[1], or whether all the acid available is used; but it is desirable to use quantities of marble and acid such that the amount of gas transferred is sufficiently great to reduce the proportionate effect of the experimental error.

Procedure.

Put the marble and water into A, the concentrated acid into B (be careful to keep the outside and the rim of the tube free from acid), and lower B into A; with E open, firmly insert the cork, which should be very slightly greased or moistened in its lower half, a combination of a gentle pushing and screwing motion being generally most effective; be careful that the glass tube, which during the course of the experiment must dip under the water, does not protrude to such an extent that in the process of pushing in the cork it breaks through the bottom of the flask; manipulation of B requires some care. Fill F as directed above, and test for the efficient working of corks, rubber connections and ground glass joint by joining at E, taking out D, connecting G to M and cautiously opening the screw clip N; if all is well, as the water runs out *very* slowly, one drop at a time, air bubbles one at a time will pass through the liquid in A, and hence through F. Disconnect, close C and E by the stoppers D and D', close F and G by J and J', and carefully weigh each piece of apparatus. It has been found in practice that carrying out the experiment with the apparatus here recommended and in the manner described entails an experimental error of a magnitude such that weighing to milligrams is legitimate and necessary, weighing to fractions of milligrams futile.

Connect the two pieces of apparatus as shown in the figure, carefully labelling and putting away D', J and J'. Start the reaction by shaking out a few drops of acid from B. When the evolution of gas has nearly ceased, shake out a little more acid, and continue to do so until all the acid

---

[1] There is a difference in this respect from the somewhat similar experiment, in which the object is the determination of the amount of carbon dioxide contained in a known weight of pure dry carbonate (*post*, p. 356). For this purpose the decomposition of the carbonate must be made *complete* by using excess of acid, the weight of carbon dioxide being determined either from the loss in weight of the apparatus in which it has been evolved (Clowes and Coleman, *Quantitative Analysis*, 10th ed., 1914, pp. 108 *et seq.*; Newth, *Manual of Chemical Analysis*, 1907, pp. 259 *et seq.*), or from the gain in weight of a suitable absorption apparatus (Clowes and Coleman, *ibid.* pp. 103 *et seq.*; Newth, *ibid.* pp. 261 *et seq.*); in either case it is essential that what passes *from* the evolution apparatus or *into* the absorption apparatus should be carbon dioxide *only*, and hence the necessity for the provision in that case of a suitable drying tube attached to and constituting part of the evolution apparatus.

has been used, being careful to produce the gas *at a very slow rate.* Note that at first a gas—which is air—bubbles through all the three compartments of *F*, but that soon none passes beyond the first. At this stage, when the evolution of gas has come to an end, *A* as well as the big bulb of *F* will be full of carbon dioxide, and if, as should be done, the apparatus is now left to itself, the potash will continue to absorb the gas and will rise, first slowly, then as the surface of contact becomes greater rapidly, into the big bulb (caution given above as to allowing for this in the amount of solution used), while air will pass in from *G*, thereby effecting automatically restitution of some of the air expelled at the beginning, the necessity for which has been pointed out above. The air thus introduced will of course not be dry nor free from carbon dioxide, and theoretical perfection would make it seem desirable to work from the beginning with the tube *L* attached to *G*, which would have the further advantage of preventing any leaking in by diffusion of moisture or carbon dioxide; the taking of this extra precaution would of course present no special difficulty, but its omission is not likely to introduce any serious error, because the volume of air sucked back through *G* is so small comparatively that the weight of moisture and carbon dioxide carried by it would be a quantity not recorded in the weighings, which do not go beyond milligrams. The purification of the air entering at *K* is a different matter, because a very considerable volume is used in order to sweep out from *A* the last traces of carbon dioxide, every 1 c.c. of this gas left representing 1 c.c. of air less than was in the apparatus at the beginning of the experiment, and hence a loss of about ·0012 gr. After the automatic suction of air into *A* through *F* has come to an end, leave the apparatus for at least half an hour, to give time for *F* to cool, the absorption of the carbon dioxide having been accompanied by the evolution of a considerable amount of heat, as can be ascertained by touching the bottom of *F*, which will feel quite hot. Next attach *K*, *L*, and *M* as shown in the figure, and, by suitable manipulation of the screw clip *N*, pass a slow current of air with the object already set forth; to secure the complete replacement of the carbon dioxide in *A* by air, at least two litres of air should be passed. Disconnect at *C*, at *E* and at *G*; insert *D*, *D'*, *J*, *J'*, and after having left the two sections of the apparatus for about half an hour in the balance case for the necessary temperature adjustment, weigh each section.

Inspection of the results summarised in Table B on p. 216 shows that in the majority of instances the loss exceeds the gain, which is in accordance with what we should expect, the chief probable error being failure to retain completely the carbon dioxide and water vapour carried over from *A*. It should prove interesting to let some members of a class try the experiment with the interposition of another weighed tube containing potash and calcium chloride between *G* and *L* to act as a check for the efficiency of *F*, *G* in the matter of complete absorption.

Record of results.

Comparison of the loss in weight due to interaction between marble and hydrochloric acid solution with the weight of the carbon dioxide evolved and absorbed by potash.

A. Specimen of note-book entry.

		(i)	(ii)
		gms.	gms.
Weight of the apparatus containing the marble and hydrochloric acid } before the reaction ...... $= a$		57·936	67·350
The same after the reaction ...... $= c$		57·088	66·232
Loss $= a - c$		**·848**	**1·118**
Weight of the potash bulbs before the reaction ... .. $= b$		87·857	47·716
The same after the reaction ...... $= d$		88·698	48·837
Gain $= d - b$		**·841**	**1·121**
Difference between loss of wt. of marble and hydrochloric acid, and weight of } carbon dioxide evolved and collected $= (a - c) - (d - b)$		− ·007	+ ·003
Difference calculated for 100 of carbon dioxide		− 0·8	+ 0·3

B. Summary of results obtained in different years by sections of the classes.

	Loss of system marble + hydrochloric acid $= A$	Gain of potash bulbs $= B$	Difference of loss and gain $= A - B$	Difference expressed as percentage of B $= \dfrac{(A - B)100}{B}$
Student: A	·2445	·2460	+ ·0015	+ 0·6
B	·588	·588	± ·000	± 0·0
C	·848	·841	− ·007	− 0·8
A′	1·132	1·133	+ ·001	+ 0·1
B′	1·179	1·151	− ·028	(− 2·5)
C′	·537	·510	− ·027	(− 5·3)
D′	·791	·782	− ·009	− 1·2
E′	·825	·815	− ·010	− 1·2
A″	1·432	1·417	− ·015	− 1·0
B″	1·069	1·067	− ·002	− 0·2
C″	·508	·637	+ ·129	(+ 20)
A‴	1·033	·941	− ·092	(− 9·8)
B‴	·867	·870	+ ·003	+ 0·3
C‴	·875	·862	− ·013	− 1·5
D‴	1·118	1·121	+ ·003	+ 0·3
E‴	1·244	1·231	− ·013	− 1·1
F‴	1·176	1·170	− ·006	− 0·5

3. Complete syntheses.

(1) *The complete synthesis of silver sulphide* is described by Vaughan Cornish in his excellent and suggestive little book, *Practical Proofs of Chemical Laws,* to which students are referred for experimental details. Known weights of silver and sulphur are heated together in a sealed tube, and after completion of the reaction the silver sulphide formed and the sulphur left in excess are weighed. The specimen results given are excellent:

	Silver taken	Sulphur taken	Sulphur left	Sulphur used	Silver +sulphur	Silver sulphide	Difference
Exp. 1.	·8301	·1233	none	·1233	·9534	·9531	− ·0003
„ 2.	1·0000	·2963	·1476	·1487	1·1487	1·1476	− ·0011

(2) *The complete synthesis of silver iodide,* illustrative of Stas' classical research on this subject. Besides the evidence for the law of conservation of mass that this experiment is intended to supply, a number of other useful objects may be accomplished incidentally, viz. the determination of the combining ratio silver to iodine, a valuable contribution to the data collected in Chapter IX (The determination of combining weights); the technique of the gravimetric estimation of iodides and the volumetric estimation of silver by thiocyanate; the manner of effecting quantitatively the change of iodine into iodide, of metallic silver into a soluble silver salt (nitrate). Hence if a student has the time at his disposal, the experiment, though a little laborious, will well repay the time and trouble spent on it.

Experiment XXVI. The complete synthesis of silver iodide.

The method is an indirect one[1], consisting in the production of the yellow insoluble silver iodide by interaction between solutions of ammonium iodide and silver nitrate, which have been obtained by suitable reactions

[1] The direct synthesis, *i.e.* burning of a known weight of silver in iodine, similar to the process employed in the direct synthesis of silver chloride (*post*, p. 231), is a difficult matter, even if the work is confined to a determination of the ratio silver : silver iodide, the iodine being obtained by difference. Ladenburg has effected such a direct synthesis.

from known weights of the two elements. The work in the form about to be described divides itself naturally into three parts :

(i) The preparation of standard solutions of silver and of iodine, *i.e.* solutions which in a known volume (or in a known weight) contain known weights of (*a*) iodine, (*b*) silver.

(ii) Determination of the equivalency factor between these two solutions, *i.e.* of the volumes of the two solutions which contain the exact quantities of iodide and silver salt required for interaction, no excess of either iodine or silver being left in solution after the precipitation of the insoluble silver iodide.

(iii) The production and weighing of the silver iodide obtained from suitable volumes of the two solutions.

The quantities of the different substances involved which previous experience has shown to be suitable for the purposes of the experiment as described, are :

Silver. The purest[1] foil obtainable—1·75 grams (about), weighed accurately to milligrams.

Iodine. The purest re-sublimed crystals, powdered, and dried by being kept for some time in a desiccator—2·0 grams (about), weighed accurately to milligrams in a glass stoppered weighing tube.

Whatever the quantities of iodine and silver taken, they must be in the ratio of *not more* than 1·14 of iodine to 1 of silver; this is to ensure the presence of excess of silver when equal aliquot parts of the solutions containing iodine and silver in this ratio are mixed.

Ammonia. 100 c.c. of a solution of ammonia of specific gravity ·967 (which according to the tables[2] contains 8 °/₀ of ammonia) diluted by the addition of 300 c.c. of water ; this should yield a solution somewhat stronger than normal[3].

[1] See *ante*, Chapter II, p. 96.

[2] Clowes and Coleman, *Quantitative Analysis*, 10th ed., 1914, pp. 512 *et seq.* Tables have been compiled which give the composition of the solutions of a number of acids, alkalis and salts as a function of the specific gravity. These tables greatly facilitate the work of preparing solutions of any desired strength. The specific gravity determination is done most quickly by means of a hydrometer (Glazebrook, *Hydrostatics*, § 58). Of course the solution taken for dilution need not have the special specific gravity above quoted ; by the use of tables any other could be made to serve, the corresponding amount of dilution required being calculated. All that matters is to produce a solution *about* normal, and even that is only a question of convenience, the volumes of solution required being of convenient magnitude if this is the strength.

[3] For meaning of terms normal, decinormal, seminormal etc., see Clowes and Coleman, *ibid.* p. 148; Newth, *Manual of Chemical Analysis*, 1907, pp. 312 *et seq.*

Sulphurous acid. About 200 c.c. of a solution of specific gravity 1·030 (which according to the tables contains 5·5 °/₀ of sulphurous anhydride) diluted to 300 c.c., should give a solution somewhat stronger than normal.

Potassium thiocyanate. Pure re-crystallised—about 2 grams dissolved in 1 litre.

Iron alum. Pure re-crystallised—a small volume of 5 °/₀ solution.

(i) (a) *Preparation of the standard solution of iodine,* by changing a known weight of iodine into ammonium iodide by the action of alkaline ammonium sulphite.

Preparation of the ammonium sulphite solution, containing twice as much ammonia as is required for the neutralisation of the acid (ante, Chapter I, p. 56). Using solutions of ammonia and sulphurous acid of the strength above recommended, make about 250 c.c. of the sulphite, either by adding to a known volume of ammonia half the volume of sulphurous acid required for complete neutralisation, or, what comes to the same thing, adding to a known volume of sulphurous acid twice the volume of ammonia required for complete neutralisation.

Record of results.

	Volume of ammonia solution taken	Volume of sulphurous acid required for neutralisation	Volume of sulphurous acid required for neutralisation of 1 c.c. of ammonia
	m	n	$\dfrac{n}{m}$
(i)	10 c.c.	27·50	2·750
(ii)	10 c.c.	27·45	2·745
(iii)	10 c.c.	27·35	2·735
	Mean	27·43	2·743

For the preparation of the ammonium sulphite solution 200 c.c. of ammonia are mixed with $100 \times \dfrac{n}{m}$ c.c. of acid. Of course it will be a saving of time if the ammonium sulphite is prepared in larger quantities for the use of a number of students.

Change of the iodine to ammonium iodide[1].

Put about 25 c.c. of the sulphite solution into a beaker of about 200 c.c. capacity. Shake into it the iodine from the carefully weighed stoppered

---

[1] The equation for the reaction occurring is:

$$(NH_4)_2SO_3 + H_2O + I_2 = (NH_4)_2SO_4 + 2HI$$
$$2HI + 2(NH_4)OH = 2(NH_4)I + 2H_2O.$$

tube, which contains about the amount required and weigh the tube again. From a burette or other apparatus suitable for the delivery of the solution in small drops, add sulphite until the last trace of colour of the iodine solution is destroyed. As the reaction proceeds, the brown solution becomes yellow and ultimately colourless. The change from pale yellow to colourless is at this concentration so sharp that there should not be any difficulty in detecting the end point with extreme precision. If too much sulphite should have been added, a further amount of carefully weighed iodine might be added by shaking a tiny crystal out of a weighing tube and adding the weight to that originally taken. The volume of sulphite solution used does not enter into the quantitative relations we wish to ascertain, and so need not be actually measured; but it will be useful to note and put on record its approximate value at any rate, in case the same solution should be used again in a repetition of the experiment.

The next step consists in making the solution of the ammonium iodide up to a known volume; the transference to the graduated flask will have to be effected with the utmost care to prevent any loss, the beaker being washed out several times in succession with small quantities of water each time. Remember to shake. Call the strength of this solution, *i.e.* the weight of iodine per c.c., M.

(b) *Change of a known weight of silver into silver nitrate.*

The amount of silver required (1·75 grams, or if the amount of iodine has been other than 2·00 grams, the corresponding proportionate amount) is accurately weighed. It is then dissolved in nitric acid (30 c.c. of acid of specific gravity 1·4 diluted to 100 c.c.) with due precautions against mechanical loss by spurting, etc.; the procedure is in every way identical with that described in Chapter II, p. 99. The solution is gently boiled until the brown acid fumes have disappeared, cooled and made up with the usual precautions to a known volume. Call the strength of this solution, *i.e.* the weight of silver per c.c., N.

The table on p. 222 summarises the measurements required for the making of the above standard solutions a and b.

(ii) *Determination of the equivalency factor between the two solutions.*

The quantities of silver and iodine used in the preparation of the two solutions have been chosen on the basis of antecedent knowledge, according to which, on mixing equal volumes of the two, the solution above the precipitated silver iodide will contain a small excess of silver nitrate. The object of this section of the experiment is to determine the amount of this excess, and thereby to ascertain the volumes of the two solutions which will be exactly equivalent, *i.e.* will form insoluble silver iodide without leaving in the solution an excess of either of the interacting substances. In research work the interaction of exactly equivalent quantities of iodine and silver would be produced by the cautious addition of a standard solution of iodide to the silver solution until no turbidity is produced as a drop falls through the clear liquid above the precipitated iodide. The technique of this process is however somewhat elaborate, and the method about to be described (see Clowes and Coleman, *Quantitative Analysis*) may be substituted, since it is of an accuracy quite commensurate with that likely to be attained by the average student in the rest of the measurements involved. The method is based on the occurrence of the following reactions :

Potassium or ammonium thiocyanate precipitates from silver nitrate white silver thiocyanate, insoluble in nitric acid ; thiocyanate gives with ferric salts a deep red colour, due to the formation of iron thiocyanate, which is not destroyed by dilute nitric acid ; red iron thiocyanate is decomposed by a soluble silver salt, with the formation of white insoluble silver thiocyanate. Hence if to a solution of a silver salt (which may contain free nitric acid) a drop of iron alum solution is added, and thiocyanate solution is then run in from a burette, the appearance of a pink tinge (due to the formation of iron thiocyanate) will indicate that all the silver has been precipitated as white insoluble silver thiocyanate, and hence the end of the reaction. The presence of precipitated silver iodide does not interfere with this reaction.

Perform qualitative tests illustrative of these reactions.

Determination of the silver equivalent of the thiocyanate solution.

Using known volumes of the standard silver solution made in (i) (*b*), find the volume of the thiocyanate solution required for the complete precipitation of all the silver; use lots of 10 c.c.

Estimation of the excess of silver nitrate solution when known volumes of the two standard solutions are mixed.

Use 25 c.c. of each solution, and determine the excess of the silver nitrate by titration with the previously standardised thiocyanate solution.

(iii) *The determination of C, the weight of silver iodide produced by the interaction between A and B*, weights of silver and iodine found from the preceding work to be strictly equivalent.

In (ii) we ascertained the volume P of silver nitrate solution that has to be added to every 1 c.c. of iodide solution in order to make the precipitation complete, just complete, and no more than complete. But since from (i) we know M and N, the strengths of each of these solutions in terms of iodine and silver, we know that by adding to Q c.c. of iodide solution $P \times Q$ c.c. of silver nitrate solution, we shall get for A the weight of iodine $Q \times M$, and for B the weight of silver $P \times Q \times N$.

The preparation of silver iodide by mixing the iodide and the silver solutions.

Use 50 c.c. of iodide solution, and add to it $50 \times P$, the calculated quantity of silver nitrate solution, using graduated flasks, burette, pipettes, as may seem best for the most simple and most accurate production of the required volume. In order to produce the silver iodide in the form best suited for filtering, it may be best to measure out the required volumes of the two solutions each in a separate beaker, warm each, and then mix, remembering the necessity for complete transference of the one solution into the other.

Determination of the weight of the precipitated silver iodide.

Filter through a Gooch crucible, wash, dry and weigh ; for details see any text book of Quantitative Analysis, e.g. Clowes and Coleman.

(i) Determination of M and N, the strengths, i.e. weight in grams per c.c. of solution, of the iodine and silver solution.

(a) *Preparation of the standard iodine solution.*

		gms.
Wt. of empty weighing tube (approx., to decigrs. only) .... R		7·8
Wt. weighing tube + iodine .... S (iodine being put into the tube until value of $S = $ a little more than $R + 2·0$ gms.)		9·9
Wt. weighing tube + iodine (accurate to mgs.) ... W_1		9·900
,, ,, after shaking out the bulk of the iodine .... W_2		8·895
,, ,, after shaking out an additional small crystal of iodine ...... W_3		8·8866
∴ Wt. of iodine taken................... $\begin{array}{l} = W_2 - W_1 \\ \text{or } W_3 - W_1 \end{array}$ a		1·0134
Vol. to which the ammonium iodide solution obtained from a grams of iodine is made up... V c.c.		250
∴ M, the strength of the solution, *i.e.* the weight of iodine contained in 1 c.c. .... $\dfrac{a}{V}$		$\dfrac{1·0134}{250}$
		$= ·00405$ gm. per c.c.

(b) *Preparation of the standard silver solution.*

		gms.
Wt. of silver (weighed on balance pan or on accurately counterpoised watch glass) .... b		8867
Vol. to which the silver nitrate solution obtained from b gms. of silver is made up.................. V_1		250
∴ N, the strength of the solution, *i.e.* the weight of silver contained in 1 c.c. .... $\dfrac{b}{V_1}$		$\dfrac{·8867}{250}$
		$= ·00355$ gm. per c.c.

(ii) Determination of P, the equivalency factor between the two solutions.

(a) *Silver equivalent of the thiocyanate solution.*

Vol. of silver nitrate solution used .... $= V_2$			10
,, thiocyanate solution required for pptn. .... }	(i)	15·25	
	(ii)	15·25	
	(iii)	15·25	
Mean $= v$		15·25	
Vol. of silver nitrate solution ppd. by 1 c.c. of thiocyanate .... } $\dfrac{V_2}{v}$			$\dfrac{10}{15\cdot25} = \cdot65$ c.c.

(b) *Estimation of excess of silver nitrate solution in inter-action between definite volumes of the two solutions.*

Vol. of iodide solution taken.......................... $= V'$		25
,, silver nitrate solution added .... $= V''$		25
,, thiocyanate solution required to ppte. the excess of silver nitrate .... } V'_s		103 c.c.

$\cdot\cdot\cdot$ 1 c.c. of thiocyanate corresponds to $\dfrac{V_2}{v}$ of silver nitrate

$\therefore$ Excess of silver nitrate............................. $= v' \cdot \dfrac{V_2}{v}$ $\quad$ $1\cdot03 \times \cdot65 = \cdot675$ c.c.

Calculation of P from the results obtained in (a) and (b).

$\therefore$ Vol. of silver nitrate required for the complete pptn. of 1 c.c. of iodide

$$= \frac{V''}{V'} - \frac{v'}{V'} \cdot \frac{V_2}{v},$$

which on substituting for the values that we may choose arbitrarily those recommended, becomes

$$\frac{25}{25} = \frac{v'}{25} \cdot \frac{10}{v} = 1 - \frac{10v'}{25v} \quad .... = P \qquad \cdot94 \text{ c.c.}$$

(iii) **Determination of C, the weight of silver iodide produced by the interaction between A and B, weights of silver and iodine found from the preceding work to be strictly equivalent.**

(a) *Preparation of the silver iodide by mixing the requisite volumes of the two solutions, which gives the values for A and B, the quantities of iodine and silver used in the synthesis.*

Vol. of iodide taken .... V $= 50$ c.c.

,, silver nitrate added .... VP

$$= 50\left(1 - \frac{10v'}{25v}\right) \quad = 48\cdot65 \text{ c.c.}$$

$$\therefore A = V \cdot M \ .... = \frac{50 \cdot a}{250} \quad = \cdot2026 \text{ gm.}$$

$$B = V \cdot P \cdot N \ .... = 50\left(1 - \frac{10v'}{25v}\right)\frac{b}{250} \quad = \cdot1725 \text{ gm.}$$

(b) *Estimation of the silver iodide obtained by synthesis from A of iodine and B of silver.*

Wt. of Gooch crucible .... (i)		12·2774 gms.
(ii)		12·2774 ,,
(iii) W'	$=$	12·2774 ,,
,, Gooch crucible + silver iodide .... (i)		12·6532 ,,
(ii)		12·6528 ,,
(iii) W''	$=$	12·6528 ,,
∴ Wt. of iodide.............................. $W'' - W'$	$= C =$	0·3754 ,,

Final result :

$$C - (A + B) \ .... = \cdot3754 - \cdot3751 \text{ ,,}$$

$$= \cdot00025$$

CHAPTER VI

I. The Nature and Scope of the Law.

In Chapter III, which dealt with the classification of complex substances, the experimental study of some simple typical cases led us to the recognition of certain fundamental differences in qualitative relations, viz. : (i) the relation of the properties of the complex substance C to the properties of its constituents A and B; (ii) the principle of the methods by which C can be resolved into $A + B$. The concomitant fundamental quantitative differences, though there referred to incidentally only, are included in the following summary of the results.

Complex substances may be :

(1) Heterogeneous structures whose composition may vary continuously from $A = 0$ and $B = 100$ to $A = 100$ and $B = 0$, and in which the properties of C are additively those of A and B.

(2) Homogeneous structures.

(a) Solutions in which the ratio between the components varies continuously, either without limit as above (complete miscibility) or with a limit such that when $A = 100$, B may have any value from 0 to m, and in which the properties of the complex substance, though not additively those of the components, are a continuous function of the composition (see diagram, Chapter I, p. 80).

(b) Complex substances, in which the properties of A and B are merged in new ones of C, and in which the composition is fixed.

Of these three classes of complex substances, (1) and (2)(a), which are heterogeneous and homogeneous respectively, agree in that their *properties* as well as their *composition* are *variable*, and that there is a definite quantitative connection between the two variables, characteristics which are summed up in the name 'mixture.' On the other hand, the class characteristics of (2) (b), which are *fixity of properties* as well as *fixity of composition*, are summed up in the name 'compound.' The expression of the quantitative relation characteristic of compounds is:

$$\frac{A}{B} \text{ or } \frac{A}{C} \text{ or } \frac{B}{C} \text{ or } \frac{C}{D} = \text{constant},$$

where D stands for the weight of another substance with which C interacts by virtue of one or other of its components being exchanged either for D itself or for a component of D. Since in the experimental work connected with the establishment of the law of fixed ratios or the testing of its accuracy, the measurement of any one of these ratios may be resorted to, it should prove useful to illustrate by a simple qualitative experiment the different forms that the problem under discussion can take.

Experiment XXVII. The synthesis and various modes of decomposition of mercury bromide.

2 to 3 c.c. of mercury placed in a porcelain dish are covered with water, and a little bromine is added ; the colour of the bromine will gradually disappear, an effect which will be accelerated if by occasional stirring fresh surfaces of the mercury are exposed. As more and more bromine is gradually added, white crystals separate from the liquid above the un-changed mercury, showing the production of a new substance which is soluble in water. The disappearance of the colour proves that the bromine must have participated in the reaction which produces these white crystals, and the same could be established for the mercury, either by continuing the addition of bromine until all the mercury has disappeared, or by showing that it decreases in weight. In this synthesis the bromine and mercury correspond to our A and B, and the white crystals to our C. Pour off some of the colourless solution from the unchanged mercury, and to a portion add a few drops of chlorine water—bromine will be liberated ; to another portion add a soluble silver salt (*e.g.* silver nitrate)—pale yellow silver bromide, insoluble in nitric acid and sparingly soluble in ammonia, will be precipitated. In these reactions the chlorine, the silver nitrate and the silver bromide stand for D.

It is evident that each of the above reactions could be used for the measurement of a ratio the constancy of whose value would demonstrate the quantitative relationship characteristic of 'compounds.' There is, as a matter of fact, no special difficulty in measuring the amounts of the various substances participating in any of the reactions shown. Starting with known weights of bromine and of mercury, and weighing the unchanged mercury, would give us $A : B$. Starting with a known weight of bromine and weighing the white residue left after careful evaporation of the solution separated from the excess of mercury would give us $A : C$. $B : C$ could be got by starting with a known weight of mercury and adding more and more bromine until all the metal has disappeared, after which cautious evaporation of the solution to dryness would remove the excess of bromine (B.P. $59°$ C.), together with the water, and leave the white solid, which is weighed. Finally, starting with a known weight of the white crystals, we could find values for $C : D$ by the measurement of the quantitative relationship between C and any one of the various substances above used for D: we could drive off the liberated bromine together with the water, and thus obtain, in a form suitable for weighing, the chloride produced through the agency of the chlorine; or we could determine the weight of silver present in the form of silver nitrate which is required for the complete transference of the bromine from its combination with mercury to a fresh combination with silver; or we could weigh the insoluble silver bromide produced.

This then is an instance of the relationship known as the *Law of Fixed Ratios*: how has this law been discovered and established?

II. Historical.

Fixity of composition as an attribute of substances possessed of fixed properties must have been recognised from the time when quantitative analyses began to be made; the large amount of such work done during the eighteenth century must be taken as proof of a general belief in the existence of a law of fixed ratios. Berzelius (1779—1848), who at the beginning of the nineteenth century initiated a new era in quantitative work by his accurate determination of the composition of hundreds of the most important and the best known of compounds, continually refers to the results obtained by a number of predecessors, Klaproth (1743—1817), Buchholz (1770—1818), Wenzel (1740—1793), Vauquelin (1763—1829), J. B. Richter (1762—1807) and others. Earlier still was the work of Bergman (1735—1784), Marggraf (1709—1782) and of Kunkel (1630—1702), whose almost startlingly accurate determination of the composition of silver chloride has been referred to in a previous chapter (*ante*, Chapter II, p. 97, footnote).

But, though evidently tacitly assumed, the law of fixity of composition was not formulated and definitely enunciated until its validity had been denied and variability of composition asserted to be the rule. The great French chemist Claude Louis Berthollet (1748—1822) at the end of the eighteenth century asserted that the composition of complex substances depends on the relative masses of the constituents present at the moment of their formation, and that the observed cases of fixity of composition represent exceptions due to what we might now-a-days call 'survival of the fittest.' According to Berthollet, when an acid acts on an alkali a theoretically unlimited number of combinations can be formed in which the composition would vary continuously and imperceptibly from acid = 0 and alkali = 100 to acid = 100 and alkali = 0, and the fact that what crystallises out from the solution is always the same neutral substance of quite fixed properties and quite fixed composition, is due to the special circumstance that of all the combinations which are possible and which actually exist in the solution, this is the least soluble which therefore separates out first; but thereby the equilibrium in the solution is disturbed, and a fresh amount of substance of this special composition is formed and is separated out again, a process which goes on until all the acid or alkali available has been used up in the production of a salt of fixed composition, formed at the cost of all other possible ones owing to a special physical property possessed by it. The same reasoning was applied to the case of gases such as water vapour, the fixity of whose composition was explained by the assumption that of all the possible combinations between the constituents, the one of fixed composition is formed at the cost of all others because it is the most volatile.

Berthollet's doctrine of variable composition was opposed and refuted by his contemporary Proust (1755—1826). For one thing, Proust showed that a great many of the cases adduced by Berthollet as instances of variable composition could be shown to be mixtures in varying ratios of two substances, each of perfectly fixed composition: *e.g.* the substance obtained by the incomplete calcination of lead in air, the composition of which may vary from 100 of lead and 0 of oxygen to 100 of lead and 7·8 of oxygen, could, by treatment with dilute acetic acid, which dissolves litharge but has no action on lead, be shown to be a mixture of varying amounts of lead and litharge. But Proust's great work consisted in the definite proof of fixity of composition by the production of experimental evidence of the most comprehensive and the most conclusive kind. He demonstrated that the composition of a substance of fixed properties was absolutely independent of the conditions operative at its formation.

(i) Minerals of different origin are identical in composition :

No differences have yet been observed between the oxides of iron from the South and those from the North. The cinnabar of Japan is constituted according to the same ratio as that of Almaden.

(ii) Native and artificially prepared substances have the same composition :

The malachites[1] of Arragon when dissolved in nitric acid lose carbonic

[1] The beautiful green mineral malachite is a 'basic' carbonate of copper. Its composition can be demonstrated by heating a little of the powdered substance in an ignition tube, when water will condense in the upper cold parts of the tube, and the production of carbon dioxide can be shown in the usual way by the lime-water test; the black residue left in the tube is qualitatively and quantitatively identical with the black substance obtained by heating copper in

acid and leave 1 per cent. of an earthy residue; precipitation reproduces 99 parts of artificial carbonate....We must conclude that nature acts not differently in the depths of the earth...than through the agency of man.

(iii) Alteration in the amount and the kind of the material from which a substance is formed does not produce a change in its composition. Antimony sulphide is found to have the same composition whether it is synthesised from antimony and sulphur or from antimony and cinnabar, and however great may be the excess of sulphur present at its formation :

"Antimony follows the law of all the metals which can unite with sulphur. It attaches to itself an invariably fixed quantity of it, which we have no power to increase or diminish." And hence : "the properties of a true compound are as invariable as the ratio of its constituents...it is a privileged product to which nature assigns fixed ratios,...a being which nature never creates, even when through the agency of man, otherwise than balance in hand, *pondere et mesura*."

Proust carried the day, and from the early years of the nineteenth century the law of fixed ratios has been accepted as one of the fundamental principles of the science of chemistry. The recognition of fixity of composition as the essential and distinguishing characteristic of compounds is however not incompatible with belief in the possibility of very small deviations from the requirements of the exact law. Do such deviations exist ?

III. Establishment of the Law as exact.

The first period of marked progress and brilliant achievement in the realm of accurate stoichimetrical measurements began about 1810, and is associated with the name of the

air (*ante*, Chapter II, p. 90). Addition of sodium carbonate to a soluble copper salt (*e.g.* nitrate, chloride, sulphate) produces a beautiful green precipitate, carbon dioxide being evolved at the same time. The evolution of the carbon dioxide indicates the formation of a basic salt, *i.e.* a salt in which the amount of the base, in this case copper oxide, is greater than that required for the complete saturation of the carbonic acid. The green precipitate can be shown qualitatively to be made up of the same constituents as malachite, whilst the identity in quantitative composition of the two substances can be demonstrated in the simple way devised by Proust and followed in the illustrative experiment in which calcium carbonate (Iceland spar) is substituted for copper carbonate.

great Swedish chemist Berzelius (*ante*, p. 228). About half a century later a fresh advance of at least equal importance was made, due again to the ingenuity, the experimental skill, the wonderful perseverance and untiring energy of one man ; it was the Belgian Stas (1813—1891) who not only devised new and most ingenious methods, but who also applied them with such signal success, that it became possible to test for the existence or non-existence of very small deviations from constancy of composition. We owe to Stas two classical researches planned specially to test the exactness of the law of fixed ratios, viz. the synthesis of silver chloride and the analysis of ammonium chloride, in which he measured with the utmost accuracy attainable the ratios $A : C$ (silver : silver chloride) and $D : C$ (silver : ammonium chloride) for specimens prepared in diverse ways, from different material, under varying conditions of temperature and pressure.

The results obtained for silver chloride may be summarised thus :

Chloride synthesised by:

		Ratio $A : C$
Burning silver in chlorine, a reaction occurring at red heat		100 : 132·841 to 132·843
Dissolving silver in nitric acid and precipitating at the ordinary temperature with	(i) hydrogen chloride (gaseous)	100 : 132·846 to 132·849
	(ii) hydrochloric acid solution	100 : 132·846
	(iii) ammonium chloride	100 : 132·842
		Mean = **132·845**

Silver chloride, though produced under very different conditions, has a composition which I am forced to regard as constant. In fact I dare not attribute the insignificant differences in the results to other causes than the inevitable errors of observation. (Stas, 1865.)

The work of Stas has been accepted by chemists as settling the question of fixity of composition for the time at any rate ; absolute fixity in a matter such as this is impossible to attain. What has been proved is, that if deviations from the requirements of the exact law do exist, they are less than the experimental error involved in the measurements—even the most accurate—of the combining ratios. But we cannot foretell what the future may bring, what may be revealed by a possibly enormously increased accuracy of weighings and a correspondingly great reduction of the experimental error.

It was not only in the above two researches, which were undertaken with the special object of testing the accuracy of the law of fixed ratios and in which everything hinged on this point, that Stas used the device of working with specimens of the same substance derived from different sources and prepared under different conditions. The example set by him in this respect has been followed as a matter of course in recent work which aims at a high degree of accuracy[1]. Thus in the researches carried out at Harvard by T. W. Richards and his collaborators, work which does not fall far short of a revision of the atomic weight of nearly every one of the more important elements, most extensive use is made of this principle :

> No single method of preparation is adequate for a case of this sort...some entirely different method must be adopted for preparing another sample, in order to be certain that constant impurity has not found its way into the first, in spite of all precautions. Stas realised this.

The constancy of the values obtained for such different specimens supplies the required proof of the purity of the material used, fixity of composition having been assumed. This case is in every way analogous to that of the complete syntheses and analyses discussed in the preceding chapter, where also the degree of agreement of the results obtained with the requirements of a law postulated as exact was used as a test of the success of the experimental work. And here as there, by an interchange of what is *assumed* and what *tested*, we can use this type of combining weight determinations as a measure of the exactness of the law of fixed ratios, thereby supplying evidence additional to that obtained by Stas when he deliberately set to work to investigate this stoichiometrical relationship. Thus in a recent re-determination

---

[1] The hydrogen chloride used in Gray and Burt's density measurements was derived from three different chlorides (*ante*, Introd., p. 23). The tellurium compounds used in Baker and Bennett's determination of the atomic weight of tellurium were treated in eight different ways (*ante*, Chapter III, p. 138); in Lord Rayleigh's measurements of the density of nitrogen, the atmospheric nitrogen was obtained by using three different absorbents for the oxygen, and the chemical nitrogen was obtained from three different compounds (*Study of Chemical Composition*, p. 90).

of the ratio sodium chloride (*C*) to silver chloride (*D*), the
following values were obtained :

Source of the sodium chloride	(Weight of sodium chloride equivalent to 100·000 parts of silver chloride)
German rock salt.................................	40·779
	40·779
	40·779
Sodium bicarbonate and pure hydrochloric acid...	40·780
	40·778
	40·780

IV. Students' illustrative experiments.

1. **The combining ratios of substances interacting in solution are independent of the concentration.**

Experiment XXVIII. To show that when a neutral salt is produced, the ratio of the weights acid : alkali is independent of the concentration of the solutions used.

Starting with a solution of an acid (*e.g.* hydrochloric or sulphuric) and a solution of an alkali (*e.g.* potash or soda) of strength *a* grams per c.c. and *b* grams per c.c. respectively, the required ratio is given by $\frac{v}{V} \cdot \frac{a}{b}$, and the point at issue is to compare a number of values

$$\frac{v'}{V'} \cdot \frac{a}{b}; \quad \frac{v''}{V''} \cdot \frac{a}{b}; \quad \frac{v'''}{V'''} \cdot \frac{a}{b}; \ldots\ldots$$

obtained in determinations in which to the carefully measured volumes *V'*, *V''*, *V'''*, we have added unknown *different* quantities of water, whereby though we have altered the concentration we have not altered *Vb* the total quantity of alkali used. Obviously the constant factor $\frac{a}{b}$ can be cancelled, and the results stated in terms of volumes only (provided always we work throughout with the same two solutions).

Try the action of potash and of soda, of hydrochloric acid and of sulphuric acid, on a number of the colouring matters litmus, methyl orange, phenol-phthalein and cochineal, which are the most important representatives of a class of substances named 'indicators[1],' and record the results in tabular form. Show that by the careful addition of the acid to the alkali, to which a few drops of indicator have been added, a point is reached at which a sharp colour change appears, *e.g.* from blue to purple (or red) in the case of litmus, from deep red to colourless in the case of phenol-phthalein ; measure the volumes required for this. Then, without the addition of the indicator, mix alkali and acid in this ratio, using at least 100 c.c. of the

---

[1] Clowes and Coleman, *Quantitative Analysis*, 1911, §§ 262 *et seq.*, give instructions concerning the preparation of solutions of suitable strength.

alkali. Evaporate the solution to about half its bulk and let it stand until crystals deposit; collect the crystals[1], drain on a filter plate, wash with the least quantity of cold water and dry on a porous plate. Compare these crystals with the acid and alkali from which they have been made[2]. Ascertain with all possible accuracy the relative quantities of acid and of alkali required for the production of the crystalline substance above studied, by adding the one solution to a known volume of the other until the colour change in an indicator shows the absence of either free acid or alkali.

Into three beakers measure by means of a pipette (preferably the same throughout the experiment, so as to diminish errors due to incorrect calibration) three different quantities of the same potash or soda solution[3], and add to each an arbitrarily chosen different quantity of water; add an amount of indicator just sufficient to colour the solution[4], and from a burette run in the acid until the colour change characteristic of the special indicator used has occurred. Read the burette at the beginning and at the end, taking all the precautions necessary for obtaining correct values (Clowes and Coleman, 1911, § 256).

Record of results.

A. Form of note-book entry.

Vol. of original undiluted potash solution taken V	Vol. of water added	Vol. of sulphuric acid solution required for neutralisation. Burette readings at			Vol. of acid required for neutralisation of 1 c.c. of potash $= \dfrac{v}{V}$
		(i) beginning	(ii) end	Difference (ii) − (i) = v	
10 c.c.	10 c.c.	0·00	6·65	6·65 (m)	·665
10 c.c.	20 c.c.	6·65	13·35	6·70 (m')	·670
20 c.c.	10 c.c.	14·20	27·40	13·20 (n)	·660
20 c.c.	20 c.c.	27·40	40·70	13·30 (n')	·665
30 c.c.	10 c.c.	0·00	20·10	20·10 (p)	·670
30 c.c.	20 c.c.	20·10	40·25	20·15 (p')	·672

[1] The procedure is the same as that followed in the recrystallisation of potassium chlorate, Chapter II, p. 91.

[2] The principle is that set forth in Chapter I.

[3] Solutions about normal will prove of a suitable strength; such solutions are obtained by making up to 1 litre the solution of about 45 grams of caustic soda, 60 grams of caustic potash, 250 c.c. of hydrochloric acid of specific gravity 1·085, 250 c.c. of sulphuric acid of specific gravity 1·155.

[4] If too much is added, it will increase the difficulty of fixing with accuracy the exact end point.

B. Summary of results obtained by five members of a class.

		$\dfrac{m+m'}{2 \times 10}$	$\dfrac{n+n'}{2 \times 20}$	$\dfrac{p+p'}{2 \times 30}$
Student	A	·660	·660	·655
	B	·670	·665	·670
	C	·655	·655	·656
	D	·670	·665	·666
	E	·650	·650	·650

As in the case of Experiment XXIV, for the majority of students it will not be worth while to do this experiment, but it has been included for two reasons, one theoretical and one practical : the fact which the experiment is intended to demonstrate, though usually taken for granted, is after all not one which can be assumed *a priori* ; it may prove useful to substitute this experiment for one of the more laborious ones following in the case of such members of the class as may require practice in accurate titration, the technique of which involves certainty in volume readings and an eye trained to the recognition of the first indication of certain colour changes.

2. **The composition of substances is independent of the mode of preparation (material used and physical conditions).**

(1) Synthesis of silver chloride (illustrative of Stas' classical synthesis).

Determination of the ratio silver : silver chloride, the silver chloride being prepared from silver

(*a*) by burning in chlorine,

(*b*) by solution in nitric acid and precipitation by hydrochloric acid.

Experiment XXIX (*a*). The gravimetric synthesis of silver chloride by burning a known weight of silver in a stream of chlorine and weighing the silver chloride formed.

A suitable apparatus is shown in fig. 49, p. 237. The silver is heated in a hard glass bulb tube, which is bent in the blowpipe and supported in a slanting position in order that the silver chloride formed should flow into the bulb, thus exposing a fresh surface of the silver to the chlorine and

minimising the danger of any of the chloride being carried away by the
stream of chlorine. The chlorine is prepared in the usual manner from
lumps of pyrolusite and pure strong hydrochloric acid ; it is dried by strong
sulphuric acid. The removal of any hydrochloric acid by water, a matter
of great importance in the density experiment (*ante*, p. 43), is in this case
immaterial. If the flask in which the chlorine is evolved is fairly large,
and a rose burner is used for heating, it is possible to regulate the rate of
evolution of chlorine quite easily and to have a steady current passing for
a considerable time. At the end of the experiment, the chlorine filling the
bulb tube and that retained by the silver chloride in the bulb is removed
by fusing the chloride in a stream of pure dry air. For this purpose the
bulb tube is detached from the rest of the apparatus, an aspirator of the
usual form (*ante*, p. 89) is connected at *N*, and tubes or wash-bottles
containing material for retaining moisture and carbon dioxide, *e.g.* soda
lime, or potash and calcium chloride, or potash and strong sulphuric acid,
are connected at *M*.

Record of Results.

Specimen of note-book entry.

Wt. of bulb tube + suspending wire .... =	40·2750 grams
,, ,, ,, ,, + silver .... =	42·0480 ,,
∴ Weight of silver................................. =	1·7730 ,,
Wt. of bulb tube + suspending wire + silver chloride =	42·6285 ,,
∴ Weight of silver chloride .... =	2·3535 ,,
∴ 100 of silver give silver chloride............. =	**132·74**

(*b*) The synthesis of the chloride from a known weight of metallic silver
dissolved in nitric acid and precipitated in hydrochloric acid has been
carried out already in Experiment X, Chapter II, p. 96, where additional
variation of condition was effected by (*a*) weighing the precipitate in the
flask in which it was produced, (*β*) weighing the precipitate collected on
a Gooch filter. The mean value of the results, summarised on p. 102,
Chapter II, was **132·25**.

Combination of results from (*a*) and (*b*) illustrating the point under investigation.

Determination of the ratio silver : silver chloride = *A* : *C*.

Chloride synthesised by:	Ratio *A* : *C*
Silver burnt in chlorine	100 : 132·74
Silver dissolved in nitric acid and precipitated by ⎫ hydrochloric acid solution .... ⎭	100 : 132·25

(2)   Analysis of potassium chloride.

Using differently prepared specimens of potassium chloride,
determine the ratios

      (i)   silver nitrate : potassium chloride,

      (ii)   silver : potassium chloride.

Apparatus for determining the weight of silver chloride obtained when a weighed quantity of silver is heated in a stream of chlorine.

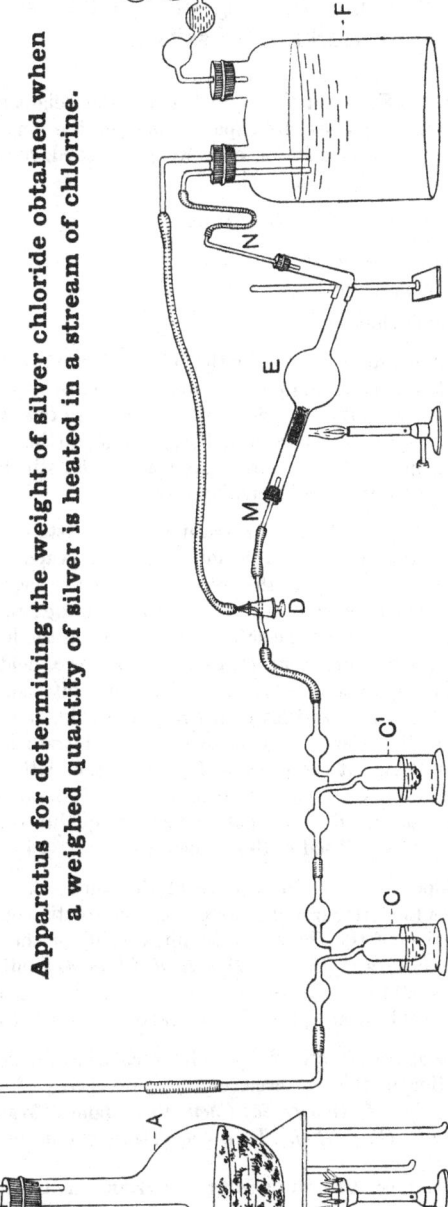

Fig. 49.

A, B, D, F, G.   Same as in apparatus, Introductory, Appendix, p. 42.

    E.   Hard glass bulb tube, with silver placed in stem to left of bulb.

  M, N.   Glass tubes passing through corks in stem of reduction tube, and connected by rubber tubes with D and F respectively.

  C, C'.   Wash-bottles filled with conc. sulphuric acid.

(i)   Determination of the ratio silver nitrate : potassium chloride, the silver nitrate being stated either as volume of solution used or as weight of the substance.

Experiment XXX.   Find the ratio between the weight of potassium chloride—using differently prepared specimens—and the amount of silver nitrate, stated as volume of solution or as weight of solid, required for the complete precipitation of the chloride.

A.   The preparation of the material from :

(*a*)   potassium bicarbonate,

(*b*)   potassium nitrate,

(*c*)   potassium chlorate.

(*a*)   Prepare potassium chloride by the action of pure hydrochloric acid on pure recrystallised potassium bicarbonate, added until there is no more effervescence and the solution is acid to litmus ; remove the excess of acid by evaporating to dryness in a porcelain dish, grinding up the solid obtained and then igniting in a crucible.   Grind up again whilst still hot, and put into a small stoppered or corked weighing tube[1].

(*b*)   Prepare potassium chloride by treating pure recrystallised potassium nitrate with concentrated pure hydrochloric acid and evaporating to dryness (draught cupboard) ; repeat this process once or twice, whereby the greater part of the nitric acid is replaced by hydrochloric ; to remove the rest of the nitric acid grind up the substance, dissolve in the least quantity of boiling water, add pure concentrated hydrochloric acid and cool ; beautiful crystals of potassium chloride will be precipitated.   Drain the crystals, wash with small quantities of hydrochloric acid, and test for the absence of nitrate by 'nitron' (see below).   If the nitrate has not been completely removed, repeat the process of precipitation with hydrochloric acid until this has been achieved ; finally remove the excess of hydrochloric acid and preserve as in (*a*).   Incidentally perform qualitative tests illustrative of the reactions utilised in this preparation.

(*a*)   Hydrochloric acid, which is of about the same volatility as nitric acid, partly expels the latter from its salts.   Test the solution obtained after the first evaporation of the chloride with nitric acid, for the presence of both chloride and nitrate (Clowes, *Practical Chemistry*, 8th ed., 1908, pp. 209 *et seq.*, pp. 214 *et seq.*).   Show also that after the second or third treatment by hydrochloric acid, the nitrate reaction becomes feebler.

(*β*)   Chlorides are much less soluble in hydrochloric acid than in water (for the explanation of this in terms of the ionic theory of solution, see Ostwald, *Principles of Inorganic Chemistry*, James Walker, *Introduction to Physical Chemistry*.   To a concentrated aqueous solution of

---

[1]   A small specimen tube fitted with a soft new cork answers the purpose well.

potassium chloride add strong hydrochloric acid, and note the precipitation of the crystalline chloride. The precipitation may be made more complete if instead of adding a solution of the acid, gaseous hydrochloric acid is led into the solution by means of a small funnel[1] which just touches the surface of the liquid.

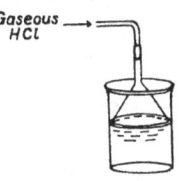

Fig. 50.

(γ) The nitrate of the organic base called 'nitron' is so very insoluble that its formation is a most sensitive test for nitrates. Dissolve a little nitron in acetic acid, and add a few drops of it to about 1 c.c. of an extremely dilute solution of nitric acid contained in a test-tube and further diluted by the addition of water until the tube is full; allow to stand, and note the formation of a white crystalline precipitate. Apply this test to the chloride made from the nitrate.

(c) Prepare potassium chloride by the ignition of pure recrystallised potassium chlorate.

Proceed as in Experiment IX, Chap. II, p. 90, and do all you can to ensure the reaction being made complete by repeated grinding up of the solid, followed by strong heating. Remove any chlorate remaining by boiling with conc. hydrochloric acid.

Show the qualitative identity of the three substances prepared under (a), (b), (c) by the following reactions:

(a) Flame test,

(β) Action of concentrated sulphuric acid on solid,

(γ) Action of concentrated sulphuric acid on solid mixed with manganese dioxide,

(δ) Action of silver nitrate on a solution.

Apply to each of the three samples the tests for purity given by Merck[2]

B. The analysis of the three specimens of potassium chloride.

(1) Determination of the ratio potassium chloride : silver nitrate. Using potassium chromate as indicator, find the volume of silver nitrate required for the precipitation of a known weight (about ·2 gram, weighed out most accurately) of the potassium chloride. The principle and practice of this method have been fully discussed in Chap. I, when an analysis in every respect analogous to the present one was made of (a) ammonium chloride and (b) the substance obtained by its sublimation. As a matter of fact, that experiment could find a place in this series if, content with the qualitative proofs for the identity of the two substances, we made the analysis a *test* of the law of fixed ratios, which in the earlier experiment was assumed. Everything that was said there as regards the following points still holds good, viz.: the sufficiency, for the purpose of the experiment,

---

[1] The use of the funnel is to prevent sucking back of the solution, a danger due to the extreme solubility of hydrogen chloride.

[2] *Chemical Reagents, their purity and tests* (1907), p. 186.

of determining the volume of silver nitrate solution; the advantage of expressing the relation in terms of weight instead of volume of silver nitrate; and the desirability of a critical examination of the differences between the value found and the standard value.

Record of Results.

A. Specimen of note-book entry.

Determination of the

ratio wt. potass. chloride : (*a*) vol. }of silver nitrate
(*b*) weight}

(Strength of silver nitrate solution used = *m* grams per c.c.)

Source of chloride	Wt. of chloride A gms.	Vol. of silver nitrate solution v c.c.	Wt. of silver nitrate solution vm gms.	Vol. of silver nitrate interacting with 1 gm. of potass. chloride $\frac{v}{A}$ c.c.	Wt. of $\frac{vm}{A}$ gms.
(*a*) Potass. bicarbonate (i)	·2116	28·6	·4862	135·2	2·298
and hydrochloric (ii)	·3890	52·65	·9090	135·4	2·301
acid .... (iii)	·4037	54·60	·9282	135·3	2·300
Mean				**135·3**	**2·300**
(*b*) Potass. chlorate ... (i)	·3220	43·50	·7395	135·1	2·296
(ii)	·2116	28·60	·4862	135·2	2·298
(iii)	·2019	27·30	·4641	135·2	2·299
Mean				**135·2**	**2·298**

B. Summary of Results.

Weight of silver nitrate required for the precipitation of 1 gram of potassium chloride.

	Potass. bicarbonate and hydrochloric acid	Potass. nitrate and hydrochloric acid	Potass. chlorate
Student A	2·31	2·28	2·34
	2·31	2·33	2·31
B	(2·25)	2·33	2·27
	2·30		
C	2·31	2·29	2·35
	2·34	2·32	(3·24)
D	2·27		(2·36)
	2·33		2·32
Mean	**2·307**	**2·310**	**2·318**

Mean of whole series .... = **2·312**

Standard value, from combination of ratios: }
Silver nitrate : silver and } = **2·2792**
Silver   : potass. chloride }

(ii) Determination of the ratio silver : potassium chloride[1], $D : C$.

Principle of the method : Determination of (i) the *relative* silver-precipitating power exerted on the same solution of silver nitrate (which is of suitable but unknown strength) by (*a*) C grams of potassium chloride and (*b*) 1 c.c. of a certain solution of thiocyanate (also of suitable but unknown strength); (ii) the *absolute* silver precipitating value of the thiocyanate solution, as measured by its interaction with a solution containing a known weight of silver, viz. b grams per c.c.

The method is analogous in principle and involves the same technique as the one used in Experiment XXVI, Chapter V, p. 217, for the measurement of the ratio silver : iodine.

Experiment XXXI. Find the ratio between the weight of potassium chloride—using differently prepared specimens—and the weight of silver (in the form of silver nitrate) required for the complete precipitation of the chloride.

Materials required :

The differently prepared specimens of potassium chloride (see above, p. 238).

Pure silver (about 1 gram) weighed very accurately, dissolved with the usual precautions against loss, and made up to a known volume, viz. 100 c.c.

A silver nitrate solution of suitable but unknown strength, containing about 10 grams of the pure recrystallised salt in 1 litre.

A thiocyanate solution of suitable but unknown strength, containing about 3 grams of the pure recrystallised salt in 1 litre.

An iron alum solution to be used as indicator.

Procedure.

The operations required are :

(1) Preparation of a silver solution of known strength. Detailed instructions concerning the preparation of such a solution have been given before (p. 220). Let the weight of silver taken be b grams and the volume of the solution made up from it V_1 c.c., then 1 c.c. contains $\dfrac{b}{V_1}$ grams of silver.

---

[1] This is a value of considerable importance in the building up of the system of combining weights (*post*, p. 334; *Study of Chemical Composition*, pp. 205 *et seq.*).

(2) The determination of the silver equivalent of the thiocyanate solution. This corresponds to (ii) (*a*) in Experiment XXVI (see p. 223). The silver solution, which is in the dish or beaker, is measured by a pipette, the thiocyanate is run in from and measured by a burette. Let it be found that V_2 c.c. of the silver solution made in (1) require for the completion of the reaction, indicated by the appearance of a pink tinge, p c.c. of thiocyanate.

∴ 1 c.c. of thiocyanate precipitates
(interacts with—corresponds to) $\dfrac{V_2 \cdot b}{p \cdot V_1}$ grams of silver.

(3) The standardising of the unknown silver nitrate solution by the thiocyanate solution. The procedure is the same as above in (2).

Let it be found that V_3 c.c. of silver nitrate solution is precipitated by q c.c. of the thiocyanate.

∴ 1 c.c. of thiocyanate interacts with $\dfrac{V_3}{q}$ of the silver nitrate solution,

and

1 c.c. of silver nitrate solution corresponds to $\dfrac{q}{V_3}$ of the thiocyanate

solution.

(4) Determination of the volume of silver nitrate precipitated by C grams of potassium chloride. This measurement, which corresponds to (ii)(*b*) in Experiment XXVI is made indirectly by adding to a known weight of the potassium chloride (dissolved in an arbitrarily chosen unknown amount of water) a volume of the silver nitrate solution which is known, and which is more than sufficient for effecting the precipitation, the excess of silver nitrate being estimated by titration with the thiocyanate solution.

Let the weight of potassium chloride taken be.................................C grams

,,   volume of silver nitrate added to the chloride solution be ......V c.c.

,,   volume of the thiocyanate solution required to precipitate
        the excess of silver nitrate be ....v c.c.

Calculation.

Thiocyanate equivalent of the V c.c. of silver nitrate used in the

precipitation of the C grams of potassium chloride ....$= \dfrac{V \cdot q}{V_3}$ c.c.
$\qquad\qquad\qquad\qquad\qquad\qquad\qquad\qquad\qquad\qquad\qquad$ (from 3)

∵ thiocyanate required to precipitate excess of silver nitrate ....$= v$ c.c.
$\qquad\qquad\qquad\qquad\qquad\qquad\qquad\qquad\qquad\qquad\qquad$ (from 4)

∴ thiocyanate equivalent in silver precipitating power to

C grams of potassium chloride .... $= \left(\dfrac{V \cdot q}{V_3} - v \right)$ c.c.

∵ 1 c.c. of thiocyanate solution precipitates $\dfrac{V_2 \cdot b}{p \cdot V_1}$ grams of silver (from 2)

∴ silver precipitated by C grams of potassium chloride $= \left(\dfrac{V \cdot q}{V_3} - v \right) \cdot \dfrac{V_2 \cdot b}{p \cdot V_1}$ gms.

∴ Weight of potassium chloride which precipitates 100

of silver .... $= \dfrac{C \times 100}{\left(\dfrac{V \cdot q}{V_3} - v \right) \cdot \dfrac{V_2 \cdot b}{p \cdot V_1}}$ gms.

Record of results.

Specimen of note-book entry.

Weight of potassium chloride which precipitates 100 of silver.

(1) Preparation of the standard silver solution.

Wt. of silver taken $= 1\cdot000$ grams
Volume of the solution $= 100$ c.c.

(2) Silver equivalent of the thiocyanate solution.

Volume of silver solution taken V_2 c.c.	Vol. of thiocyanate required for precipitation p c.c.	Silver equivalent of thiocyanate $\dfrac{V_2 \cdot b}{pV_1}$ grams per c.c.
5 c.c.	5·70 c.c.	
,, $+15$ c.c.	23·2 c.c. $= 4 \times 5\cdot80$ c.c.	
,, ,, $+10$ c.c.	34·95 c.c. $= 6 \times 5\cdot82$ c.c.	
10 c.c.	11·65 c.c. $= 2 \times \mathbf{5\cdot82}$ c.c.	0·00859 grams per c.c.

(3) Thiocyanate equivalent of unknown silver nitrate solution.

Vol. of silver nitrate $= V_3$ c.c.	Vol. of thiocyanate required $= q$ c.c.	Vol. of thiocyanate which precipitates 1 c.c. of silver nitrate $= \dfrac{q}{V_3}$
5 c.c.	12·4 c.c.	
,, $+5$ c.c.	25 c.c. $= 2 \times 12\cdot5$ c.c.	
,, ,, $+5$ c.c.	37·5 c.c. $= 3 \times \mathbf{12\cdot5}$ c.c.	$\mathbf{2\cdot50}$ c.c.

(4) Thiocyanate equivalent of the potassium chloride.

Source of the chlorides	Potass. bicarbonate		Potass. nitrate		Potass. chlorate	
	(i)	(ii)	(i)	(ii)	(i)	(ii)
Wt. of potass. chloride $= C$ grams	·451	·354	·366	·387	·314	·382
Vol. of silver nitrate solution added... $\Big\} = V$ c.c.	40	40	40	40	40	40
Thiocyanate equivalent of silver nitrate solution added $\Big. = \dfrac{V \cdot q}{V_3}$ c.c.	100	100	100	100	100	100
Thiocyanate required for precipitating excess of silver $\Big. = v$ c.c.	24·10	40·15	38·90	35·00	47·65	35·70

Source of the chlorides	Potass. bicarbonate		Potass. nitrate		Potass. chlorate	
	(i)	(ii)	(i)	(ii)	(i)	(ii)

Calculation.

Thiocyanate equivalent of C grams of potass. chloride $\Big\} = \dfrac{V \cdot q}{V_3} - v$ c.c.	75·9	59·85	61·10	65·00	52·35	64·30
Silver precipitated by C grams of potass. chloride $= \left(\dfrac{V \cdot q}{V_3} - v\right)\dfrac{V_2 \cdot b}{p \cdot V_1}$ grams	·6519	·5141	·5248	·5584	·4496	·5524
Wt. of potass. chloride required to precipitate 100 grams silver $= \dfrac{C \times 100}{\left(\dfrac{V \cdot q}{V_3} - v\right)\dfrac{V_2 \cdot b}{p \cdot V_1}}$ grams	69·18	68·85	69·74	69·31	69·82	69·16
Mean =	**69·02**		**69·53**		**69·49**	

Two students who used the more laborious but no doubt more satisfactory method of making a special standard silver solution for each experiment obtained the following results:

Weight of potassium chloride which precipitates 100 of silver.

	Student A		Student B			
Source of potass. chloride	Potassium bicarbonate	Potassium chlorate			Potassium perchlorate	
Weight of chloride $= a$ grams	·2769	·4358	·5147	1·4773	2·9692	2·9606
Wt. of silver added $= b$ grams	·4040	·7244	·8415	2·1674	4·4712	4·4460
Excess of silver estimated by titration with thiocyanate $= c$ grams	·0061	·0975	·1005	·0271	·1701	·1584
Wt. of silver used in the precipitation of the chloride $= b - c$ grams	·3979	·6269	·7410	2·1403	4·3011	4·2876
Wt. of potassium chloride which precipitates 100 of silver $= \dfrac{a \times 100}{b - c}$ grams	**69·6**	**69·5**	**69·5**	**69·02**	**69·03**	**69·05**

Standard Value (Richards and Staehler) **69·1073**.

3. **The composition of the compound is independent of the relative amounts of the constituents present at its formation.**

The composition of water shown to be independent of the original amounts of oxygen and hydrogen present.

Experiment XXXII. To find the volume of gaseous hydrogen combining with 1 of oxygen in the formation of liquid water, the volume of which is assumed to be negligible[1]. This experiment is illustrative of Gay-Lussac and Humboldt's classical determination in 1805, and of more recent exact measurements by Scott in 1893 (*Study of Chemical Composition*, p. 312).

Fig. 51 shows an arrangement of apparatus suitable for

   (i)   the preparation and the purification of the two gases,

   (ii)   the bringing together of measured quantities of the gases,

   (iii)   the use of an electric spark in bringing about combination in a mixture of the gases confined over water.

(i) The hydrogen is made by the action of dilute sulphuric acid on zinc, and is purified by passing it through an alkaline solution of potassium permanganate, which retains hydrogen sulphide, phosphide and arsenide[2] and carbon dioxide; the measurement of the gas being done over water, there is no need to dry it. It is very important to ensure the hydrogen being free from air by letting a slow stream of the gas pass for at least 20 minutes before collecting.

The oxygen is made from sodium peroxide, and stored in the manner described before (Chapter III, p. 114).

(ii) It is recommended to use for a 'eudiometer[3]' a U-tube the closed

---

[1] It will be instructive to do the simple calculation required to show the percentage error introduced by this assumption. The data are:

        1 c.c. hydrogen   weighs ·00009 gram

        1 c.c. oxygen   ,,   ·00143 ,,

        1 c.c. liquid water ,,   1   ,,

2 vols. gaseous hydrogen unite with 1 vol. gaseous oxygen.

∴ 3 litres of gaseous mixture weighing $(2 \times \cdot 09) + 1\cdot43 = 1\cdot61$ grams form an equal weight of liquid water, which since its density is 1 corresponds to 1·61 c.c.

The ratio of the volume neglected to the original volume

$$= \frac{1\cdot61}{3000} = \cdot05 \text{ per cent. approx.}$$

In ordinary volume measurements the experimental error is generally much greater than ·05 per cent.

[2] The interaction between these hydrides and the permanganate is one of oxidation of the hydride accompanied by reduction of the permanganate to the green manganate. Students should note the difference in the reduction of the alkaline permanganate in this case from that occurring in acid solution, when the reduced substance is colourless manganous salt.

[3] Perkin and Lean, *Introd. to Chem. and Physics*, II, p. 100, 1901.

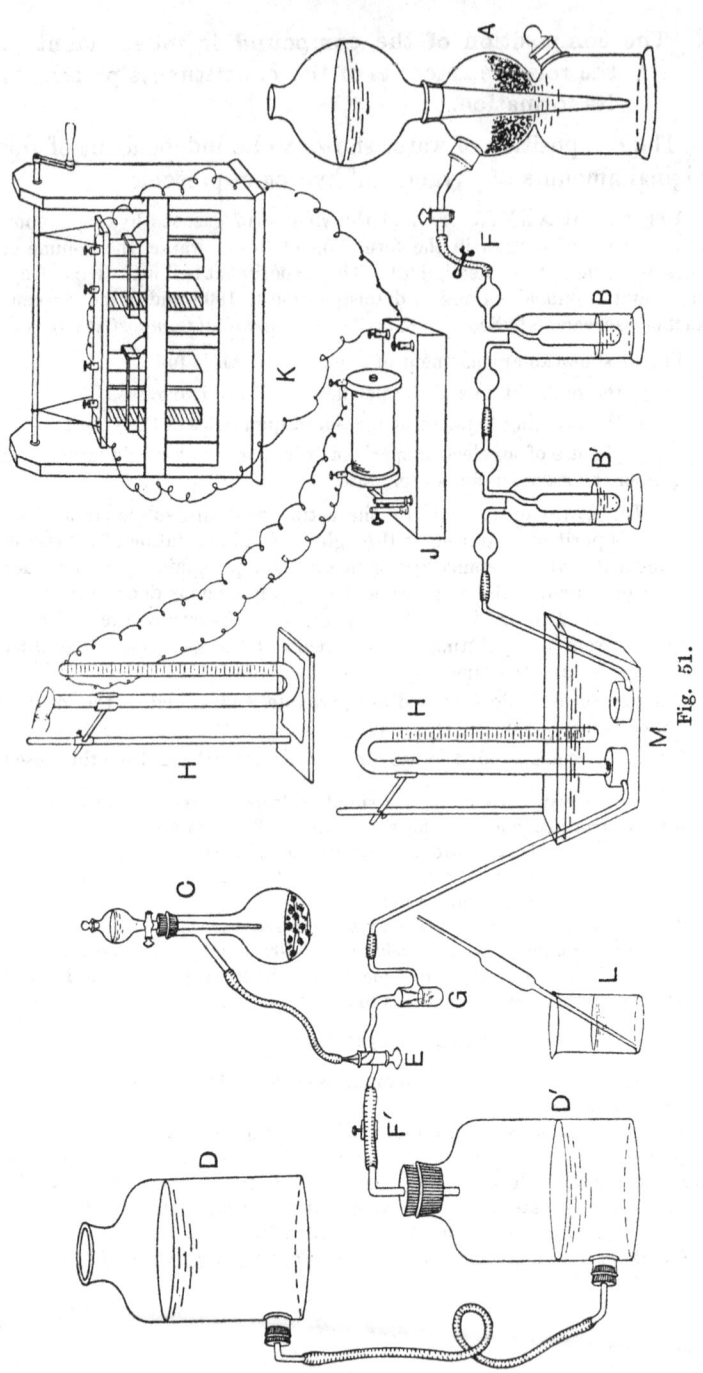

Fig. 51.

A. Kipp for the production of hydrogen from zinc and dilute sulphuric acid.

B, B'. Wash-bottles filled with alkaline solution of permanganate.

C. Apparatus for the production of oxygen by the action of water on sodium peroxide.

D, D'. Aspirators for the storing and delivering of the oxygen.

E. Three-way glass stop-cock.

F, F'. Screw clips.

G. Small wash-bottle containing a few drops of water, for showing the rate at which the oxygen is delivered.

H. U-tube arranged to serve as eudiometer by being closed at one end, fitted with platinum wires for sparking, and graduated in $\frac{1}{5}$ c.c.

J. Small induction coil.

K. Battery.

L. Pipette to be used in the addition or withdrawal of water required for bringing the pressure of the gas in the closed limb of the U-tube to atmospheric.

M. Trough with two beehives for the introduction of the two gases into the eudiometer tube.

limb of which is graduated in 1/5 c.c. and fitted with platinum wires for sparking. The advantage of this arrangement is that the pressure of the gas in the closed limb can be easily brought to that of the atmosphere by the addition or withdrawal of water in the open limb. A certain knack, which can only be acquired by experience, is required for completely filling the closed limb of the tube with water, as well as for transferring the gas collected in the bend of the tube to the closed end. In the measurement of the gaseous volumes, made with the object of finding a ratio, no corrections are made for temperature, pressure and tension of aqueous vapour ; the assumption being that temperature and pressure have remained the same during the course of the experiment; this also applies to the partial pressure of the aqueous vapour present in the gas, which is that of complete saturation at one and the same temperature.

(iii)　A battery and induction coil[1] are used for producing a spark.

Procedure.

By the joint action of the glass tap and the screw clip *F*, adjust the rate of production of the hydrogen so that the bubbles pass slowly, about 1 in 2 seconds ; when full time has been allowed for sweeping out all the air from *B* and *B'*, collect some hydrogen in the U-tube, which has been completely filled with water ; transfer the hydrogen from the bend to the closed end, equalise the levels by the addition or withdrawal of water, and take a reading ; this gives *A*, the volume of hydrogen.

Start the delivery of oxygen from the aspirator at a steady slow rate by suitable manipulation of the screw clips and the three-way stop-cock. Completely fill the open limb of the eudiometer with water, invert it in the trough, and add some oxygen to the hydrogen already present. It will be convenient to ascertain beforehand, by means of a straight graduated gas tube, the approximate volume value of each bubble of oxygen and hydrogen

[1] Glazebrook, *Electricity*, Chapter XIII, pp. 389 *et seq.*

respectively as they come off at the adjusted rate of about 1 in 2 seconds. By counting the number of bubbles that have passed, the volumes of the gases can be made approximately equal to any desired values. Measure the volume of the mixed gases oxygen and hydrogen, after adjusting the pressure to atmospheric as before; this gives B, the volume of oxygen + hydrogen, which it is convenient to make about 20 c.c., and which should not be more than 25 c.c., so as not to get too violent an explosion when the gases combine.

Reduce the pressure by withdrawing some of the water from the open tube, clamp the tube tightly, with its bend resting on a soft duster several times folded, hook the wires from the induction coil into the platinum loops, close the open end of the U-tube with the thumb, and pass a spark. The object of the thumb is to make the air in the open limb of the U-tube act as a kind of elastic cushion, which prevents the water closing the bend of the tube from being hurled out by the explosion, and outside air from passing up into the closed limb. Adjust the pressure and take another reading; this gives C, the volume of the residual gas. Fill up the open end of the U-tube with water, close it with the thumb, transfer the bubble of residual gas from the closed to the open limb, and test to ascertain whether it is oxygen or hydrogen.

Record of results.

The Volumetric Synthesis of Water.

	Vol. of hydrogen	Vol. of hydrogen + oxygen	Vol. of oxygen	Vol. and nature of residual gas		Vol. hydrogen mixed with 1 vol. oxygen	Vol. hydrogen combined with 1 vol. oxygen
	A	B	$B-A$	C		$\dfrac{A}{B-A}$	$\dfrac{A}{B-A-C}$ or $\dfrac{A-C}{B-A}$
				hydrogen	oxygen		
Exp. 1	15·0	35·2	20·2		12·8	0·78	2·02
,, 2	6·5	21·4	14·9		11·6	0·44	1·97
,, 3	10·8	25·8	15·0		9·6	0·72	2·00
,, 4	6·5	18·3	11·8		8·7	0·55	(2·09)
,, 5	12·8	27·8	15·0		8·6	0·85	2·00
,, 6	14·4	28·8	14·4		7·2	1·00	2·00
,, 7	10·4	19·4	9·0		3·8	1·16	2·00
,, 8	10·2	18·8	8·6		3·7	1·19	(2·08)
,, 9	9·8	18·4	8·6		3·7	1·14	2·00
,, 10	8·6	13·8	5·2		1·0	1·65	2·05
,, 11	14·6	22·2	7·6		0·2	1·92	2·02
,, 12	10·4	14·9	4·5	1·5		2·31	1·98
,, 13	21·6	28·5	6·9	8·3		3·13	(1·93)
,, 14	23·2	27·7	4·5	14·3		5·15	1·98

Mean = 2·00

The above 14 results were obtained in two different years by students working in pairs.

4. The composition of substances is independent of the differences in the conditions operative in their natural and artificial formation.

The composition of native Iceland spar shown to be the same as that of artificially prepared precipitated calcium carbonate.

Experiment XXXIII. Determination of the weight of calcium carbonate obtained by precipitating a solution of a known weight of Iceland spar in hydrochloric acid by addition of ammonium carbonate. (This experiment is illustrative of Proust's classical research on malachite and artificial copper carbonate.)

The experimental work involved in the production, collection and weighing of the precipitate could be made identical with that described in every text book of quantitative analysis under the heading " Estimation of calcium as carbonate" (*e.g.* Clowes and Coleman, *Quantitative Analysis*, 1911, §§ 145, 146), and if they work on these lines, students can incidentally add an important item to their knowledge of analytical work. But on the other hand, the special purpose of the experiment makes it desirable to carry it out in as simple a manner as possible, so as not to obscure the point at issue by a number of incidental chemical changes arranged so as to counteract one another. Thus the difficulty of complete separation of the filter paper from the precipitate is generally overcome by heating the paper, together with that portion of the precipitate which clings to it, to a temperature so high that the organic matter (cellulose) composing the paper is destroyed, and only an extremely small weight of ' ash ' (inorganic substances, stable at high temperatures) is left. But in this process, known as the 'burning of the filter paper,' the high temperature and the presence of the reducing agent carbon (separated out in the charring and only burnt away later) generally cause some chemical change in the adhering precipitate, *e.g.* as in this case, change of calcium carbonate into oxide. Hence the solid mixed with the ash must be made to undergo some further change, which is to bring it back to its original condition, *e.g.* heating of the calcium oxide with ammonium carbonate. To avoid the necessity of carrying out a number of processes which involve the application of a considerable amount of previous knowledge of the properties of calcium carbonate, it will be better to use either a Gooch crucible (see p. 100) or a so-called 'weighed filter paper.' But owing to the difficulty of getting a filter paper to constant weight, students are recommended to use a 'counterpoised filter paper.' If we start with two filter papers dried under identical conditions, and make their weights equal by snipping off bits from the heavier one, and if then throughout we treat the counterpoise filter in exactly the same way as that holding the precipitate (dry both in the same oven for the same length of time, keep both in the same desiccator, etc., etc.), and if then finally we place the filter and precipitate in one balance pan, and the counterpoise paper with the weights in the other pan, there is every chance that any change produced in the weight of the one paper (such as that due to absorption of moisture) will be produced to the same extent in the other, and therefore the weight of the filter paper does not enter into the calculation at all.

Iceland spar, a very pure variety of native calcium carbonate[1], is a substance interesting from a great many points of view. It is found in a cave in Iceland which when it was discovered was filled with this form of clear crystallised calcite; its cleavage is perfect along the rhombohedral faces, a sharp blow producing a number of absolutely perfect small rhombohedra; its special optical properties are strongly marked and are used in the instrument known as Nicol's prism[2].

A known weight of the spar, assumed to be quite pure calcium carbonate, is dissolved in dilute acid, when carbon dioxide is evolved and a new calcium salt remains in solution. The calcium is precipitated as carbonate by the addition in excess of soluble carbonate; the precipitate is collected and weighed. Concordance between the weight of spar used and the weight of precipitated carbonate is taken as indicating identity of composition. It should be realised that in so doing the assumption is made that all the lime originally present in the spar is also present in the precipitated carbonate, *i.e.* that the solution of the spar in acid has not involved the removal of any calcium; also that no calcium has remained behind in the liquid from which the precipitate has been removed. It would take us too far if at this stage we were to attempt to demonstrate the validity of these two assumptions, but it is very important that students should realise that an assumption most important and fundamental has been made. But if, as is assumed, the amount of lime present in the Iceland spar and in the precipitated carbonate is the same, and if we find that the carbonates themselves have the same weight, then the amount of carbon dioxide must also be the same, and hence the ratio between lime and carbon dioxide is constant.

Procedure.

Put some filter papers to dry, keeping them at a temperature between $110°$ and $120°$ for at least an hour. Counterpoise two of these papers against each other, and cut the heavier until the weight of the two is accurately the same. Carefully weigh a small empty porcelain dish which has been heated and then cooled in the desiccator. In this dish weigh a few clear crystals of Iceland spar (about 1 gram); then transfer them to a clean beaker of about 100 c.c. capacity, add about 25 c.c. of hydrochloric acid—the bench reagent will do—cover with a watch-glass and heat very gently on a sand bath. When solution is completed, rinse the underside of the watch-glass into the beaker, add ammonia until the solution just smells

---

[1] The two minerals calcite and aragonite, crystallising in the rhombohedral and rhombic systems respectively, were shown as far back as 1788 to be chemically identical, both of them being calcium carbonate.

[2] See Miers, *Mineralogy*, 1902, pp. 128, 130, 392 *et seq.*

of it, heat to boiling and gradually add a solution of ammonium carbonate; the reason for the neutralisation by ammonia preceding the addition of ammonium carbonate is the necessity of preventing violent effervescence with its attendant danger of mechanical loss. Let the precipitate settle, pour in a little more ammonium carbonate to see whether the precipitation has been complete. Filter whilst still hot through one of the counterpoised filter papers; wash with hot water till the washings are free from chloride; cover up the funnel and set it to dry in an oven whose temperature should be well above 100° (to remove all moisture), and which must not be above 120°, lest the filter paper should be charred and the experiment spoiled. When the filter paper and the precipitate are fairly dry, remove them to the previously weighed dish; place the dish and its contents, to which the counterpoised filter paper has been added, in the oven, and leave them there for two or three hours longer. Cool in desiccator, and when quite cold, weigh, placing the dish and its contents on one scale pan, and the counterpoised filter paper together with the requisite weights on the other. Put back into the oven for an hour or so, cool and weigh as before, and repeat the process until two successive weighings do not differ by more than 1 milligram.

Record of results.

A. Specimen of note-book entry.

The weight of precipitated calcium carbonate obtained from a known weight of Iceland spar.

	(i)	(ii)	(iii)
Wt. of empty porcelain dish, previously heated and cooled in desiccator	24·678	26·879	22·345
Wt. of dish + spar	25·748	27·898	24·200
∴ Wt. of spar $= a$	**1·070**	**1·019**	**1·855**
Wt. of dish + carbonate (i)	+ 25·752	+ 27·904	+ 24·208
(ii)	+ 25·749	+ 27·902	+ 24·203
(iii)	+ 25·749	+ 27·897	+ 24·202
(iv)		+ 27·896	
∴ Wt. of precipitated carbonate $= b$	1·071	1·017	1·857
Difference between weight of spar and of precipitated carbonate $= a - b$	+ 0·001 gms.	− 0·002 gms.	+ 0·002 gms.
This difference expressed as percentage of weight of spar $= \dfrac{a-b}{a} \times 100$	**+ 0·1**	**− 0·2**	**+ 0·1**

B. Summary of results obtained by a class of five students.

In the experiments marked (P), the precipitated carbonate was weighed on a counterpoised filter paper; in those marked (G) a Gooch crucible was used.

			Weight of Iceland spar $= a$	Weight of precipitated carbonate obtained $= b$	Difference $= a - b$	Difference expressed as percentage of a $= \dfrac{(a-b)100}{a}$
Student	A					
		(P)	1·002 gms.	1·006 gms.	+0·004 gms.	+0·4
		(G)	1·158　,,	1·155　,,	−0·003　,,	−0·3
,,	B					
		(P)	0·925　,,	0·922　,,	−0·003　,,	−0·3
		(G)	1·230　,,	1·228　,,	−0·002　,,	−0·2
,,	C					
		(P)	0·991　,,	0·988　,,	−0·003　,,	−0·3
,,	D					
		(G)	0·758　,,	0·750　,,	−0·008　,,	−1·0
,,	E					
		(P)	0·950　,,	0·953　,,	+0·003　,,	+0·3
		(G)	0·489　,,	0·494　,,	+0·005　,,	+1·0
Total of 8 experiments			7·503	7·496	−0·007	**−0·1**

CHAPTER VII

THE LAW OF MULTIPLE RATIOS

I. Nature and Scope of the Law.

1. **Existence of more than one compound of the same constituents A and B, and characteristics of such substances C_1, C_2, C_3.**

Proust, the successful champion of the doctrine of fixity of composition, had in the wide range of his stoichiometrical measurements come across this class of substances which at first sight seems an exception to the law. Having worked with the two sulphides of iron, the two oxides of tin, and the two oxides of copper, he recognised and stated clearly and definitely the differences between substances of this type and mixtures. In the case of *mixtures* of the components[1] A and B, a theoretically infinite number of combinations is possible, composition as well as properties varying continuously; in the case under consideration C_1, C_2, C_3 are *compounds* of the same constituents A and B, because for each of them properties and composition are fixed, and the variation in both these respects from C_1 to C_2, from C_2 to C_3, etc. is sudden and considerable, with which goes the further fact that the number of compounds containing the same constituents cannot be indefinitely great and is in fact small, 2 or 3 being the most common number, whilst 4 or 5 occur but rarely[2].

[1] For the sake of simplicity the argument assumes 2 constituents only, but of course everything said is equally applicable to more complex substances in which the number of constituents is 3, 4, 5 or even more.

[2] This is exclusive of carbon compounds, in which, owing to special circumstances, viz. the high valency of the carbon atom and the power possessed by it to satisfy some of these valencies by combination with other carbon atoms, the number of combinations between the same constituents is very large.

It should be useful to introduce at this point a number of students' experiments illustrative of the characteristics of the class of compounds under consideration, in which (a) the constituents are the same; (b) the properties are different; (c) the ratio between the constituents is different.

Experiment XXXIV.
Synthesis of two iodides of mercury.

Five grams of mercury rubbed in a mortar with about 3 grams of iodine yield a green powder; the same amount of mercury rubbed with about twice the former amount of iodine yields a brilliant red powder. The difference in colour, together with the mode of production, show the existence in this case of the relations above enumerated under (a), (b), (c). Moreover, (c) is easily demonstrated in a striking manner, by converting one compound into the other by adding one or other of the constituents: the green powder when rubbed with more iodine becomes red; the red when rubbed with more mercury becomes green. The original weighings need not be done at all accurately; as a matter of fact we can dispense with all weighings by starting with a *very few* crystals of iodine and rubbing these with mercury added in very fine drops (see fig. 52), until first a red and then a green powder is obtained, and afterwards reproducing the red powder by the addition of more iodine. The combination is greatly facilitated by the addition of a few drops of alcohol (or methylated spirit), which by dissolving the iodine increases the surface of contact between the iodine and mercury, and which, owing to its easy volatility, is removed automatically in the course of the process.

Fig. 52.

a. Bottle containing mercury.

b. Glass tube drawn out to fine nozzle.

Experiment XXXV.
Synthesis of two chlorides of copper.

The process consists in preparing a new and different substance by the addition of more copper to the green salt known to be a chloride of copper. Either start with commercial cupric chloride, the composition of which

Thus in Richter's *Lexicon der Kohlenstoffverbindungen* (1900) about 650 hydrocarbons—combinations between the elements carbon and hydrogen—are dealt with. This number of course does not represent all that *can* exist, only those that so far have been isolated and studied, and additions are continually being made.

may be assumed as known, or prepare a solution of this substance[1] by :
(i) synthesising the black oxide of copper by either heating copper in a stream of air (*ante*, p. 89) or dissolving copper in nitric acid, evaporating to dryness and igniting[2]; (ii) dissolving the black substance in hydrochloric acid. This procedure establishes the green substance as a compound of copper, but unless we assume previous knowledge of the composition of hydrochloric acid, and of the nature of the double decomposition which occurs when an acid interacts with a metallic oxide, we shall not really know anything definite about the nature of the constituent B combined with A, the copper; the production of an identical substance by burning copper in a stream of chlorine would give the required additional information. For the right interpretation of the next process, which consists in the heating of the green chloride with more copper, it is essential to remember that in the absence of air copper does not dissolve in hydrochloric acid, and hence that such reaction as occurs is one between the green chloride and the copper.

Fig. 53.

In a round-bottomed flask of about 200 c.c. capacity, fitted with a cork and glass tube drawn out to a fine nozzle, place about 2 grams of crystalline green chloride of copper (or the solution obtained from about 1 gram of copper) with about 100 c.c. of concentrated hydrochloric acid and about

---

[1] In the current system of chemical nomenclature the termination ‘-ide,’ distinctive of a binary combination, is affixed to the more electronegative of the two components, ‘copper *oxide*,’ ‘copper *chloride*’ being the names given to any combination between the two elements copper and oxygen, copper and chlorine respectively. When there are several combinations between the same constituents, these are differentiated by the addition of the suffixes ‘-ous’ and ‘-ic’ to the other component, cupr*ic* oxide being the combination containing comparatively more oxygen than the cupr*ous* oxide; or ‘oxide’ is qualified by the prefixes ‘mon-,’ ‘di-,’ etc., as carbon *mon*oxide and carbon *di*oxide; or ‘sub-,’ ‘hypo-,’ ‘per-’ are used to indicate the gradation in the relative amount of the constituent to which they are prefixed, as copper *sub*oxide and copper oxide, lead oxide and lead *per*oxide, *hypo*chlorite and chlorite.

[2] Students are recommended to consider what form they would give to a simple experiment intended to demonstrate the identity of the substance thus obtained with the oxide of copper produced by direct combination of the copper and oxygen.

5 grams of copper turnings (copper wool would do very well); heat gently on a sand bath in the draught cupboard. Note the change of the green colour to opaque brown on the addition of the concentrated acid[1], the gradual disappearance of the colour and its reappearance when, on stopping the heating, air enters through the capillary nozzle. When the liquid has become quite colourless, pour some of it whilst still hot into a tall cylinder filled with water, when a crystalline white precipitate will be formed, showing that the substance produced from the green chloride by the action of copper, whilst soluble in concentrated hydrochloric acid, is insoluble in water. Remove excess of hydrochloric acid by washing by decantation, place some of the white crystals in a small beaker and add a little chlorine water, noting

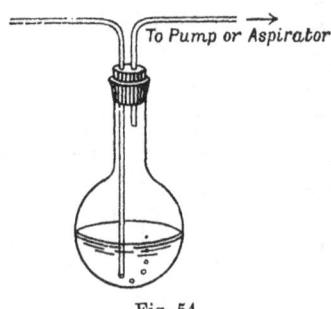

the almost instantaneous production of a green solution which, as far as appearance goes, is the same as the original green chloride. Put another lot of the crystals into a flask with dilute hydrochloric acid, or take some of the colourless solution somewhat diluted with water, and draw a stream of air through the solution (fig. 54). The white crystals will disappear and a green solution be formed, hydrochloric acid and air (or rather the oxygen contained in the air) producing

Fig. 54.

an effect equivalent to that of chlorine.

Hence, though the proof has not been complete, what has been done establishes a strong likelihood that the green crystals soluble in water and the white crystals insoluble in water are both compounds of copper with chlorine, and that the difference between the two substances is due to a difference in the relative amounts of the two constituents, addition of copper to the green substance producing the white, and addition of chlorine to the white producing the green, a cycle of changes which can be repeated an indefinite number of times at will.

Experiment XXXVI.

Synthesis of carbon dioxide from carbon monoxide and oxygen, the composition of the carbon dioxide itself being either assumed as known or determined as part of the experiment by the direct combination of carbon and oxygen.

[1] Solid crystalline cupric chloride (which contains water of crystallisation) as well as strong aqueous solutions of the chloride, are green; the anhydrous salt and solutions in concentrated hydrochloric acid are brown; dilute aqueous solutions are blue. For an explanation of these colour changes in terms of the ionic theory of solution, see Ostwald, *Foundations of Analytical Chemistry*, 3rd ed. 1908, pp. 129 *et seq.*; Walker, *Introduction to Physical Chemistry*, 6th ed. 1910, pp. 354 *et seq.*

Though the composition of carbon dioxide (for summary of properties see *ante*, p. 210) may be demonstrated by analysis—a brightly burning strip of magnesium ribbon, when plunged into a jar of the gas, continues to burn, and black carbon is deposited—synthesis supplies a simpler and more conclusive proof.—The simple apparatus depicted in fig. 55, in which some pure carbon in the form of lampblack is burned in a stream of oxygen (or air), and the gas formed tested by lime water, shows the principle of the method, which in order to supply real proof would have to include a number of other tests, qualitative and quantitative, of the gas whose identity with that obtained by the action of hydrochloric acid on marble we wish to establish.

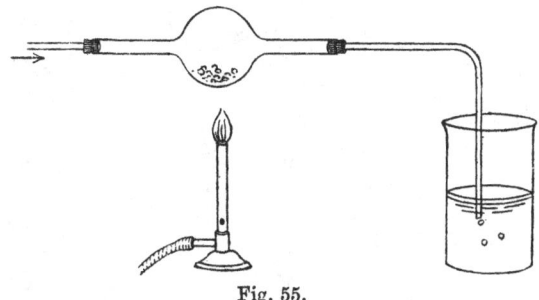

Fig. 55.

The object of the present experiment is to show that carbon dioxide can also be produced by the addition of oxygen to an inflammable gas obtainable in a variety of ways. The following method of preparing this gas consists in the decomposition of oxalic acid by concentrated sulphuric acid, and the storing in an aspirator of the portion of the gaseous mixture produced which is not absorbed by potash (fig. 56).

Procedure.

With G, the glass stop-cock placed so as to connect F and H, and with L open, raise K and let water flow into J until all air has been expelled from it and the delivery tube attached to it, after which close L. From the dropping funnel let enough concentrated sulphuric acid fall into A to cover the solid oxalic acid, and heat with a *very* small flame; when the reaction has been started it will probably be necessary to take the burner away, and perhaps even to cool the bottom of the flask by immersion in a trough with cold water, so as to prevent the evolution of the gas occurring at a rate too fast for the complete removal of the portion absorbed by potash. At this stage it may be desirable to get some rough idea concerning the quantity of gas rejected before collection in J is begun. This can be done by placing H under the beehive in the trough, and collecting the gas in cylinders, the volume of which is approximately known and allows of a comparison with that of the whole apparatus $A—F$, itself approximately evaluated. Two cylinders of about 150 c.c. capacity each make a useful combination; whilst the second is being filled, the contents of the first are

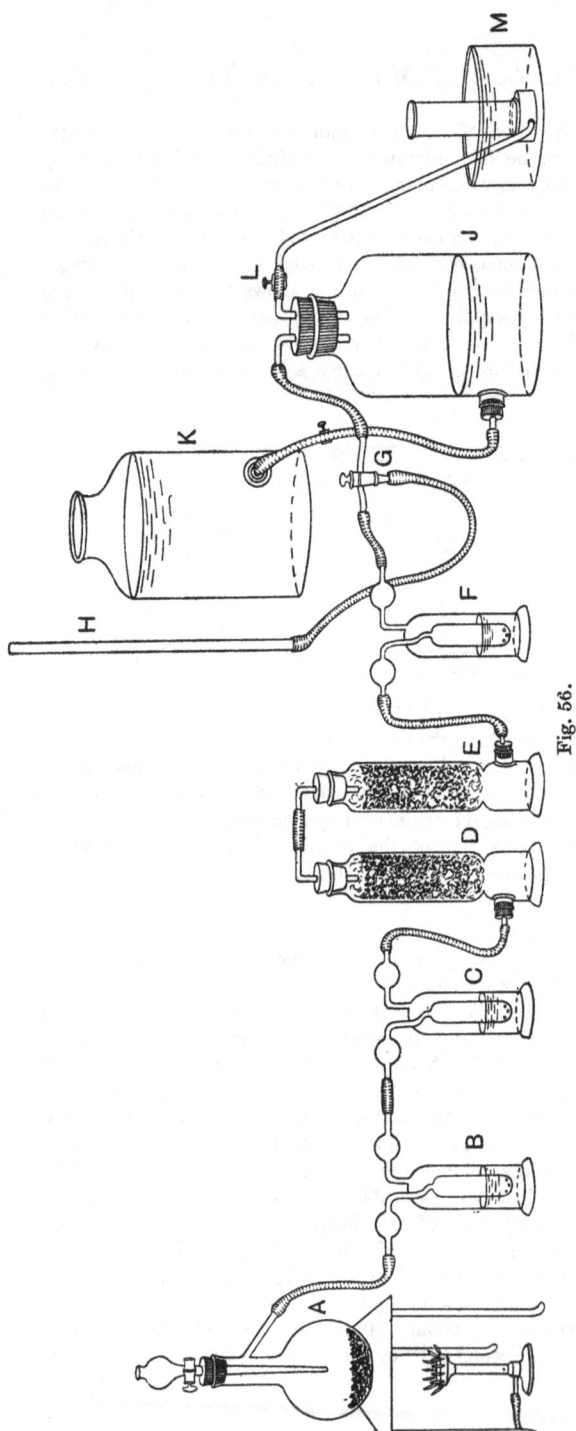

Fig. 56.

A. Round-bottomed flask of about 200 c.c. capacity, holding about 50 grams of oxalic acid (preferably anhydrous) and covered by the concentrated sulphuric acid delivered through the dropping funnel.

B. Wash-bottle containing baryta water, to show the presence in the gas of carbon dioxide.

C, D, E. Wash-bottle and towers containing potash (solution and solid) to absorb the carbon dioxide.

F. Wash-bottle containing lime water, to test the completeness of the removal of the carbon dioxide.

G. Three-way glass stop-cock.

H. Tube for carrying off gas not stored into a flue or burner.

J, K. Aspirators arranged for the storing and delivering of the portion of gas not absorbed by potash.

L. Pinch cock.

M. Trough and beehive for the transference of easily regulated volumes of gas from J into cylinders or tubes.

tested with a lighted taper, after which it is filled again with water—preferably by immersion in the trough if this is of suitable size—and got ready for collecting the gas by the time the second is full, which then has its contents tested in turn. This process is continued until from the volume collected and the behaviour of the last cylinder full of gas tested, it *seems probable* that all the air has been expelled.—When ready to collect the gas in *J*, turn the stop-cock *G* so as to connect *F* with *J*, *L* being closed and the aspirator *K* lowered.—After collecting a sufficient amount of gas in *J*, again make connection between *F* and *H*, the open end of which is placed under a draught chimney or over a Bunsen flame so as to dispose of the gas, which is highly poisonous.

Test the gas stored in *J* with the special object of showing its difference from carbon dioxide:

Action of potash. The mode of collection shows that it is not absorbed.

Action of lime water. The absence of a precipitate in *F* shows that there is no action similar to that of carbon dioxide.

Inflammability. A taper plunged into a cylinder filled with the gas is extinguished, but the gas burns at the mouth of the cylinder with a characteristically coloured flame, like that sometimes seen over the clear glowing coal of a fire which has stopped emitting jets of luminous flame. The flame will show specially well if a tall, rather narrow cylinder is used, the gas lighted and some water poured rapidly into the cylinder down its side.—The product of combustion of the gas gives the lime water test for carbon dioxide, as can be easily shown by pouring a few c.c. of clear lime water into a cylinder full of the gas[1], igniting it and shaking up with the lime water when combustion has ceased. Hence the gas formed *may* be carbon dioxide; students are recommended to consider what further work would be required to change probability into certainty.—Supposing this necessary additional work to have been done, we should be justified in inferring that since the inflammable gas under consideration is by burning (combination with oxygen) changed into carbon dioxide, known from its synthesis to be a compound of carbon and oxygen, *both* gases are oxides of carbon, and the difference in their properties is due to the presence in the carbon dioxide of a relatively greater amount of oxygen. The validity of this view can be demonstrated by making the inflammable gas from carbon dioxide by the addition of more carbon, which can be accomplished by leading it over heated charcoal (see fig. 57).

Procedure.

D and *E* being disconnected, the charcoal in *E* is made red-hot, the current of gas from *A* is started so as to expel all air from *B*, *C*, and *D*, and connection is then made; the rate at which the gas is allowed to pass should be a *very* slow one, so as to ensure completeness of the action which occurs when it comes into contact with the strongly heated charcoal in *E*.

---

[1] Remember the necessity of using boiled-out water in any work involving the lime water test for carbon dioxide (*ante*, p. 172).

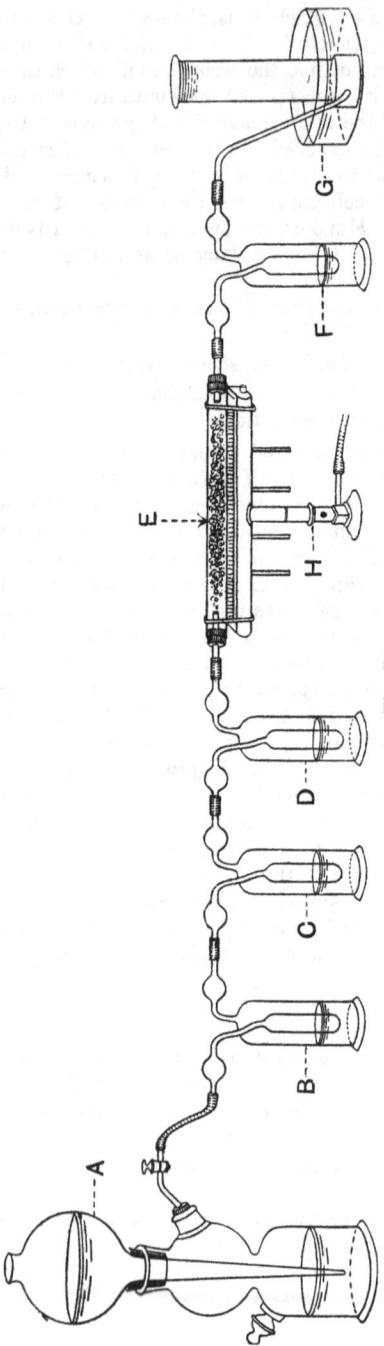

Fig. 57.

A. Kipp for the generation of carbon dioxide from marble and hydrochloric acid.
B. Wash-bottle containing water, to retain hydrochloric acid spray.
C, D. Wash-bottles containing concentrated sulphuric acid, to dry the gas.
E. Tube of combustion (hard potash) glass, filled with charcoal broken up small and heated strongly, either by a small combustion furnace or by a 'Ramsay tube burner H[1].'
F. Wash-bottle containing lime water.
G. Trough with beehive, etc., to collect the gas.

[1] Some asbestos paper in the form of a pent roof put over the tube greatly increases the efficiency of a Ramsay burner.

Experiment XXXVII.

Analysis of the two carbonates of potassium.

The existence of two different salts, both produced from potash and carbon dioxide, has long been known. The following table summarises for these two substances the various properties pertinent to our present enquiry.

	Normal Carbonate or Monocarbonate	Acid Carbonate or Bicarbonate
(a) The constituents are the same, as is proved by:		
(i) Synthesis	Made directly by leading carbon dioxide into potash solution	
(ii) Analysis	Treated with dilute acid, both evolve carbon dioxide: both give the flame coloration and all other reactions characteristic of potassium compounds (Clowes, *Qualitative Analysis*, 8th ed., p. 115); both turn litmus blue	
(b) The two substances have different properties:		
(i) State of hydration	Crystallises with water	Crystallises without water
(ii) Solubility in water	Very soluble	Sparingly soluble, easily recrystallised from hot water
Wt. dissolved by 100 grams water at 20° at 60°	112 grams 127 ,,	26·9 grams 41·3 ,,
(iii) Action of heat	Solid fuses and then remains unchanged; solution not changed	Both solution[1] and solid lose carbon dioxide and pass into the normal salt[2]
(iv) Action of carbon dioxide	The gas is absorbed and the salt changed into the acid carbonate	No action
(v) Magnesium sulphate solution added to the solution of the salt	Immediate precipitate	No precipitate until heated
(vi) Mercuric chloride solution added to the solution of the salt	Red precipitate	White precipitate changed by heat to red
(c) The observed differences in properties are due to a difference in the relative amounts of carbon dioxide and alkali	Evidence for this is supplied by (iii) and (iv) above, which show how the addition of more carbon dioxide to the normal salt produces the acid salt, whilst conversely, removal of carbon dioxide from the acid salt produces the normal salt	

For footnotes 1 and 2 see p. 262.

No special directions are needed for performing the greater number of the above experiments. To do (*a*) (i) and (*b*) (ii) would require a disproportionate amount of time. Experiment (*b*) (iv) is instructive as showing how the bicarbonate can be actually prepared; various plans are recommended in different text books[3], the starting-point being always a saturated solution of the normal carbonate; and it will be a saving of time to prepare this beforehand by making a hot saturated solution, allowing it to stand, and

[1] The decomposing effect of heat on the solution must be taken into account when trying to recrystallise the acid carbonate; heating to too high a temperature or for too long a time would produce so much decomposition as to transform a great portion of the acid salt into the normal salt, separation from which may be the very object of the process. This difficulty can be overcome by leading carbon dioxide into the hot filtered solution (see fig. 58) contained in the

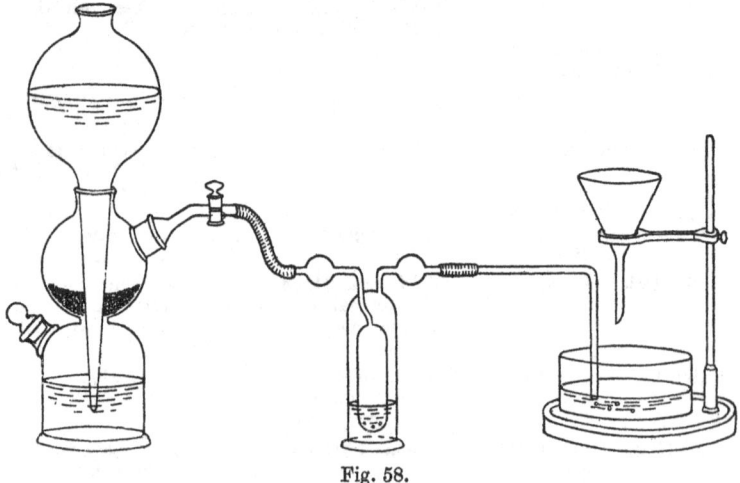

Fig. 58.

crystallising dish. In performing the differentiating tests (v) and (vi) between the two salts, and in any other case when bicarbonate free from carbonate is wanted, the solution must be effected in the cold (by shaking up the finely-powdered substance with cold water); and the further precaution of a preliminary 'washing' of the solid with small quantities of cold water, to remove any outer layer of changed salt, is desirable whenever practicable.

[2] This property is made use of in cooking, when a dough into which some acid carbonate (of sodium) has been incorporated whilst cold is made 'light' by the bubbles of carbonic acid gas produced on heating, which are retained in the elastic structure provided by the gluten of the flour.

[3] Emerson Reynolds, *Experimental Chemistry*, Part III, Metals.

pouring off the liquid from the precipitated crystal cake as wanted.

Experiment XXXVIII.

Analysis of the two sulphates of potassium, usually called normal or monosulphate and acid or bisulphate respectively.

(*a*) Both salts give all the analytical tests for potassium and for sulphate (Clowes, *Qualitative Analysis*, 8th ed., 1908, pp. 115, 198).

(*b*) The salts differ in (i) appearance—examine the crystals with a lens, or watch through a low-power microscope the deposition of crystals from drops of saturated solution placed on a slide; (ii) solubility; (iii) action on moist litmus, one being neutral and the other acid; (iv) behaviour when heated, the nórmal salt remaining solid and apparently unchanged, whilst the acid salt fuses easily (M.P. 197°) and at a red heat is decomposed with the evolution of sulphuric acid.

(*c*) The differences observed under (*b*) are due to the relatively different amounts of acid present: Some powdered crystals of the acid salt contained in a small crucible, when heated very strongly—either by the foot-blowpipe flame, a Mecker, Teklu, or any other specially effective burner—yield fumes of sulphuric acid[1], which can be condensed for purposes of subsequent testing by stopping the heating and placing a cold lid on the crucible. The residual substance can be identified as the normal neutral salt only if the decomposition has been made complete, which requires a high temperature but can be easily achieved by the direct heating in the blowpipe flame of a bead of the acid sulphate held in a loop of platinum wire.

Experiment XXXIX.

The relationship between potassium chlorate and potassium perchlorate.

Students are recommended to read carefully the directions for the laboratory preparation of these two salts (Bruce and Harper, *Practical Chemistry*, p. 32; Biltz, *Laboratory Methods of Inorganic Chemistry* p. 118) which in the case of the chlorate consists in the action of chlorine on hot concentrated potash solution, followed by the separation of the less soluble chlorate from the more soluble chloride formed simultaneously; whilst the perchlorate is made by heating the chlorate to a definite temperature, followed by separation from the simultaneously formed chloride and the undecomposed chlorate (*ante*, Chapter II, p. 91).

The preparation of the perchlorate, which does not require the setting up of any special apparatus, is an experiment which well repays the time

---

[1] This is the property utilised when the acid potassium sulphate is used to supply the free sulphuric acid required for the occurrence of certain reactions; *e.g.* in testing for a fluoride or fluosilicate by the green coloration imparted to the non-luminous Bunsen flame by volatile boron fluoride, a bead is made from the solid obtained by mixing the substance to be tested with borax and acid potassium sulphate (Newth, *Manual of Chemical Analysis*, 1909, p. 135).

spent on it. Starting with about 15 grams of chlorate—which of course need not be freed from chloride present as an impurity—heat in a small dish or large crucible, finding out by trial the conditions effective for the production and maintenance of the required temperature, viz. between 380° and 400°, recognised by the slow continuous evolution of a gas from the fused mass. At intervals withdraw a little of the fused mass by dipping in the end of a very thin glass rod, and test for chlorate by boiling with dilute hydrochloric acid (evolution of chlorine); when this test shows that little, if any, chlorate is left, decompose the residual chlorate by evaporating to dryness with hydrochloric acid. Separate the perchlorate from the chloride by fractional crystallisation; inspection of the solubility diagram given in Chapter II, p. 90, will show the desirability of removing at the outset as much chloride as possible by repeatedly shaking up the finely ground solid with small quantities of cold water, after which the residue is dissolved in the least possible quantity of boiling water, and the crystals which separate out on cooling are drained, washed with small quantities of cold water and tested for the absence of chloride and of chlorate.

It will be desirable to invert the order followed in the previous experiments and to begin the comparative study of the chlorate and perchlorate by the consideration of the differences between the two substances. The following table summarises these differences, giving alike those most suitable for rapid testing, which should be done by students not already familiar with them, and those others which, since they are quantitative, take too much time to be used in ordinary classes.

	Potassium chlorate	Potassium perchlorate
(b) Differences in properties:		
(i) Solubility	Both sparingly soluble in cold water, but perchlorate so much so as to be almost insoluble; hence its use as a test for potassium	
100 gms. water dissolve at 20° 100°	8 grams 56 ,,	1·7 grams 18 ,,
(ii) Formation	By the action of chlorine on hot strong solution of potash, and separation from the chloride by crystallisation	By heating the chlorate to about 400° C., when perchlorate and chloride are formed, and oxygen is set free
(iii) Action of heat	Melts at 359° C. and begins to decompose at 372° C. into perchlorate, chloride and oxygen	Decomposes above 400° C. into chloride and oxygen. M.P. 610
(iv) Action of concentrated sulphuric acid on the solid	The chloric acid first liberated is decomposed into oxygen and an orange-coloured explosive gas which is an oxide of chlorine	The perchloric acid liberated is so stable that it can be distilled

	Potassium chlorate	Potassium perchlorate
(v) Action of dilute hydrochloric acid on the solid	No action in the cold (difference from hypochlorite); on heating, chlorine is evolved, the chlorate being reduced to chloride	No action

(*a*) Identity of the constituents.

That the chlorate and perchlorate contain the same constituent elements can be shown by analysis, the products of decomposition by heat being the same for both, viz. oxygen and potassium chloride. It is a simple matter to show that the gas evolved from either salt is possessed of the qualitative properties of oxygen (colourless, non-inflammable, excellent supporter of combustion), and that the residual solids give all the qualitative tests for chloride and for potassium; but this leaves open the possibility of the gas being a *mixture* of oxygen with some other colourless gas, elementary or complex, and of the solid being a mixture containing potassium chloride as a constituent. Hence without quantitative work there can be no certainty that the products of decomposition are in both cases the same, viz. oxygen and potassium chloride only. This quantitative work may take the form of the determination of the density of the gas evolved or of the silver nitrate reaction ratio of the solid residue which must contain no undecomposed chlorate or perchlorate.

(i) Determination of the silver nitrate reaction ratio for the solid residue.

The work is in principle and in experimental procedure identical with that done when we demonstrated the identity of ammonium chloride with the sublimate obtained from it (*ante*, Chapter I, p. 67), and again, when we showed the identity of variously prepared specimens of potassium chloride (*ante*, Chapter VI, p. 239).

Record of results.

Specimen of note-book entry.

Determination of the volume of a certain silver nitrate solution required for the complete precipitation of 1 gram of the solid residue obtained from:

	I.		II.	
	Potass. chlorate	Potass. perchlorate	Potass. chlorate	Potass. perchlorate
Weight of sample of residue ...	·536 gm.	·365 gm.	·477 gm.	·454 gm.
Vol. of silver nitrate solution (not standard) required for complete precipitation[1]	9·35 c.c.	6·35 c.c.	9·95 c.c.	9·50 c.c.
∴ Volume of silver nitrate solution required for the precipitation of 1 gram of solid residue	17·45 c.c.	17·40 c.c.	20·85 c.c.	20·92 c.c.

[1] Different silver nitrate solutions were used in Experiments I and II.

(ii) Determination of the density of the gas evolved. The principle of the method consists in finding the volume occupied by the gas, the weight of which is given by the loss in weight of the previously carefully dried chlorate (or perchlorate) from which it has been produced (*ante*, Chapter IV, p. 191).

If the two density measurements are done in quick succession, temperature and pressure may be assumed to have remained the same, and the *comparison* of the two densities, which is our real object, can be made without multiplying both values by the same factor $\dfrac{273 + T}{273} \cdot \dfrac{760}{B - t}$; but the advantage of making this calculation is that it gives us values which we can compare with the standard ones for the different gases known, and by which we can identify our gases and be able to show, not only that both gases are the same, but also that both are oxygen.

Record of results.

Specimen of note-book entry.

Determination of the density of the gas obtained by heating.

	Potassium chlorate	Potassium perchlorate
Wt. of tube + asbestos plugs + suspending wire + salt		
(i) before evolution of gas.................$= a$	24·3425 gms.	24·3518 gms.
(ii) after evolution of gas ....$= b$	24·2610 ,,	24·2445 ,,
∴ Weight of gas W $= a - b$	0·0815 ,,	0·1073 ,,
Reading of the gas burette in which the gas is collected over water or over mercury		
(i) before the reaction ....$= c$	2·2 c.c.	3·4 c.c.
(ii) after the reaction........................$= d$	61·2 ,,	80·4 ,,
∴ Volume of gas V $= d - c$	59·0 ,,	77·0 ,,
Temperature...................................$= T$	14° C.	14° C.
Barometer..$= B$	769 mm.	769 mm.
Tension of aqueous vapour at temperature T° ....$= t$	$= 0$, gas having been collected over mercury	
Pressure ....$= B - t$	769 mm.	769 mm.
Density = wt. of 1 litre of the gas calculated to 0° and 760 mm. $$= \dfrac{W \times 1000}{V \times \dfrac{273}{273 + T} \times \dfrac{B - t}{760}}$$	**1·435** gms. per litre	**1·448** gms. per litre

2. Simple numerical relation between $\dfrac{A_1}{100 - A_1}$, $\dfrac{A_2}{100 - A_2}$, $\dfrac{A_3}{100 - A_3}$, the ratios of the same constituents A and B present in the different compounds C_1, C_2, C_3.

Experiments XXXVI—XXXIX had for object the demonstration of the class-characteristics of compounds whose qualitative and quantitative composition is represented by $\dfrac{A_1}{B_1} \gtrless \dfrac{A_2}{B_2} \lessgtr \dfrac{A_3}{B_3}$ where A_1, A_2, A_3 represent different quantities of the same constituent A, and B_1, B_2, B_3 different quantities of the same constituent B. By the simple device of expressing the composition of C_1, C_2, C_3 in the usual way as percentages, *e.g.* $\dfrac{A_1}{100 - A_1}$, but with A_1, A_2, A_3 all referred to the *same* weight of B, a simple numerical relationship becomes apparent

$$\frac{A_1}{B_1} : \frac{A_2}{B_2} = \frac{A_1}{100 - A_1} : \frac{A_2}{100 - A_2} = m : n$$

when it is found that the ratio $m : n$ closely approximates to that of simple whole numbers. Thus in the case of the two oxides of carbon, dealt with in Experiment XXXVI, the composition by weight has been found to be:

Percentage		Carbon monoxide		Carbon dioxide	
Oxygen $=$	A	$57{\cdot}2$	$= A_1$	$72{\cdot}7$	$= A_2$
Carbon $= B = 100 - A$		$42{\cdot}8 = B_1 = 100 - A_1$		$27{\cdot}3 = B_2 = 100 - A_2$	

$$m : n = \frac{A_1}{100 - A_1} : \frac{A_2}{100 - A_2}$$
$$= \frac{57{\cdot}2}{42{\cdot}8} : \frac{72{\cdot}7}{27{\cdot}3} = 1{\cdot}34 : 2{\cdot}66 = 1 : 1{\cdot}99$$
$$= 1 : 2 \text{ very nearly.}$$

Therefore in carbon dioxide gas for the *same* amount of carbon there is twice as much oxygen as in carbon monoxide. This fundamental stoichiometrical relationship, which is known as the "law of multiple ratios" can be formulated thus:

"If two substances A and B unite in more than one ratio, the various masses of A which combine with a fixed mass of B bear a simple ratio to each other."

How has this law been discovered and established?

II. Historical.

1. John Dalton (1766—1844), by whose name the law is known, made his first statement on the subject in 1802, but it was not until 1807 that the discovery reached a wider public through the account given of it by the Glasgow Professor Thomas Thomson in his *System of Chemistry*. The very important part played by this law in the foundation of the Atomic Theory accounts for the great interest taken in the actual mode of its discovery. Painstaking historical research has shown that Dalton did not arrive at the law inductively by a generalisation from a number of data collected in the course of analytical work, but that it represented a deductive inference from theoretical speculations concerning the ultimate constitution of matter; these speculations he proceeded to test by an investigation of all the pertinent facts he could find, using for the purpose data already available and supplying new ones of his own. The substances the composition of which gave data showing the newly discovered relationship were:

(i) The two oxides of carbon—analyses by Lavoisier and other French chemists.

(ii) The two oxides of nitrogen, now named nitrous oxide (laughing gas, see *ante*, Chapter III, p. 120) and nitric oxide (*ante*, Chapter III, p. 140)—analyses by Sir Humphry Davy.

(iii) The two hydrides of carbon, then named 'marsh gas' or 'light carburetted hydrogen' and 'olefiant gas' or 'heavy carburetted hydrogen' (our methane and ethylene). These two substances were investigated by Dalton himself, who found that when decomposed by sparking, each yields a volume of hydrogen practically twice that of the original gas; and that when exploded with excess of oxygen, carbon dioxide gas is obtained, the volume of which, when compared with that of the gas from which it is derived, is as 1/1 in the case of the marsh gas, 2/1 in the case of the olefiant gas[1].

2. W. H. Wollaston (1766—1828) published in 1808 results obtained prior to his cognisance of Dalton's discovery. His experiments were done with the alkali salts of the polybasic acids carbonic, sulphuric and oxalic, and were specially noteworthy for the ingenious and simple methods by which he showed that the quantities of acid combined with a certain amount of base in the so-called 'acid[2] salt' is just twice or four times that combined with the *same* amount of base in the so-called 'normal salt.' In Section IV of this chapter Experiments XLI and XLII are illustrative of Wollaston's classical work on the carbonates and the oxalates. In the case of the sulphates he synthesised the two potassium salts from potassium carbonate and sulphuric acid; he first made the acid salt by treating a known weight of carbonate with excess of sulphuric acid and heating to a temperature at which the free acid volatilises; he then showed that in order to change this acid salt to the normal—in this case neutral— salt, he had to add an amount of potassium carbonate equal to that originally used in the preparation of the acid salt. Note that as regards the sulphuric acid all that needed to be known and was known was, that the two salts contained the *same* though unknown amount of acid.

3. J. J. Berzelius (1779—1848) in a series of papers published between 1811 and 1812[3] gave to the world the results

---

[1] The actual numbers obtained were :

	Hydrogen	Carbon dioxide
100 volumes marsh gas yield	200 vols.	100 to 105 vols.
	(almost exactly)	
100 volumes olefiant gas yield	195 vols.	185 to 190 vols.

[2] This term is used to designate that one of the two salts which contains the relatively greater amount of acid, irrespective of whether it has an acid reaction or not: thus, whilst the acid sulphate or acid oxalate of potassium turn blue litmus red, the acid carbonate does the opposite, *i.e.* exhibits an alkaline reaction.

[3] *Essay to ascertain the Fixed and Simple Ratios in which the Constituents of Inorganic Nature are combined*, reprinted in *Klassiker der Exakten Wissenschaften*, No. 35.

of stoichiometrical work which greatly surpassed in accuracy that of any of his predecessors or contemporaries, and by which he may be said to have *established*[1] the Law of Multiple Ratios. Many years later, Berzelius himself clearly characterised the relative share and merit of Dalton and himself in this matter :

> It seems as if in this investigation the illustrious scientist had not at the outset been provided with a sufficiently firm experimental basis. To me it has seemed as if the small number of analyses given showed the desire of the experimenter for certain results[2]; but this is the very attitude we should avoid when seeking proofs for or against a preconceived theory. Notwithstanding all this, it is to Dalton that belongs the honour of the discovery of that part of the doctrine of chemical combination termed the *law of multiple ratios*, which none of his predecessors had observed.

Berzelius' accurate analyses covered a wide field, comprising as they did the composition of (*a*) the two oxides of sulphur, sulphurous and sulphuric anhydride (*ante*, Chapter III, p. 114); (*b*) the two sulphides of iron, the artificially prepared substance obtained by fusing together iron and sulphur (*ante*, Chapter III, p. 113) and the mineral iron pyrites; (*c*) the two oxides of iron ; (*d*) the two oxides and chlorides of copper (p. 255); (*e*) the three oxides of lead. Moreover later on, when the immediate object was no longer the establishment of the fundamental stoichiometrical laws, but the determination of the accurate combining ratios of the elements and of the criteria for the fixing of the relation between these and the atomic weights (*post*, Chapter IX), he produced numbers which showed the existence of multiple ratios in the case of the two oxides of chromium, of phosphorus and of arsenic. The following example illustrates the method he employed in the case of the two oxides of lead:

(i) The synthesis of litharge : 10 grams of lead, when changed in a variety of ways to the yellow oxide, yielded in

---

[1] The fact that before he was cognisant of Dalton's and Wollaston's results he had found that "in the basic chloride of lead and in the basic chloride of copper the acid is saturated by 4 times as much of the base as in the neutral salts" entitles him to be considered an independent discoverer of the law.

[2] Explicable by the fact that Dalton had not arrived at the law inductively, but that he was on the look-out for experimental verification of a theoretically established deduction.

the mean 10·780 grams of litharge, which makes the percentage of oxygen 7·24.

(ii) Analysis of the brown oxide : 5 grams of brown oxide ignited left 4·545 grams litharge mixed with 0·130 grams of impurities (lead sulphate and silica); the loss of weight of 0·325 gram represents the oxygen given off by the brown oxide in passing into 4·545 grams of yellow oxide, which, according to (i), contains 0·330 gram of oxygen and 4·215 grams of lead.

Hence the quantities of oxygen combined with the same 4·215 grams of lead in the yellow and the brown oxide of lead are 0·330 gram and (0·330 + 0·325) gram respectively, quantities which are in the ratio of 1 : 1·985, or very nearly 1 : 2.

III. Classification of the Law as Exact.

Marvellous as was the degree of accuracy attained by Berzelius considering the stage of development reached before him, his work does not lend itself to testing the existence or non-existence of small deviations from the requirements of the law if exact. In fact, there is no work on record the object of which has been the determination, under conditions more favourable to accuracy, of the ratio $m : n$ for compounds $C_1, C_2 \ldots$, whose composition is represented by

$$\frac{A_1}{B_1} : \frac{A_2}{B_2} = \frac{mA}{nB} : \frac{A}{B}.$$

But indirectly the degree of approximation of $m : n$ to simple whole numbers can be found from data obtained in accurate work on C_1 and C_2 separately, undertaken for a different purpose and by different investigators ; and thus it is possible to judge of the deviations from whole numbers by the magnitude of the probable experimental error attaching to the various measurements involved.

Thus we have a number of values for the ratio carbon to carbon dioxide, and carbon monoxide to carbon dioxide, according to the most trustworthy of these the weight of oxygen combined with the *same* weight of carbon in the two oxides is in the ratio of $1 : 1.99995$. (*Study of Chemical Composition*, pp. 167 *et seq.*)

Another set of pertinent data has been supplied by the recent very accurate analyses of nitric and nitrous oxides.

(i) Nitrous oxide : Guye and Bogdan (1904) found as a total of 5 experiments :

Wt. of nitrous oxide decomposed	Wt. of oxygen withdrawn by heated iron spiral	Wt. of nitrogen (by difference)	Wt. of nitrogen combined with 1·0000 gram of oxygen
5·6269 gms.	2·0454 gms.	3·5815 gms.	1·7510

(ii) Nitric oxide : Gray (1905) found as a total of 5 experiments (*ante*, Chapter V, p. 201) :

Wt. of nitric oxide decomposed	Wt. of oxygen removed by finely divided nickel	Wt. of nitrogen (a) by difference	(b) by weighing of residual gas	Wt. of nitrogen combined with 1·00000 gram of oxygen
2·93057	1·56229	1·36828	1·36819	0·87579
		Mean = 1·36824		

∴ ratio of weights of nitrogen combined with the same weight of oxygen in nitrous and nitric oxide respectively $\Big\} = 1.7510 : .87579 = \mathbf{1.9993 : 1}$.

The deviation from the whole number 2 is $\dfrac{7}{20000}$, and well within the experimental error of the two researches involved.

IV. Students' Illustrative Experiments.

The special feature of those about to be described is that they are either simple adaptations of classical experiments, or that they incidentally supply data of fundamental importance in the establishment of combining weight values directly referred to the standard oxygen (*post*, Chapter IX, p. 315).

1. **The composition of the two carbonates of potassium (or sodium).** Experiment illustrative of Wollaston's classical experiment (Alembic Club Reprints, No. 2 ; *Study of Chemical Composition*, p. 160). The qualitative relations between these two salts have been dealt with earlier in this chapter (p. 261).

Experiment XL. Determination of the ratio between the quantities of carbon dioxide (measured as volumes) combined with the same (though unknown) weight of potash in the two carbonates.

It is assumed as proved that when the carbonate which contains the relatively greater amount of carbon dioxide (and which is also the less soluble of the two) is heated, it loses part of the carbon dioxide, but retains all the base (potash or soda) originally present in it, and that the other carbonate is formed which contains the same amount of base—whatever that amount may have been—as the original salt. Let that unknown weight of potash contained in a grams of the higher—so-called acid— carbonate be X, and let it be found that by suitable treatment it yields a volume of carbon dioxide which at the temperature and pressure of the experiment occupies v c.c. Then if we heat ma grams of the higher carbonate, carbon dioxide (and water) will be given up and Y grams of the lower carbonate will be left. With the actual values of Y we are not concerned, but on the above suppositions we know it to contain mX of potash, and we can, by treating it as we did the parent salt, find the carbon dioxide left; let this volume be v' c.c. Then if the temperature and pressure in the two volume measurements had been the same, $\dfrac{v}{X} : \dfrac{v'}{mX}$ *i.e.* $\dfrac{vm}{v'}$ is the ratio between the amounts of carbon dioxide combined in the two salts with the same weight X of potash which is what we require.

In practice it is convenient to make m equal to 2, because experience has shown that then v and v' become practically equal, and the value of the ratio will not be altered in spite of the fact that owing to the solubility of the carbon dioxide in the solution from which under the conditions of the experiment it is liberated, the apparent values of v and v' are smaller than the true values. For if we arrange so as to have about the same volume of solution in each case, g, the volume of gas retained, will be about the same also, and the true ratio would be $\dfrac{(v+g)}{v'+g}$, in which, since g is comparatively small, if v is practically equal to v' we may neglect g.

For the success of the experiment it is essential that the acid carbonate used should be quite free from normal carbonate, for the latter being unchanged by heat, would yield the same volume of gas before and after the ignition; on the other hand, the presence of other impurities does not matter, provided only that they are evenly distributed through the sample used. Hence: (i) it is necessary to recrystallise the acid salt, and to do so just before performing the experiment, because the salt does not keep well, a slow change to the normal salt occurring spontaneously on the air-exposed surfaces; (ii) it is not necessary to dry the recrystallised salt completely, provided uniformity is ensured by grinding or any other form of thorough mixing.

Fig. 59, p. 274, gives a simple apparatus suitable for the quick performance of the experiment by groups of students. The carbon dioxide is liberated by dilute hydrochloric acid, and is measured by the volume of water it expels from an aspirator. Since carbon dioxide is fairly easily soluble in water (at 15° and 760 mm. 1 litre of water dissolves 1·002 litres of the gas), it is essential that the water in the aspirator should be saturated with the gas[1].

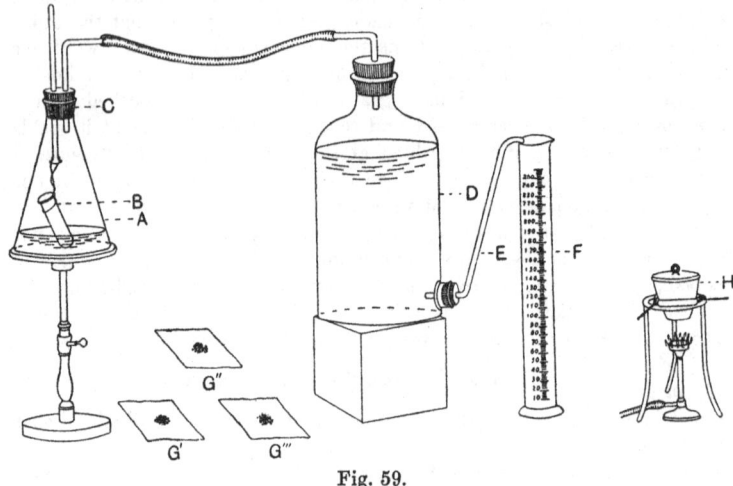

Fig. 59.

A. Erlenmeyer flask of about 150 c.c. capacity, into which the carbonates are put, together with an amount of water (about 10 c.c.) which should be the same in both cases.

B. Small tube of 8 to 10 c.c. capacity for holding the concentrated hydrochloric acid, carried by a copper wire attached to a glass rod, which is greased so as to fit well but slide easily in the rubber cork C.

D. Aspirator filled with water saturated with carbon dioxide and coloured by methyl orange.

E. Delivery tube which can be tilted so as to make the pressure of the gas in D equal to that of the atmosphere (*ante*, p. 89, fig. 90, C_2).

F. Graduated cylinder to measure the water expelled, holding 250 c.c. and graduated to 2 c.c.

G′, G″, G‴. Three *equal* lots of about 1 gram each of the freshly recrystallised acid carbonate on black glazed paper.

H. Small crucible with lid, for the careful heating of the $2a$ grams of the acid salt.

---

[1] Considering the great effect on gaseous solubility of change of temperature and pressure, the use of this form of apparatus might seem to be open to the objection that the large volume of carbon dioxide solution in the aspirator may, owing to its having been supersaturated to start with or owing to a rise in temperature, itself evolve a volume of gas sufficiently large to vitiate the measurements made; this is certainly true theoretically, but in practice the parts of the experiment involving the volume measurements are performed so quickly that there is not enough time for an appreciable change to occur in the equilibrium condition of the solution. 'Appreciable change' means a change greater than what is covered by the other experimental errors involved.

Procedure.

(i) Recrystallise some of the acid salt, following the directions given in the footnote on p. 262; place on a porous plate to dry, and mix well. For the reason already given, the recrystallisation must be done quite shortly before the performance of the experiment; in any case, whatever the salt used, the absence of any appreciable amount of the normal carbonate should be ascertained by the mercuric chloride test[1].

(ii) Prepare the saturated solution for D by filling it with water, adding a few drops of methyl orange, then a small handful of common washing soda and finally hydrochloric or sulphuric acid in excess, *i.e.* until the colour of the liquid is deep pink. The presence of undecomposed normal carbonate in the solution would be fatal, because owing to the tendency of this salt to pass into the acid carbonate, its power of absorbing carbon dioxide is considerably greater than that of water.

(iii) Weigh out three *equal* lots of about 1 gram each of the recrystallised salt. A quick way of doing this consists in preparing two pieces of black glazed paper which exactly counterpoise one another, placing one together with a 1 gram weight on the right balance pan, and shovelling salt on to the paper in the other balance pan until equilibrium is established. Considering the probable error of the volume measurement—at least 1/250— an accuracy of 2 to 3 mgs. is all that is required. When one lot of salt has been weighed, it is transferred with the help of a camel-hair brush to another piece of glazed paper (size and weight immaterial), a second lot is weighed, and so on.

(iv) Unite two lots of the salt and set them to heat in the covered crucible, taking all possible precautions against loss by spurting, *e.g.* raising the temperature very gradually. Ten minutes' heating over a Bunsen burner—which should at first be used with a rose—will complete the reaction, but a *proof* of this having been accomplished would involve weighing the crucible with its contents after it had been left to cool in a desiccator; heating, cooling and weighing a second time, and repeating these processes until the weight had become constant.

(v) Introduce the a grams of the salt into the flask A; add a roughly measured volume of water (say 10 c.c.), carefully washing down any particles of solid clinging to the side of the flask; fill B with concentrated acid, taking the usual precautions against spilling any; connect the apparatus as shown in the figure, being very careful to push all the corks well in; adjust E until its open end is at exactly the level of the coloured water in D; push down B, spilling the acid into the carbonate solution, and collect the water expelled from D in F; tilt E at the end so as to equalise the levels again. As the gas comes off in A with a great rush, there is

---

[1] Mercuric chloride gives with bicarbonate a pale yellow, with carbonate an orange red precipitate.

some danger of the corks in A and in D being forced out, and the water spurting out from E with such force as to be projected over the opening of F; to prevent any of these calamities the use of two pairs of hands is desirable. When the evolution of gas has apparently ceased, shaking A will probably disengage a little more; continue to collect water in F until further shaking of A produces practically no more effect. Measure v, the volume of water expelled, remembering the probable error of this measurement (see p. 38). Pour back into the aspirator the water collected in F; wash out A, being specially careful to remove all acid. Repeat the same process with the product obtained in (iv) by the heating of $2a$ grams of the acid salt, and measure v'.

Record of results.

Summary of results obtained by classes of 8 and of 12 students working in pairs.

Wollaston's Experiment on the two carbonates of potash.

	Volume of carbon dioxide, measured at the temperature and pressure of the room (supposed to have remained the same during the time of the experiment) evolved from:		Ratio of volumes, and hence of weights, of the carbon dioxide combined with the *same* amount of potash in the two salts
	(1) a grams of the less soluble salt (a = about 1 gram)	(2) $2a$ grams of the less soluble salt changed by heating and consequent loss of carbon dioxide into the more soluble salt	
	$= v$	$= v'$	$= v : \dfrac{v'}{2}$
Experimenters			
Pair A	217 c.c.	216 c.c.	2·01 : 1
B	218 ,,	218 ,,	2·00 : 1
C	220 ,,	222 ,,	1·98 : 1
D	216 ,,	216 ,,	2·00 : 1
A′	220 ,,	212 ,,	(2·07) : 1
B′	228 ,,	224 ,,	2·04 : 1
C′	218 ,,	218 ,,	2·00 : 1
D′	216 ,,	218 ,,	1·98 : 1
E′	230 ,,	230 ,,	2·00 : 1
F′	253 ,,	256 ,,	1·98 : 1

Mean = **2·00 : 1**

2. The composition of the three oxalates of potassium. Experiment illustrative of Wollaston's classical experiment (Alembic Club Reprints, No. 2 ; *Study of Chemical Composition*, p. 161).

Experiment XLI. Determination of the ratio between the quantities of potash combined with the same (though unknown) amount of oxalic acid in the three oxalates.

Of these salts, the one containing the smallest amount of the acid is neutral to litmus, whilst the other two have an acid reaction ; when ignited, any one of these salts is changed into potassium carbonate, which contains all the base originally present in the oxalate[1] from which it was derived ; addition of more base to either of the acid salts changes it to the neutral salt. It is therefore possible to change a known amount of either of the acid salts into the neutral salt by the addition of potassium carbonate, which latter can be obtained by ignition of a known amount of one of the acid salts.

Procedure and calculation.

Use *either* of the two acid salts sold under the names of *binoxalate* or salts of sorrel, and of *quadroxalate* or acid oxalate respectively. Weigh out 2 quantities a and b of the finely ground salt ; dissolve the one lot in water, making the solution up to m c.c. (about 8 grams of salt in 250 c.c. are suitable amounts) ; change the other lot, viz. the b grams, into carbonate by ignition in a covered crucible[2], the temperature of the foot-blowpipe flame or that produced by two large Bunsen burners being required to ensure the completion of the reaction, and the use of a nickel crucible being conducive to the saving of time ; dissolve the carbonate, separating it by filtration from the simultaneously formed carbon, and carefully wash the crucible and filter paper containing carbon, adding the washings to the bulk of the filtrate and making up to a known volume n ; find the volumes v' and v of the two solutions—acid oxalate and carbonate—which must interact to produce neutrality. Neither methyl orange (which cannot be used for 'weak[3]' acids, *e.g.* oxalic) nor phenol phthalein (which does not act with carbonate) being available, litmus must be used as an indicator, which involves the necessity of having the carbonate in the dish and the acid oxalate in the burette, and of boiling away the carbon dioxide as fast as it is liberated[4].

---

[1] Students are recommended to think out and describe in writing some simple experiment by which they would demonstrate the validity of this statement.

[2] Guard against spurting, which is very apt to occur at the beginning when the crystals 'decrepitate.'

[3] See J. Walker, *Introduction to Physical Chemistry*, for an explanation of this based on the Ionic Theory.

[4] Clowes and Coleman, *Quantitative Analysis*, 10th ed., 1914, p. 154.

(i)   Solution of the acid oxalate.

Wt. of salt taken $=a$ grams ; volume of solution ....$=m$ c.c.

Let the weight of potash in a grams of the acid oxalate be...... X grams

$\therefore$ wt. of potash in 1 c.c. of the acid oxalate solution ....$=\dfrac{X}{m}$,,

(ii)   Solution of the carbonate.

Wt. of the acid oxalate used in the production of the carbonate $=b$ grams

Volume of the solution ....$=n$ c.c.

and $\because$ wt. of potash in b grams of the acid oxalate, and hence also in the unknown weight of carbonate obtained by the ignition of that amount of acid oxalate ....$=\dfrac{bX}{a}$ grams

$\therefore$ wt. of potash in 1 c.c. of the carbonate solution ....$=\dfrac{bX}{an}$,,

(iii)   Determination of neutralisation equivalent.

Neutral oxalate is produced by adding to v' c.c. of acid oxalate, containing $\dfrac{v'X}{m}$ grams of potash, v c.c. of carbonate containing $\dfrac{vbX}{an}$ grams of potash.

But since the weight of oxalic acid in the acid oxalate taken is the same as that contained in the neutral oxalate produced from it by addition of potassium carbonate,

$\therefore$ the weight of oxalic acid which

in the acid oxalate is combined with .... $\dfrac{v'X}{m}$ of potash,

in the neutral salt is combined with...... $\dfrac{v'X}{m}+\dfrac{vbX}{an}$,,   ,,

$\therefore$ the ratio of the quantities of potash combined with the *same* amount of acid in the two salts is

$$\frac{v'X}{m} : \left(\frac{v'X}{m}+\frac{vbX}{an}\right)$$

$$= 1 : \left(1+\frac{vbm}{v'an}\right),$$

and if, as is most simple in working, we make $a=b$, *i.e.* take two exactly equal weights of the said oxalate, and if we make $m=n$, *i.e.* make the two solutions up to the same volume, and if we put for $\dfrac{v}{v'}$ (*i.e.* the volume of carbonate required to neutralise 1 c.c. of acid oxalate) p, the required ratio becomes

$$1 : (1+p).$$

Record of results

A. Specimen of note-book entry.

Determination of the ratio between the quantities of potash combined with the same amount of acid in the acid oxalate (binoxalate).

	I	II	III
Wt. of salt taken for acid oxalate solution[1] .... $= a$	8 grams	8 grams	8 grams
Wt. of salt taken for ignition to carbonates.......................... $= b$	8 grams	8 grams	8 grams
Volume of acid oxalate solution $= m$	250 c.c.	250 c.c.	250 c.c.
,, ,, carbonate ,, $= n$	250 ,,	250 ,,	250 ,,
Volume of carbonate solution taken for titration .... $= v$	25 ,,	25 ,,	25 ,,
Volume of acid oxalate solution required for neutralising v c.c. of carbonate (i) $= v'$	25·05 c.c.	24·80 c.c.	25·00 c.c.
(ii)	25·15 ,,	24·80 ,,	24·95 ,,
(iii)	25·10 ,,	24·95 ,,	24·95 ,,
Mean	25·10 ,,	24·85 ,,	24·95 ,,
Volume of carbonate required for neutralising 1 c.c. of acid oxalate............................ $= \dfrac{v}{v'} = p$	·996 c.c.	1·006 c.c.	1·002 c.c.
Ratio between quantities of potash combined in the acid and in the neutral oxalate with the **same** amount of acid............ $= 1 : 1 + p$	**1 : 1·996**	**1 : 2·006**	**1 : 2·002**
In a corresponding experiment with the quadroxalate, the ratios found were	**1 : 4·05**	**1 : 3·98**	**1 : 4·03**

---

[1] The quantities used in the experiment quoted will be found convenient; the actual values given to a and b, and m and n may of course vary, so long as a is made equal to b and m to n.

3. The two chlorides of copper. Experiment illustrative of Berzelius' classical experiment (*Klassiker der Exakten Wissenschaften*, No. 35, pp. 27 *et seq.*). The qualitative relations between these two salts have been dealt with earlier in this chapter (Experiment XXXV, p. 254).

Experiment XLII. Determination of the ratio between the weights of copper combined with the same amount of chlorine in the two chlorides.

Out of contact with air, copper is not acted upon by concentrated hydrochloric acid; a solution of the green chloride in concentrated hydrochloric acid, when digested with copper out of contact with air, is changed into the white chloride by taking up more copper. The utilisation of these facts constitutes the principle of Berzelius' experiment. A certain unknown amount of the green chloride, which has been made from a known weight of copper—or from a known weight of copper oxide, whose composition is assumed as an antecedent datum—is digested out of contact with air with a weighed strip of copper until the solution has become quite colourless, when the copper strip is withdrawn and weighed; the loss in weight of the copper strip gives the weight of copper that had to be added to that already present in the green salt to change it to the white salt.

Procedure.

(i) Change of a known weight *a* grams of pure metallic copper, first to the black oxide and then to the corresponding green chloride. The metal is dissolved in nitric acid (which should not be too concentrated). The solution is next evaporated, and then ignited, when the nitrate decomposes easily, yielding the black oxide (*ante*, Chapter III, p. 129). In all these processes great care must be taken to prevent mechanical loss, and all the precautions recommended in the analogous case of silver (*ante*, Chapter II, p. 99) should be observed. The insertion into the neck of the flask of a bulb of the kind used in ammonia distillations (Clowes and Coleman, 10th ed. 1914, p. 101; see fig. 60) proves very useful, as the spray carried off by the escaping gases and vapours is in great part retained in the bulb, where it can be changed by heat to the oxide, dissolved in concentrated hydrochloric acid and added to the contents of the flask; the colour of the solution makes it easy to see how far the precautions taken against mechanical loss have been effective. The black oxide is then dissolved by the addition of concentrated hydrochloric acid, whereby we obtain a certain (unknown) weight of the green

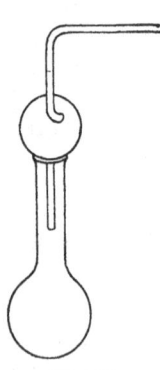

Fig. 60.

chloride soluble in both water and hydrochloric acid, which from its mode of preparation we know to contain a grams of copper[1].

(ii) Change of the green chloride containing a grams of copper into the white chloride, by digestion out of contact with air with a weighed strip of copper. The success of the experiment depends on carrying out the change in the complete absence of air. No doubt this condition could be best secured by using a sealed tube, as in the determination of halogens in organic compounds by Carius' method (Perkin and Kipping, *Organic Chemistry*, 1911, pp. 27 *et seq.*), but with students not versed in the technique of sealing off glass tubes, the same object can be attained fairly well by using the same small round-bottomed Jena glass flask in which the solution and ignition were effected—which has the advantage of avoiding loss in transference—and by closing it as shown in the figure. Concentrated hydrochloric acid is added to the solution up to within a short distance from the mouth of the flask, and the liquid is then heated by means of a rose burner; when hot, the copper strip weighing b grams is dropped in and the well-fitting rubber cork with its glass tube and open rubber tube pressed in tight, after which the screw clip is closed, the glass rod slipped in and the cork wired in securely. The copper should be used in the form of thick foil, as even so there is a good deal of crumbling of the edges during solution, and consequent difficulty in the next stage, the collection for weighing of all the residual copper. The flask is heated by a very small flame[2] at a considerable distance, until the solution, which at first was opaque and almost black, has become practically colourless, which may take as long as two or three days. Note that the sequence of operations is arranged so as to close up the flask when the liquid contained in it has been brought to a temperature such as to make the volume at least as great as it is likely to be in the course of the subsequent digestion.

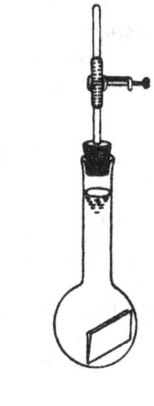

Fig. 61.

[1] It being essential that all the nitrate should have been destroyed before the next stage—the nitric acid liberated by the hydrochloric acid would dissolve some of the copper—it might seem as if at this point we ought to evaporate once or twice to dryness with hydrochloric acid. But the additional expenditure of time would not really be justified, considering how easily the nitrate decomposes, and how much may be done in the way of removing nitrous fumes by a continuous removal of the gas in the flask (by suction from water pump), and how greatly we should be adding to the danger of mechanical loss by increasing the number of evaporations.

[2] A convenient way of obtaining the right amount of heat is to use the luminous flame of an ordinary Bunsen burner from which the air tube has been removed (see fig. 61).

(iii) Determination of the loss of weight of the copper strip. Remove the residual copper from the hydrochloric acid solution of the white chloride; wash it with water, and finally, to ensure quick drying—so as to minimise the danger of surface oxidation—with a mixture of alcohol and ether; determine its weight c. Remember the tendency of the white chloride, when in contact with air and hydrochloric acid, to re-form the green chloride, and hence the necessity of performing quickly the process of removal and washing, lest the change

white chloride + hydrochloric acid + oxygen = green chloride

should go on, thereby making the weight of copper dissolved no longer a measure of its action on the original weight of green chloride. The use of a quickly acting siphon, made from a glass tube of fairly wide bore, greatly facilitates the withdrawal from the copper of the chloride solution.

Record of results of a demonstration Experiment.

	I	II
Weight of copper used in the production of the green chloride (by solution in nitric acid, evaporation to dryness and solution in hydrochloric acid) .... $=a$	4·231 gms.	3·059 gms.
Weight of strip of copper placed in the solution of the green chloride .... $=b$	8·010 ,,	5·768 ,,
Weight of strip of copper after the complete decolorisation of the green chloride $=c$	3·824 ,,	2·678 ,,
Loss in wt. of strip of copper .... $=b-c$	4·186 ,,	3·090 ,,
Ratio of wts. of copper combined with the *same* amount of acid radicle in the two chlorides .... $=a:(a+b-c)$ $=1:\left(1+\dfrac{b-c}{a}\right)$	**1·989**	**2·010**

Results obtained in other similar experiments were:

2·060   2·040   2·018

Standard Value: Berzelius (1811—1812)   2·0025

4. The chlorate and the perchlorate of potassium. Both the qualitative and the quantitative relations between these two salts have been dealt with in detail on previous occasions (*ante*, Chapter II, Experiment IX, p. 90; Chapter VII, Experiment XXXIX, p. 263). The following results were obtained in the manner described (Chapter II, p. 91) for the value of the ratio residual chloride to oxygen:

Ratio between the weights of oxygen combined with the same weight of potassium chloride in the chlorate and the perchlorate of potassium respectively[1].

	I		II		III	
	Chlorate	Per-chlorate	Chlorate	Per-chlorate	Chlorate	Per-chlorate
Wt. of salt taken ...$=a$	1·1233	1·1093	5·1961	3·9918	4·5087	6·6672
Wt. of chloride left $=b$	·6831	·5966	3·1589	2·1515	2·7379	3·5860
Wt. of oxygen ......$=a-b$	·4402	·5127	2·0372	1·8403	1·7708	3·0812
Ratio between the wts. of oxygen evolved from wts. of the two salts which yields the same weight of residual chloride $=\dfrac{a'-b'}{b'}:\dfrac{a''-b''}{b''}$	$\dfrac{·4402}{·6831}$	$:\dfrac{·5127}{·5966}$				
	$=1:1·334$		$1:1·326$		$1:1·329$	
	$=3:4·002$		$=3:3·978$		$=3:3·987$	

[1] This experiment, which in its entirety comprises the determination of the silver nitrate equivalent of the residual chloride and of the density of the gas evolved (see *ante*, pp. 265, 266), has in the writer's laboratory always been done by a few picked students only, and hence no "class results" are available.

CHAPTER VIII

PERMANENT RATIOS

I. Nature and scope of the law.

In Chapter III we recognised as the characteristic property of the complex substances termed compounds the possession of absolutely fixed properties which stand in no definite relation to the properties of the components; in Chapters V—VII the study of the gravimetric composition of compounds showed the existence of simple relations which, if C, C', C'' stand for the weight of the various compounds containing the same two constituents, and A, A', A'', and B, B', B'' for the weights of these constituents, may be represented by:

$$A + B = C \; ; \; A' + B' = C' \; ; \; A'' + B'' = C'' \text{ (Law of Conservation of Mass),}$$

$$\frac{A}{B}, \frac{A}{C}, \frac{B}{C}; \; \frac{A'}{B'}, \frac{A'}{C'}, \frac{B'}{C'}; \; \frac{A''}{B''}, \frac{A''}{C''}, \frac{B''}{C''} = \text{constant} \text{ (Law of Fixed Ratios),}$$

$$\frac{A}{B} : \frac{A'}{B'} : \frac{A''}{B''} = \frac{A}{B} : \frac{mA}{nB} : \frac{pA}{qB} \text{ (Law of Multiple Ratios),}$$

where m, n, p, q are simple whole numbers.

But admitting the constancy of the ratio $\frac{A}{B}$ in the compound C, what is the relation between A and B in other combinations into which these same components enter?

The constancy of the composition of compounds does not prove that the ratio between the weights of the constituent elements is exactly the same in their combinations with other substances. Thus the composition of the sulphide and the sulphate of barium may each be constant, and yet the ratio between the weights of sulphur and barium in the sulphide need not be absolutely identical with the ratio of the same elements in the sulphate (Stas).

There is nothing *a priori* for or against the identity of the two constant ratios, but the comprehensive study of gravimetric composition has shown that the ratio $A : B$ of the constituents in a compound C is not only fixed for that compound, but regulates the relation between the quantities of these constituents in any combination whatever, including that of equivalency in the power of combination with the quantities N, M, P, Q of other elements. Thus, designating by the symbols A, B, M, N, definite experimentally found quantities of the four elements sulphur, oxygen, lead and silver, the scheme given below shows how the ratio $A : B$, which is the constant regulating the combinations between sulphur and oxygen, is also the ratio between the quantities of sulphur and oxygen which combine with the same amount M of lead, or the same amount N of silver in the sulphides and oxides of these two metals, and that the ratio between lead and sulphur is the same in the sulphide as in the sulphate.

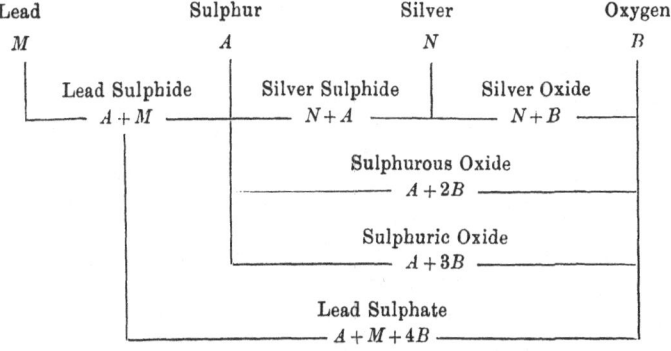

It might conduce to a more vivid appreciation of the quantitative relations embodied in the above scheme if students not already familiar with the appearance, properties and relationships of the substances involved were to do some qualitative experiments with them.

Experiment XLIII. Preparation and properties of some simple compounds of sulphur, oxygen, lead and silver.

(i) Sulphurous and sulphuric oxide: The production of sulphurous oxide by the burning of sulphur in oxygen or in air is an experiment described in all text books, as also is the preparation of the same gas by the action of dilute acid on sulphites, or by the reduction of sulphuric acid by copper or carbon (Newth, 1907 ed., p. 416 ; Ostwald, *Principles of Inorganic Chemistry*, 4th ed., 1914, pp. 306 *et seq.* ; Perkin and Lean, *Introduction to Chemistry and Physics*, Vol. II, pp. 14, 79). The change from sulphurous to sulphuric oxide by the addition of more oxygen and the concomitant change of properties has been made the subject of study in Experiment XII (p. 114).

(ii) **Lead oxide (litharge).** This substance can be made from lead by the action of nitric acid, and subsequent heating of the nitrate thus obtained; and from red lead or brown oxide of lead (by the action of heat). Its power to change into red lead by taking up oxygen from the air, has been dealt with already (Experiment XXIII, p. 177).

Try the action on (*a*) litharge, (*b*) red lead, (*c*) lead peroxide, of: (*a*) acetic acid, (*β*) hydrochloric acid (heating), (*γ*) nitric acid (dilute), (*δ*) sulphuretted hydrogen water.

(iii)   **Lead sulphide :** Found native as the mineral galena, which crystallises in beautiful cubes, grey in colour and with a high metallic lustre. In the laboratory the sulphide is made either by direct synthesis from lead and sulphur or by the action of sulphuretted hydrogen on any soluble or certain insoluble lead salts.

(*a*)   Mix 7 grams of finely divided lead with 1 gram of flowers of sulphur in a hard glass tube and heat. The two elements combine with evolution of heat and light.

(*b*)   To sulphuretted hydrogen water contained in a set of test-tubes add : solutions of lead acetate, or lead nitrate ; scraps of solid white lead chloride, lead oxalate, lead phosphate, basic lead carbonate (the substance used as a paint), lead sulphate. It is on the formation of the black lead sulphide that is based the usual very sensitive test for sulphuretted hydrogen by means of lead acetate (Clowes, *Qualitative Analysis*, 8th ed., 1908, p. 42).

Test the effect on freshly precipitated lead sulphide of hydrochloric acid, and nitric acid both dilute and concentrated.

(iv)   **Lead sulphate :** Produced by :

(*a*)   Oxidation of the sulphide. (*a*) Digest a very little finely powdered galena with concentrated nitric acid added in successive small quantities ; (*β*) to some precipitated lead sulphide suspended in water add a solution of hydrogen peroxide.

(*b*)   Precipitation from a soluble lead salt, or decomposition of lead salts insoluble in water by sulphuric acid or a soluble sulphate. Add a relatively large volume of dilute sulphuric acid to a small volume (about 1 c.c.) of a dilute solution of lead acetate, add methylated spirit (see below, p. 287) and allow to stand.

Warm a little lead oxalate with (*a*) water, (*β*) dilute sulphuric acid; filter (preferably by decantation) and test the filtrate for oxalic acid (Clowes, *Qualitative Analysis*, 8th ed., 1908, pp. 263 *et seq.*.) ; do the same, using lead phosphate and testing for phosphoric acid (Clowes, *ibid.*, p. 226). The white lead sulphate is very similar to barium sulphate, but differs in being more soluble in water and more easily decomposed. Demonstrate (*a*) the insolubility of the lead sulphate in : hydrochloric acid dilute or concentrated ; nitric acid dilute or concentrated ; sulphuric acid dilute ; (*b*) the solubility in sodium acetate; (c) the fact that by boiling for some time with ammonium carbonate there occurs double decomposition into soluble ammonium sulphate and into lead carbonate insoluble in water but easily soluble in acetic or nitric acid. The turbidity produced on diluting commercial sulphuric acid made by the lead chamber process (Newth, 1907, pp. 428 *et seq.*; Ostwald, *ibid.*, 4th ed., 1914, pp. 314, 315) is due to the precipitation of lead sulphate which is soluble in concentrated but insoluble in dilute sulphuric acid. Lead sulphate is more soluble in water than is barium sulphate, but like all sulphates it is practically insoluble in alcohol. Add dilute sulphuric acid to a small volume of lead acetate solution until no further precipitation occurs, withdraw some of the clear liquid above the precipitate, add a large volume of methylated spirit and note the formation of a further amount of precipitate in appearance like the first— what further experiments would be necessary to establish the identity of the two substances? When oil paintings darken in the course of time and are 'restored' by the use of hydrogen peroxide, the reactions occurring are as follows: Through the agency of the sulphuretted hydrogen contained in the air, the white basic lead carbonate which constitutes the paint is converted into black lead sulphide; the hydrogen peroxide in acting on the black sulphide gives up oxygen, transforming the black sulphide into white sulphate and is itself changed into water.

(v) Silver oxide: A brown powder precipitated by the addition to a soluble silver salt of any alkaline hydroxide other than ammonia. Prepare a specimen of the oxide by filtering a solution of barium hydroxide into excess of silver nitrate (Ramsay, *Exp. Proofs of Chem. Theory*, 1884, p. 97); wash by decantation, dry on the water bath, and show that by gentle heating the substance is decomposed into oxygen and metallic silver.

(vi) Silver sulphide: A black substance with metallic lustre, produced by direct synthesis from the elements (*ante*, Chapter V, p. 217) or by the usual method of double decomposition between a silver salt and a soluble sulphide. To sulphuretted hydrogen water contained in a series of small test-tubes add: solution of silver nitrate; scraps of solid silver phosphate; silver chloride; scraps of metallic silver. The tarnishing of silver is due to the action of the sulphuretted hydrogen contained in the air on the metal, and the formation of black silver sulphide in the form of a stain on a coin is used as a test for sulphur, whatever its form of combination. Test for sulphur in iron pyrites, barium sulphate, and the organic substances sulphonal and veronal. In each case strongly heat on charcoal an intimate mixture of the substance with excess of sodium carbonate, place the fused mass on a silver coin and moisten with a drop of water [1].

---

[1] The reaction occurring if the sulphur compound is a sulphate is represented by:

$$\begin{cases} X''SO_4 + Na_2CO_3 = Na_2SO_4 + X''CO_3 \\ \qquad\qquad\qquad\qquad\qquad\quad X''O + CO_2 \\ Na_2SO_4 + 4C \quad = Na_2S \quad + 4CO \end{cases}$$

Whatever the sulphur compound present, sodium sulphide is formed with or

Of the various names given to the constancy of the ratio $A : B$ in any combination whatever, viz. 'law of equivalent ratios, 'law of reciprocal ratios,' 'law of permanent ratios,' 'law of definite ratios,' since the first two refer only to certain limited aspects of the law, one or other of the last two is preferable. The law may be formulated thus :

The ratio of the masses of two substances—elementary or compound—which combine with one another in a binary compound is also the ratio (or a simple whole multiple or submultiple of the ratio) of the masses of these substances present in any more complex compound, and is also that of the masses equivalent as regards power of combination with the same amount of any third substance. (See scheme on p. 285.)

It is the operation of this law which, being at the basis of the conception of combining weights, *i.e.* of constants characteristic for each element which constitute the combining units (*post*, p. 314), introduces simplicity into the study and representation of the composition of the enormous number of compounds dealt with by the chemist.

II. Historical : discovery and establishment of the law.

It is with the name of J. B. Richter (1762—1807) that the establishment, and above all, the deductive application of this most important of all the stoichiometrical laws, is associated, and justly so, considering the extent of his contribution to the subject.

1. But the introduction into the science of the term 'equivalent' and the recognition of the permanency of the neutralisation equivalent of acids independent of the base neutralised, goes back to an earlier time and is due to Cavendish. In a paper written in 1767 he says :

without simultaneous reduction (withdrawal of oxygen) by means of the heated carbon.

$$Na_2S + 2H_2O = 2NaOH + H_2S$$
$$H_2S + 2Ag = Ag_2S + H_2$$
dark stain

...into the other part I put as much fixed alkali (potassium carbonate) as was *equivalent* to $46\frac{8}{10}$ grains of calcareous earth, *i.e.* which would saturate as much acid,

and in 1788, when he gave the record of experiments on the freezing of nitrous (nitric) and vitriolic (sulphuric) acids, in which he compared the strength of the solid freezing out with that of the liquid from which it separates[1], he did so by the application of the principle of the permanency of the neutralisation equivalent. In the case of the nitric acid the relative strengths of the two portions were determined by measuring the quantities of marble required to saturate them ; in the case of vitriolic acid, though for the sake of uniformity the strengths were expressed in terms of the same unit, the formation of insoluble calcium sulphate (gypsum) which coats the surface and 'in good measure defends the marble from the action of the acid' made it impossible to use a similar method. The procedure consisted in finding the weight of the *plumbum vitriolatum* [lead sulphate] formed by the addition of sugar of lead to the sulphuric acid; the strength of the acid was calculated from this value on the supposition that a quantity of oil of vitriol sufficient to produce 100 parts of *plumbum vitriolatum* will dissolve 33 of marble.

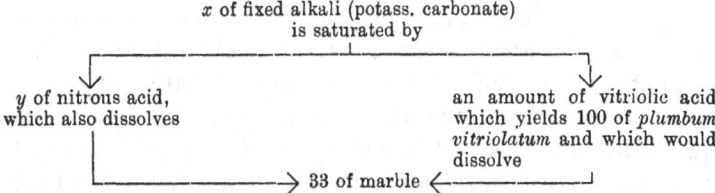

x of fixed alkali (potass. carbonate) is saturated by

y of nitrous acid, which also dissolves

an amount of vitriolic acid which yields 100 of *plumbum vitriolatum* and which would dissolve

→ 33 of marble ←

Here, then, Cavendish assumed as established that the quantities of fixed alkali and marble equivalent as regards the power of neutralising nitric acid are also the quantities equivalent as regards the power of neutralising sulphuric acid.

2. Richter between 1792 and 1799 supplied a comprehensive experimental basis for the relationship the existence of

---

[1] For the production of a sufficiently low temperature, Cavendish was reduced to sending his solutions to Henley House, a fort of the Hudson Bay Company situated about 100 miles from the mouth of the Albany River, the person in charge decanting the liquid from the solid and bringing back to Cavendish two liquids, one the decanted liquid, the other the molten solid.

which had been recognised by Cavendish, and like Cavendish used it deductively for calculating the quantities of B and D, which took part in a certain reaction using as data A and C, A and D, B and C, the experimentally found values for three other reactions. The phenomena dealt with by him were :

(1) The permanence of the neutralisation equivalents of different acids (and bases) independent of the base (or acid) neutralised. This is the relationship which had been recognised and utilised by Cavendish, and is best demonstrated by the students' illustrative experiment XLIV (*post*, p. 299). The principle is that if a certain quantity, say A, of any one acid, say hydrochloric, is neutralised by m of potash, n of soda, p of ammonia etc., and if B of another acid, say sulphuric, is neutralised by r of potash, s of soda, t of ammonia etc., then

$$m : n : p : \ldots = r : s : t : \ldots$$

and s can be calculated from m and n and r, or from n and p and t

and t can be calculated from n and p and s, or from m and p and r.

(For actual data see *Study of Chemical Composition*, p. 175.)

(2) The maintenance of neutrality after double decomposition between neutral salts, such as barium chloride and sodium sulphate, which yield insoluble barium sulphate and sodium chloride. Representing each of the four salts involved by the quantities of acid A and B, and of base C and D, from which they can be formed, the amount of barium chloride used will be $(C + A)$, and the amount of sodium sulphate containing the amount of sulphate radicle required for the change of C of baryta into insoluble sulphate will be $(D + B)$, the amount of neutral salt, barium sulphate, formed being $(C + B)$; along with the barium sulphate is formed sodium chloride, and since the resulting solution is neutral to indicators, there cannot be present excess of either hydrochloric acid or of soda, and the total quantities A and D of these originally involved must therefore have gone to the formation of $(D + A)$ of sodium chloride, and we have the equation :

Barium chloride + sodium sulphate = barium sulphate + sodium chloride

$$\left(\begin{smallmatrix}\text{baryta} \\ \text{alkali}\end{smallmatrix} + \begin{smallmatrix}\text{hydro-} \\ \text{chloric} \\ \text{acid}\end{smallmatrix}\right) + \left(\begin{smallmatrix}\text{soda} \\ \text{alkali}\end{smallmatrix} + \begin{smallmatrix}\text{sul-} \\ \text{phuric} \\ \text{acid}\end{smallmatrix}\right) = \left(\begin{smallmatrix}\text{baryta} \\ \text{alkali}\end{smallmatrix} + \begin{smallmatrix}\text{sul-} \\ \text{phuric} \\ \text{acid}\end{smallmatrix}\right) + \left(\begin{smallmatrix}\text{soda} \\ \text{alkali}\end{smallmatrix} + \begin{smallmatrix}\text{hydro-} \\ \text{chloric} \\ \text{acid}\end{smallmatrix}\right)$$

$$(C+A) \qquad + \qquad (D+B) \qquad = \qquad (C+B) \qquad + \qquad (D+A)$$

Hence the ratio of the quantities of the two acids which in barium chloride $(C+A)$ and barium sulphate $(C+B)$ neutralise the same amount of baryta is the same as the ratio of the quantities of the two acids which in sodium chloride $(D+A)$ and sodium sulphate $(D+B)$ neutralise the same amount of soda and is equal to $A:B$.

(3) Permanency of the ratio between the weights of the 15 metals experimented with which dissolve in 1000 parts of sulphuric acid (or another equal weight m of hydrochloric or n of nitric acid) and constancy of the weight of oxygen combined with these weights of the different metals in their oxides. The first part of this statement, which asserts that the ratio of the weights of the different metals which dissolve in the same weight of acid is independent of the acid, is of the same nature as that dealt with under (1), only that the acid-neutralising substance is in this case the metal instead of the base, which is the metallic oxide. Moreover a connection between these two, the metal and the base, is brought to light in the fact that throughout, the difference between the quantity of base and the quantity of metal equivalent in neutralising power is the same; which, put into present-day language, amounts to saying that *since the weights of the various metallic oxides which neutralise a constant weight of any one acid contain the same weight of oxygen, therefore the quantities of metal equivalent as regards hydrogen*

Name of metal	1000 of sulphuric acid are neutralised by either of the following :		∴ Wt. of oxygen $B=(M-C)$ combined with the different weights of metal C which represent the quantities required for the neutralisation of the same weight, viz. 1000, of sulphuric acid
	Wt. of metal $= C$	Wt. of metallic oxide $= M$	
Iron	841	1280	439
Lead	3153	3592	439
Zinc	909	1348	439
Silver	3289	3728	439
Copper	1317	1756	439
Mercury	5465	5904	439

substituting power are also equivalent as regards oxygen binding power. The table on p. 291 reproduces some of the numbers given by Richter in proof of the existence of this relationship[1]. It is obvious that these numbers have been rounded off to give them the uniformity *expected* ; this was a common practice in those days—Dalton, Wollaston, Thomas Thomson did likewise (*Study of Chemical Composition,* pp. 156, 158, 165) —but the licence taken by Richter in this respect exceeded all limits, and was due to the fact that not only did he wish to demonstrate really existing stoichiometrical regularities, but also mathematical relationships[2] between the equivalents of the individual acids and bases which experience has shown to have been purely imaginary. That the low value of Richter's quantitative work, however, did not vitiate the great importance of his discovery was recognised by Berzelius :

> On reading Richter's work on chemical ratios, we are amazed that the study of this subject could ever have been neglected.

3. Berzelius' own contributions to the establishment of Richter's law are no less important than those he made to Dalton's law (*ante,* Chapter VII). Here again it may be claimed for him that he was an independent discoverer.

> I made the discovery that in all chlorides the quantities of bases saturated by the same amount of hydrochloric acid contained the same amount of oxygen. This was also found to be the case with sulphates. This discovery is really due to...J. B. Richter...who has tried to substantiate it by ingenious though not always sufficiently accurate experiments.

Berzelius' own analyses, which he made ' as accurate as possible' and the most important of which he 'repeated several times before [he] ventured to trust to them,' brought out the following different partial aspects of the law :

(1) The data obtained in the analysis of the oxides and sulphides gave for the ratio of the weights of sulphur and of oxygen combined with 100 of lead, 100 of copper, 100 of iron, 2·02, 2·08, 1·98, numbers sufficiently near to show that the quantities A and B equivalent in combining power towards one metal are so towards all.

[1] Data taken from Muir, *History of Chemical Theories and Laws,* p. 275.

[2] Such as that whilst the bases form an arithmetical progression, the acids form a geometrical progression.

(2) Lead sulphide treated with nitric acid under conditions which ensured the retention within the system of all the lead and sulphur originally found in the sulphide was changed to white insoluble lead sulphate (*ante*, p. 286); it was shown that in the clear liquid above the insoluble sulphate there was present neither lead nitrate nor sulphuric acid, which would have been the compounds formed under the action of the nitric acid from any lead or sulphur not used in the production of the lead sulphate[1]. Hence the ratio $A : B$ of lead to sulphur is the same in the binary sulphide AB as in the ternary sulphate ABC.

(3) The data obtained in the analysis of the three compounds, lead oxide, lead sulphide and lead sulphate, together with the proof of the identity of the ratio lead : sulphur in the sulphide and sulphate (see above), demonstrate the existence of another important relation. If we consider the sulphate as made up of the basic oxide litharge and the acidic oxide sulphuric anhydride, then $A : B$ represents not only the ratio of the quantities of sulphur and oxygen equivalent as regards power of combining with M of lead, but also the combining ratio of these two elements in the sulphuric anhydride, the composition of which is given by $A + 3B$.

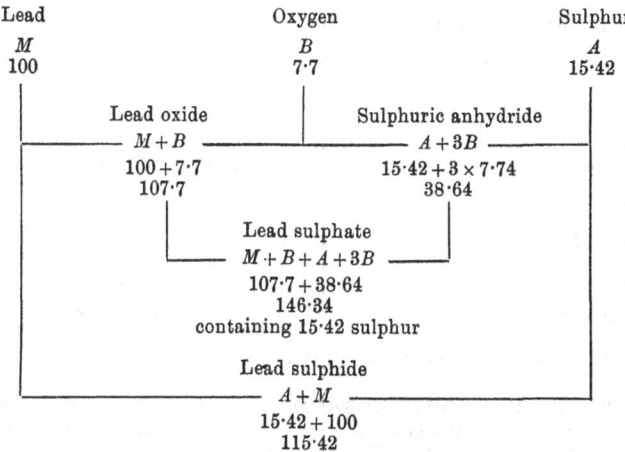

---

[1] Lead dissolves in nitric acid to form soluble lead nitrate; sulphur heated for some time with nitric acid yields sulphuric acid.

III. Deductive application of Richter's law.

In the 1811—1812 Memoir Berzelius wrote :

It is the merit of Richter to have shown long ago that the composition of salts can be ascertained by calculation.

1. We have seen in a previous part of this chapter (p. 289) how before Richter's work was published Cavendish had *calculated* the quantity of marble saturating a definite amount of sulphuric acid from the following three neutralisation ratios : (i) potass. carbonate : nitric acid, (ii) potass. carbonate : sulphuric acid, (iii) marble : nitric acid.

2. Richter tested the validity of his stoichiometrical generalisation by the calculation of neutralisation ratios and the comparison of these calculated values with those experimentally found. He considered that this afforded a true touchstone for the correctness of the experimental results, saying that if the numbers experimentally found are not in keeping with those required by the application of the law, they must be rejected as erroneous. The following is an example of the manner in which he carried out this comparison between calculated and experimental values, the substance involved being silver nitrate :

Found experimentally.

1000 of silver dissolved in nitric acid, evaporated and dried gave   1567   of silver nitrate

1000 of silver dissolved in acid, ppd. by alkali and dried gave ....   1133·5 of base

∴ 1000 of silver combine with (1567 – 1133·5) of nitric acid[1] ....   =**433·5**

Calculated.

1000 of soda are neutralised by   1164·7 of sulphuric acid
,, ,, ,, ,,   1636·6 of nitric acid
1000 of silver ,, ,,   305·8 of sulphuric acid

∴ 1000 of silver ,, ,, $\dfrac{305\cdot8 \times 1636\cdot6}{1164\cdot7}$ of nitric acid = **429·7**

which agrees well with **433·5**, the experimentally found number.

[1] What Richter and his contemporaries and all the chemists during the first decades of the nineteenth century designate as acid is the acidic oxide, which we now term the acid anhydride, whilst our acids were called hydrated acids. Salts were looked upon as binary compounds of two oxides, the one a metallic oxide termed the base, and the other the acid.

3. Berzelius followed up a precise formulation of the law by an equally definite statement concerning its deductive application and the importance of such application.

When two substances A and B have affinity for two others C and D, the ratio of the quantities C and D which saturate the same amount of A is the same as that between the quantities C and D which saturate the same amount of B. If for instance 100 parts of lead combine in the lowest oxide with 7·8 of oxygen and in the sulphide with 15·6 of sulphur, and further if, according to an analysis I shall have occasion to quote later, 100 of iron combine in the lowest sulphide with 58·8 of sulphur, it becomes possible to calculate the composition of the oxide of iron from the simple proportion :

$$15\cdot6 : 7\cdot8 = 58\cdot8 : 29\cdot4,$$

and in the oxide of iron 100 of iron must be combined with 29·4 of oxygen. The experiments which I am going to describe confirm this result.

Berzelius as a matter of fact found that 100 of iron combine in the lowest oxide with 29·6 of oxygen.

...In this manner we can calculate the composition of all binary compounds....It must be evident that when the data involved are firmly established, the result of such calculations must be much more reliable than ordinary analyses. With this object I have endeavoured to make the analyses about to be described as accurate as possible...but nevertheless they are not yet sufficiently accurate to give by calculation more than approximations.

IV.   Is the law exact ?

The above estimation of the magnitude of the experimental error involved shows that Berzelius' analyses and syntheses do not lend themselves to deciding this question. Half a century later Stas wrote :

All the analyses and syntheses performed [hitherto] are equally inadequate for demonstrating the mathematical exactness of the law of definite ratios.   As a matter of fact, however great may be the skill of a chemist, it is impossible that he should perform an analytical or synthetical operation without incurring errors in the measurements.   And so far there is nothing to prove that the differences between the experimental results obtained in analyses and the corresponding values calculated on the basis of a certain assumption should be attributed entirely to the experimental errors ; there is nothing to show that a certain portion of the error is not due to the inexactness of the law of definite ratios, considered as a mathematical law.

When thus stated, it would seem that the difficulty in the way of finding an answer to our question is insurmountable. But it all depends on what we really are concerned to prove or disprove—if it is a case of absolute mathematical exactness, we must admit that the problem is insoluble, since the complete elimination of all experimental error is an absolute impossibility. If, however, the problem is considered from the practical point of view, it assumes a very definite and manageable form; it becomes an investigation of what is the effect of a reduction of the experimental error on the differences between values *found* and values *calculated*, and if the two decrease proportionally, *i.e.* if with reduction of the value of the experimental error the two values for the composition of a compound approximate more and more closely, and if such deviation as is found is covered by the magnitude of the experimental error we shall be justified in classing the law *provisionally* as exact, and in considering that for all practical purposes we may treat it as such. This being the logical basis for the interpretation of the facts, what are the facts and what is the inference to be drawn from them?

(i) The most accurate determinations of a combining weight ratio $B : A$, where A is the standard, viz. $8\cdot000$ of oxygen (see *post*, p. 316) show that within the limits of the experimental error B is the same whether determined directly or indirectly, and is quite independent of the special elements it is combined with in the special compounds analysed. Thus, in the determination of the value silver : oxygen, Stas obtained:

From the ratios	$\dfrac{\text{potass. chlorate}}{\text{potass. chloride}}$ and	$\dfrac{\text{potass. chloride}}{\text{silver}}$	$107\cdot943 : 16$
” ”	$\dfrac{\text{silver sulphate}}{\text{silver}}$ ”	$\dfrac{\text{silver sulphide}}{\text{silver}}$	$107\cdot927 : 16$
” ”	$\dfrac{\text{silver iodate}}{\text{silver iodide}}$ ”	$\dfrac{\text{silver iodide}}{\text{silver}}$	$107\cdot937 : 16$
” ”	$\dfrac{\text{silver chlorate}}{\text{silver chloride}}$ ”	$\dfrac{\text{silver chloride}}{\text{silver}}$	$107\cdot941 : 16$
” ”	$\dfrac{\text{silver bromate}}{\text{silver bromide}}$ ”	$\dfrac{\text{silver bromide}}{\text{silver}}$	$107\cdot923 : 16$

(ii) In a research undertaken with the definite object of testing the accuracy of the law, Stas used an ingenious method for reducing the experimental error by doing without an analysis proper, that is, without a process which practically always involves transference of matter from one vessel to another. In the rare cases where this can be obviated the chief experimental error involved is that of the inevitable weighings. What he did was to compare the values of the ratio $A : B$ in the binary and ternary combinations AB and ABC, and to test for identity in the values by ascertaining whether the ternary substance could be brought to the state of the binary compound without liberation of a fraction, however minute, of the common constituents ; or conversely, whether the binary substance could be transformed into the ternary combination without the exclusion of a portion of the elements present in the binary compound from the ternary compound formed. The method had been used by Berzelius in his transformation of binary lead sulphide to ternary lead sulphate, nitric acid having been the oxidising agent employed (*ante*, p. 293) ; Stas, using sulphurous acid as reducing agent, changed ternary silver iodate (or bromate or chlorate) into binary silver iodide (or bromide or chloride), and tested for silver and iodine (or bromine or chlorine) in the clear liquid above the precipitated insoluble silver halide. But since sooner or later a limit is reached at which it becomes impossible to detect small quantities of A and B by analytical means, such as those employed by Stas in testing for silver and halogen, the method, superior though it is to those criticised and rejected by Stas, is yet equally inadequate for proving the mathematical exactness of the law. As to the degree of approximation to the requirements of the exact law that may be established, this depends entirely on the sensitiveness of the tests for A and B, and the value of a superior limit for possible deviations can be fixed numerically if we know the minimum quantities of A and B which under the special conditions of the experiment can be detected, and the amount of either the binary compound AB or the ternary compound ABC involved. In Berzelius' experiment the question was considered as settled by the statement that the ordinary qualitative tests for lead and for sulphuric acid failed to reveal the presence of lead or of sulphur in the clear liquid above the precipitated lead sulphate, and nothing was said about the quantities of these elements which it would have been possible to detect under these circumstances. On the other hand, Stas tells us that under the conditions of his experiments 1/100 of a milligram of silver or of halogen could be detected in the solution above the precipitated insoluble silver halide ; hence in one special experiment in which 260 grams of silver chlorate were reduced, the absence of any reaction for silver or for chlorine indicates that if either of these elements had been present in the solution, the amount must have been *less* than $\dfrac{1}{10,000,000}$ of the quantity of the silver and the chlorine[1] present in the compounds AB and ABC.

Hence the final inference is that though in the case of the law of permanent ratios, as in the case of all other empirical laws, we have not and cannot expect to have conclusive proof of mathematical exactness, yet we may for all practical purposes class the law as exact, at any rate treat it as such ; what we have been able to accomplish has been to fix a superior limit for the value of possible deviations from the requirements of the exact law, and the value of this is so small that it is greatly exceeded, not only by the probable experimental error of Stas' work, but also by the much smaller error of recent most accurate work ; and so far as we can judge at present, whatever the future may bring in the way of decrease of the experimental error, there is not much likelihood of its falling in analytical work to below 1 in 10,000,000. In any case it is not until this stage has been reached that possible deviations from the law of permanent ratios will call for consideration as a practical problem.

---

[1] 260 grams of silver chlorate contain something like 50 grams of chlorine and 150 grams of silver; the possibility of detecting an excess of ·00001 gram of silver shows that in the change in which 150 grams of silver participated, the amount present in the binary compound cannot differ by more than $\dfrac{·00001}{150} = \dfrac{1}{15,000,000}$ of the original weight from that present in the ternary compound, which makes Stas' estimate of $\dfrac{1}{10,000,000}$ well within the mark. For the chlorine the value would work out to $\dfrac{1}{5,000,000}$.

V. Students' illustrative experiments.

1. **To show that the quantities A and B equivalent in one reaction are so in all reactions.**

(1) The ratio between the quantities of different acids required to neutralise a certain definite amount of any one base is permanent, *i.e.* independent of the nature of the base.

Experiment XLIV. Comparison of the ratio between the weights A and B of sulphuric and hydrochloric acids which neutralise equal amounts of different alkalis, viz. : (i) C' of potash, (ii) C''' of soda, (iii) C'''' of ammonia. N.B. The amounts of one and the same alkali—*e.g.* potash—with respect to which the neutralising powers of the two acids are compared must be the same, but of course the amounts C''' of soda and C'''' of ammonia need not be identical with C', the amount of potash.

Being provided with solutions of potash, soda, ammonia, hydrochloric acid and sulphuric acid of suitable strength [1], which may or may not be known, find by means of six *sets* of titrations (i–vi) the values for the ratios required. Choose a suitable indicator. The volume of alkali taken for titration must be sufficiently great to make the volume of acid required for the neutralisation so large that the probable experimental error shall not exceed $\frac{1}{2}$ per cent. of the quantity affected by it ; the experimental error, due to the difficulty of detecting the exact point of neutrality and to want of exactness in the reading of the burette, will amount at least to ·05 c.c., possibly to ·1 c.c., and may even be larger ; taking ·1 as the probable value, this means that the volume of acid used should not be less than 20 c.c. [2]

[1] A convenient strength for the solutions is approximately twice normal; which means that they contain per litre twice the equivalent weight in grams of the solute, which in the case of acids is the amount containing two gram-equivalents of replaceable hydrogen ($2 \times 1·0076$), and in the case of bases is twice the weight which will neutralise 1 gram-equivalent of acid.

[2] Considering the important place which volumetric analysis takes in the scheme of chemistry teaching, it must be a matter of great regret that a fundamental and simple consideration of this kind is almost invariably neglected. The writer has been very much struck by this in looking through students' note-books, where cases like the following have been met with again and again: In the analysis of a sample of a crushed and evenly mixed ore, the percentage of ferrous iron was found in the usual way by oxidation with standard bichromate. The quantities of solution taken for titration were such that the volume of the bichromate solution required was only about 3 c.c.; considering the difficulty of getting the *exact* end point in this reaction, the experimental error was probably greater than ·1 c.c., but taking it at that, it would exceed 3 °/₀, making the value for the percentage of iron, which was about 25, uncertain to 1 unit in the second significant figure. The red ink comment of the responsible teacher was to the effect that the official determination having yielded a value differing by about ·6 °/₀ from that of the student, the experiment ought to be repeated ; no comment whatever was made about the modifications required in order to

Obviously the required comparison of the relative weights of two acids equivalent in their neutralising power towards different alkalis can, provided that the *same* solution of the two acids is used throughout, be accomplished by simply comparing the *volumes* equivalent in neutralising power; but calculation of the *weight* equivalency ratio presents the advantage that on the supposition of the accuracy of the value used for the strength of the solutions, *i.e.* weight of solute per 1 c.c. of solution, it becomes possible to compare the results with the *standard* values.

obtain a value to which the standard of accuracy expected could legitimately have been applied!

And when we turn to text books, things are no better. Even in a standard text book which deals with quantitative analysis in a remarkably successful manner, in the determination of the ammonium contained in ferrous ammonium sulphate the example given to illustrate the process and the method of calculation exhibits the mistake here complained of in a most glaring form. The ammonium is determined in the usual manner by expelling the volatile base ammonia by boiling with potash, absorbing the ammonia in a known volume of standard sulphuric acid (100 c.c. of normal acid), and titrating the residual acid in the solution made up to 200 c.c. by standard-normal-soda. It is specially stated that in the determination of the residual acid, two titrations should not differ by more than ·1 c.c.; yet in the statement of the result we are told that 21·06 c.c. of this liquid neutralise 10 c.c. of normal soda. If, as it seems justifiable to assume, nothing having been said to the contrary, an ordinary burette was used, graduated to 1/10 c.c. and delivering drops of which about two correspond to a graduation, and therefore not allowing of measurements accurate to more than 1/20 c.c., the critical reader may be allowed to wonder how the volume measurement can be given to 1/100 c.c., as seems to be indicated by the number 21·06. As a matter of fact the probable experimental error, put at its lowest, would amount to ·05 c.c. (and this is evidently recognised when, as mentioned above, in the directions given for experimental procedure it is said that two titrations should not differ by more than ·1 c.c.). But if the error in about 21 c.c. is ·05 c.c., for the total volume of 200 c.c. it will be approximately ·5 c.c., which is equal to about 1 in 400; yet the directly correlated volume of normal soda required for the neutralisation of the residual acid is given as 94·969 c.c. $\left(= \dfrac{200 \times 10}{21 \cdot 06} \right)$, that is, a value pretending to an accuracy of 1 in 95,000. Moreover, the error of ·5 c.c. in the 200 c.c. of partially neutralised acid, and that of about ·25 c.c. in the corresponding 95·0 c.c. of residual normal acid, would throw itself completely on to the *difference* between the volume of original and the volume of residual acid, which supplies the measure for the ammonia absorbed and which in the special case quoted is given as 5·031 c.c., *i.e.* to four significant figures, whereas in reality the second place is uncertain to $2\frac{1}{2}$ units, making the probable error of the result $5°/_0$. It can of course only be due to chance that 9·16, the experimentally found value for the percentage of ammonium, should be in such close agreement with 9·18, the value calculated from the formula. The mistake committed has been: (i) using too little of the ferrous ammonium sulphate, so making the volume of acid neutralised too small compared with the experimental error involved in the titration of the residual acid; (ii) multiplying the effect of the experimental error of the titration through using too small an aliquot part of the liquid containing the residual acid.

Record of results.

A. Suggested form of note-book entry.

Determination of the neutralisation equivalents of certain acids towards different alkalis.

	Acid	gms. of acid contained in 1 c.c. of solution	Alkali	Volume of alkali taken $= p$ c.c.	Volume of acid required for neutral-isation $= q$ c.c.	Volume of acid required for neutral-isation of 1 c.c. of alkali $= q/p$ c.c.	Ratio of weights of hydrochloric and sulphuric acids equivalent in power of neutralisation towards the following alkalis:
(i)	Hydrochloric	b	Potash			m	$\text{Potash} = \dfrac{mb}{nc}$
(ii)	Sulphuric	c	,,			n	
(iii)	Hydrochloric	b	Soda			m'	$\text{Soda} = \dfrac{m'b}{n'c}$
(iv)	Sulphuric	c	,,			n'	
(v)	Hydrochloric	b	Ammonia			m''	$\text{Ammonia} = \dfrac{m''b}{n''c}$
(vi)	Sulphuric	c	,,			n''	

B. Summary of results obtained by a class of students.

Ratio between the weights of sulphuric and of hydrochloric acids required for the neutralisation of the same quantity of

	Potash $= \dfrac{mb}{nc}$	Soda $= \dfrac{m'b}{n'c}$	Ammonia $= \dfrac{m''b}{n''c}$
Student A	1·35	1·36	1·34
,, B	(1·29)	1·30	1·33
,, C	(1·37)	1·33	1·30
., D	1·36	1·33	1·34
,, E	1·32	1·36	(1·43)
,, F	1·31	1·32	1·32
,, G	1·31	1·35	1·34
,, H	1·35	1·35	1·35
,, I	1·34	1·33	1·33
Mean	**1·335**	**1·337**	**1·331**
Standard Value	**1·342**		

(2) To show that the equivalency ratio between a metal and the hydrogen it replaces in an acid is independent of the nature of the acid.

Magnesium (or zinc, aluminium, tin etc.) acts on aqueous

solutions of the substances named hydrochloric acid and sulphuric acid, evolving hydrogen and producing (aqueous solutions of) the salts magnesium chloride and magnesium sulphate respectively. These salts differ in composition from the parent acid only in that magnesium has taken the place of hydrogen; a portion of the constituent elements of the acid has moved as a whole to another compound, termed salt, and the identity of certain analytical reactions of the acid and the salt shows the independent existence of this same constituent group in solutions of the two substances. Thus solution of barium chloride produces with either sulphuric acid or magnesium sulphate precipitates identical in properties and in composition. This group, consisting of all the constituents of the acid except the hydrogen replaceable by metal, is termed the 'acid radicle.' Hence we have the following relations :

The weight a of magnesium (or some other metal) acting on some (in our experiment *unknown*) weight of sulphuric acid containing an unknown weight x of the radicle X, and the weight b of hydrogen expelled in the replacement effected, are *equivalent* as regards power of combination with x of the radicle X of sulphuric acid. Similarly if a' of magnesium expels b' of hydrogen from hydrochloric acid, a' and b' are the weights of magnesium and of hydrogen equivalent as regards the power of combination with the same (*also unknown*) weight y of the radicle Y.

The quickest and most accurate method of finding the ratios b/a and b'/a' consists in determining indirectly the *weights* b and b' of the hydrogen evolved by a and a', the accurately weighed quantities of magnesium, by measuring v and v', the *volumes* of hydrogen evolved, and multiplying these by d and d', the densities of dry hydrogen measured under the conditions of the experiment, that is, at temperatures t and t' and pressures p and p'. The use of the values d and d' involves a knowledge of d_0, the density of hydrogen at normal temperature and pressure, a quantity which has been determined with great accuracy by various experimenters, the standard value accepted being ·0000900 gram per c.c. for all calculations except those in which the accuracy of all the other measurements involved has been carried to a very high degree ; in this latter case the following values are available :

·000089979	Rayleigh, 1892
·000089951 and ·000089870	Morley, 1892
·000089947	Thomson, 1895

By the application of the gaseous laws, d and d' can be calculated from d_0 if the temperature and pressure are known.

Experiment XLV. Find the quantities of hydrogen expelled by 1 gram of magnesium from (i) hydrochloric acid, (ii) sulphuric acid; measure the volumes of hydrogen evolved.

According to what has been said above, the ratios required are

$$\frac{b}{a} \text{ and } \frac{b'}{a'} = \frac{vd}{a} \text{ and } \frac{v'd'}{a'},$$

but if the volume measurements are made under the same conditions of temperature and pressure, d and d', the values for the density of the hydrogen, will be identical, from which it follows that the object of our experiment can be achieved by simply comparing the volumes v and v' of hydrogen liberated by unit weight of magnesium from different acids, provided that these volume measurements are made under the same physical conditions. At least two determinations should be made with each of the acids, and the difference between the mean values for the two acids should be compared with the difference between the two values obtained in each series, in order to see how far experimental error can be made to account for the deviations from identity of the values $\frac{v}{a}$ and $\frac{v'}{a'}$.

The tables on p. 307 give a scheme for the record of the results referred (i) to the volume, (ii) to the weight of hydrogen replaced.

This being a very popular experiment, there is great variety in the apparatus used for collecting and measuring the hydrogen evolved. As suggested in the Introductory chapter, provided that the conditions are favourable, it will be advantageous to allow the learners to consult different text books and each to choose the kind of apparatus that appeals to him most

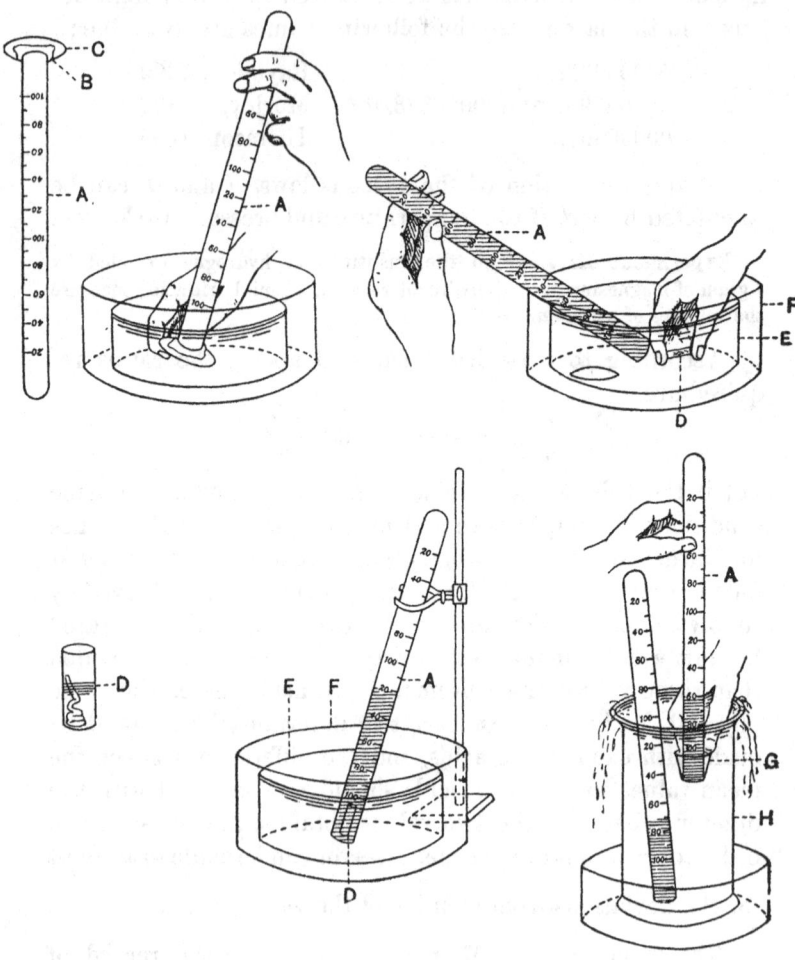

Fig. 62.

A. Gas measuring tube of capacity 200 c.c., graduated to 1/1 c.c.
   (An ordinary glass tube, one end of which has been closed by the blow-pipe, can be used, the volume of hydrogen collected being indicated by a rubber band, and its value subsequently determined with the help of a graduated cylinder or a burette.)

B. Piece of blotting paper placed over the mouth of the tube, which has been nearly filled with the dilute acid, after which water is carefully poured in until a little overflows.

C. Small watch-glass pressed with its convex side on to the blotting paper.

D. Small specimen tube for holding the weighed metal, drawn on larger scale, filled with water and with a little asbestos fibre in the mouth, which latter is to postpone the beginning of the action of the acid on the metal, thereby facilitating the introduction of D into A.

E. Dish about 10 cm. across and 7 cm. high, *completely* filled with the dilute acid.

F. Large dish to catch the overflow of acid from E.

G. Crucible for the transference of A into the cylinder,

H. Cylinder deep enough to allow of the equalisation of the levels of water inside and outside A.

(Perkin and Lean, *Introduction to the Study of Chemistry,* 1909, pp. 70 *et seq.*; Fenton, *Outlines of Chemistry,* pp. 55

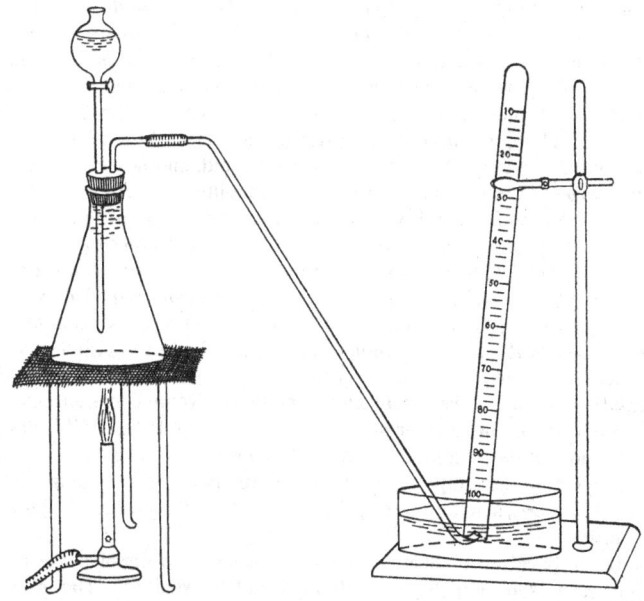

Fig. 63.

et seq.). The apparatus depicted in fig. 62, which has the advantage of great simplicity, can be used in all cases where the action occurs at ordinary temperature, as happens with the metals magnesium, zinc, aluminium ; when it is necessary to raise the temperature, *e.g.* in finding the volume of hydrogen evolved from hydrochloric acid by a known weight of tin, the apparatus depicted in fig. 63, p. 305, will be found suitable.

Procedure.

Suitable quantities of pure specimens of the metal are weighed accurately ; in the case of magnesium it is recommended to use the metal in the form of ribbon, or better still, of wire, as this makes it possible to get with ease a convenient weight by roughly weighing a long piece and then calculating and cutting off the length required which is then weighed accurately. It is useful to remember that 1 c.c. of hydrogen is evolved by (about) 1 mg. of magnesium, ·75 mg. of aluminium, or 2·7 mg. of zinc. Very dilute acid only is required in the case of magnesium (1 volume of the commercial hydrochloric or sulphuric acid to about 10 volumes of water); zinc and aluminium require stronger acid (1 : 3). The manner of filling and inverting the graduated tube and of introducing the metal is shown in the figure. Of course if the tube used should be so narrow as to make it possible to close it with one's thumb, or so flat as to make possible the use of a ground glass plate, the filter paper watch-glass device will not be necessary. The action in the case of magnesium is very rapid, and fairly stout wire is on this account preferable, since if the magnesium is used in the form of ribbon, it is apt to fall quickly to pieces which rise to the top of the liquid —carried up by the hydrogen attached to them—these cling to the sides of the tube and are left behind by the receding acid ; the process must therefore be carefully watched, and if necessary the metal washed down by a sudden upward jerk of the liquid the instant that such sticking is noticed. Considering this, it will be desirable to have the tube loose in the clip whilst the reaction is going on. The transference to a deep vessel where, after equalisation of the levels inside and outside the volume measurement is made, is effected in the manner previously described (p. 164). If it is desired to state the results in *weights* of hydrogen evolved, it becomes of course necessary to read the temperature and pressure; in this case it is important that all the liquids, acid and water, should be at the temperature of the room (*ante*, p. 164).

If the application of heat is necessary in order to dissolve the metal, the apparatus shown in fig. 63 will be found convenient. The weighed piece of metal is placed in the flask, which together with the dropping funnel delivery tube and graduated tube is entirely filled with water. When all air has been removed from the apparatus, concentrated acid is allowed to flow from the dropping funnel into the flask which is now heated, and the gas is collected in the graduated tube. When the metal is entirely dissolved, water is run into the flask until all the gas has been driven into the graduated tube. The transference and measurement of the gas is conducted as described on p. 164.

Record of results.

(i) Results stated in terms of **volumes** of hydrogen.

Specimen record of a demonstration experiment.

Volumes of hydrogen, measured at the same temperature and pressure, evolved by the same weight, viz. 1·000 gram, of magnesium from :

	Hydrochloric acid		Sulphuric acid	
	Exp. I	Exp. II	Exp. I	Exp. II
Wt. of magnesium ...	·163 gm.	·166 gm.	·167 gm.	·163 gm.
Vol. of hydrogen ...	160 c.c.	163 c.c.	165 c.c.	160 c.c.
Vol. of hydrogen evolved by 1·000 gm. of magnesium	982 c.c.	982 c.c.	988 c.c.	982 c.c.
	Mean = 982 c.c.		**Mean = 985** c.c.	

(ii) Results stated in terms of **weights** of hydrogen.

A. Specimen of students' note-book entry.

Weight of hydrogen evolved by the same weight, viz. 1·000 gram, of magnesium from :

	Hydrochloric acid		Sulphuric acid	
	Exp. I	Exp. II	Exp. I	Exp. II
Wt. of magnesium used = a	·165 gm.	·157 gm.	·158 gm.	·163 gm.
Vol. of the moist hydrogen evolved, measured at temperature of room and atmospheric pressure = v	163 c.c.	155 c.c.	156 c.c.	160 c.c.
Temperature = t	11° C.	11° C.	11° C.	11° C.
Barometric pressure = B	762 mm.	762 mm.	762 mm.	762 mm.
Tension of aqueous vapour at temp. of room (from tables) = τ	10 mm.	10 mm.	10 mm.	10 mm.
Normal density of hydrogen, *i.e.* wt. of 1 c.c. at 0° and 760 mm. (antecedent datum) = d_0 = ·0000900				
Density of hydrogen at temp. and pressure of experiment $= \dfrac{d_0 (B - \tau) 273}{760 (273 + t)} = d$	·0000856 gm. per c.c.	·0000856 gm. per c.c.	·0000856 gm. per c.c.	·0000856 gm. per c.c.
Wt. of hydrogen evolved $= v \times d$	·001395 gm.	·001326 gm.	·001335 gm.	·001370 gm.
Wt. of hydrogen evolved by 1 gram of magnesium $= \dfrac{v \times d}{a}$	·0846 gm.	·0845 gm.	·0845 gm.	·0841 gm.
	Mean = ·0846 gram.		**Mean = ·0843** gram.	
	Standard Value = ·0828 gram.			

20—2

B.   Summary of results obtained by students (small section of class in different years)[1].

Weight of magnesium required to evolve 1·01 grams of hydrogen[2] from (*a*) hydrochloric acid, (*b*) sulphuric acid:

Year	Hydrochloric acid series	Sulphuric acid series
1903	(11·8), 12·0, 12·3, (11·2), (12·8)	(11·7), 11·9, 12·1, (10·9)
1904	12·2, 12·2, 12·2, 12·2	12·2, 12·3, 12·4
1905	12·2, 12·2	12·2, 12·2
1906	12·2, 12·2	12·2, 12·2
1907	11·9, 11·9	11·9, 11·9
1909	12·0, 12·0	12·0, 12·0
1911	12·1, 12·1	12·0, 12·1
	Mean = 12·1	**Mean = 12·1**

Standard Value = 12·18

(3)   To show that the ratio between the hydrogen substituting and oxygen binding power of metals is the same for all metals.

The manner of finding the hydrogen substituting power has been dealt with in the previous experiment, and the results obtained there can be directly utilised here. The determination of the oxygen binding power is most easily made by synthesising the oxide from the metal in the manner described for copper (*ante*, p. 280), but it must be remembered that the temperature required for the decomposition of the nitrates varies with the nature of the metal (*ante*, p. 129). Moreover there is a relation between the hydrogen expelling power of a metal and the stability of its nitrate : thus magnesium acts most vigorously on even dilute acids, but its nitrate requires a much higher temperature for decomposition than does the nitrate of zinc, a metal the action of

---

[1] This experiment was never done by more than one or two members of the class, owing to the fact that it was used only incidentally as an illustration of one method of combining weight determination, various other methods being used by the rest of the class.

[2] Obviously the point we are concerned with here, viz. the permanence of the equivalency factor—weight of magnesium : weight of hydrogen independent of the acid radicle involved, is not affected by the standard of reference we may use, whether we refer the weight of hydrogen to some constant weight of magnesium, or *vice versa*. If, as is done above, we use as our standard of reference the combining weight of hydrogen 1·01 (O = 8·00), the values found will incidentally give us the combining weight of magnesium (*post*, p. 347).

which on hydrochloric or sulphuric acid is much less vigorous than that of magnesium.

Experiment XLVI. Find the ratio between the weight of metal combining with unit weight of oxygen, and that evolving unit weight of hydrogen for (*a*) magnesium, (*b*) zinc, (*c*) tin, and record your results in some tabular form (cp. p. 347).

2. **To show that the ratio $A : B$ is the same in the binary compound AB as in the ternary compound ABC.**

Berzelius oxidised lead sulphide to lead sulphate, and the resulting substance was tested for a soluble lead compound and for sulphate, these being the substances which would be formed along with the insoluble lead sulphate if in this latter ternary substance there were less of lead or less of sulphur than in the binary parent substance, the lead sulphide.

Experiment XLVII. Oxidation of lead sulphide to lead sulphate and testing of the resulting substance for the relative amounts of lead oxide and sulphate produced.

To a solution of hydrogen peroxide, which has been freed from the sulphuric acid which is often present in the commercial substance by adding saturated baryta solution till the reaction to litmus is neutral, add freshly precipitated moist lead sulphide in small quantities at a time. Bubbles of oxygen are given off and the black sulphide is gradually oxidised to the white sulphate. When there is no further action add to the liquid twice its volume of methylated spirit and leave to stand for some time. Test the supernatant liquid for lead by passing hydrogen sulphide into part of it and observing whether any black precipitate is formed, and for sulphate by adding to a part of it, a solution of barium chloride and noting whether any white sulphate is thrown down.

3. **To show that the ratio $A : B$ is the same in the combinations AC, BC and AB, i.e. that the quantities of A and B which are equivalent in the power of combination with a third substance C are also the quantities which will combine together and vice versa.**

Demonstration of the existence of the above relationship for the three elements carbon, hydrogen and oxygen.

Experiment XLVIII. Determination of the volumetric combining ratios for the substances marsh gas (hydride of carbon), carbon dioxide (oxide of carbon) and water (oxide of hydrogen), which shows that the quantities *A* of hydrogen and B of oxygen which are united with the same weight of carbon in the hydride and the oxide, are also the quantities which unite to form the oxide of hydrogen.

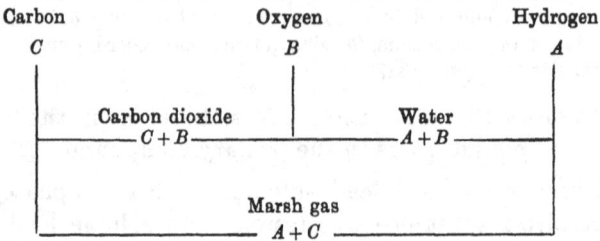

The experimental work required comprises the following volume measurements:

(i)   Ratio hydrogen : carbon dioxide, obtained in the decomposition of marsh gas.

(ii)   Ratio carbon dioxide : oxygen used in its production.

(iii)   Ratio hydrogen : oxygen which unite to form water.

(i)   **Methane.** The methane (marsh gas, Dalton's light carburetted hydrogen) is prepared in the manner already referred to (*ante*, p. 124), by heating a mixture of fused dehydrated sodium acetate with about 3 times its weight of soda lime in either a hard glass test-tube or a Florence flask—placed so that the condensed moisture may not run back upon the hot glass—and either delivering the gas directly into the vessel in which it is to be tested, or storing it in an aspirator over water, in which it is not very soluble (at 20° 1 volume of water absorbs about ·035 volume of the gas); if required dry, it must on its way to the decomposing apparatus be bubbled through concentrated sulphuric acid.

(1)   Determination of the ratio methane : hydrogen by decomposing a known volume of the gas into gaseous hydrogen and a negligible volume of solid carbon. The decomposition is effected by passing electric sparks until no further volume change occurs.

Let the ratio found be 1 : a (the standard value being 1 : 2).

(2)   Determination of the ratio methane : carbon dioxide. This is found by exploding in a suitable apparatus a known volume of the gas mixed with excess of oxygen, and measuring the volume of carbon dioxide formed by absorption with potash. The explosion must be carried out over mercury, because if done over water at the high pressure produced, a very large proportion of the carbon dioxide would dissolve in the water.

Let the ratio of methane to carbon dioxide be found to be 1 : b (the standard value being 1 : 1). Since the ratio of methane to hydrogen = 1 : a, the required ratio of the volume of hydrogen to that of carbon dioxide producible from the hydrogen and carbon united in methane = a : b.

(ii) The ratio carbon dioxide : oxygen used in its formation.

This experiment is described in most of the standard text books (Newth, 1907 ed., pp. 307 *et seq.*; Roscoe and Schorlemmer, *Treatise on Chemistry*, vol. I, p. 714). The principle of the experiment consists in synthesising the carbon dioxide by burning some carbon in oxygen or in air, and showing that no volume change occurs, from which the inference is drawn that the volume of the oxide formed is equal to that of the oxygen participating in

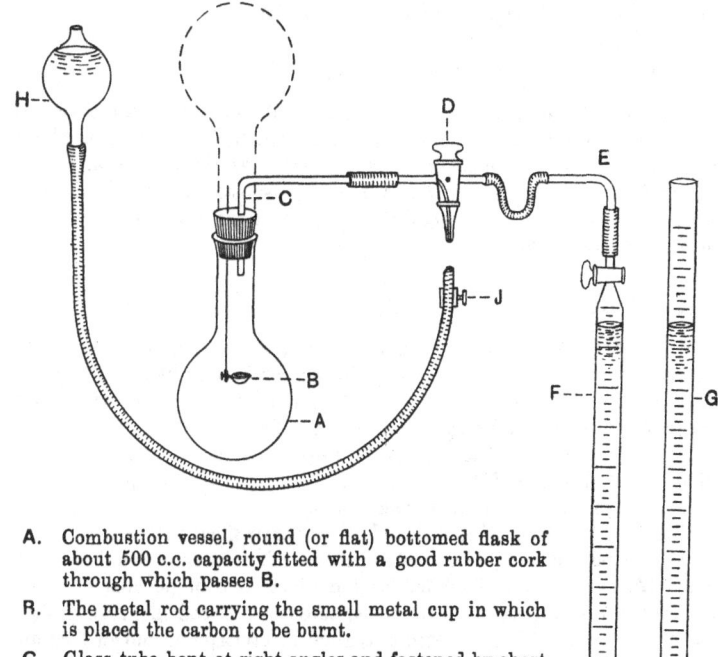
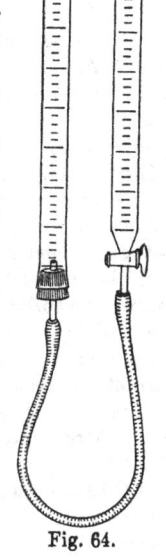

A. Combustion vessel, round (or flat) bottomed flask of about 500 c.c. capacity fitted with a good rubber cork through which passes B.

R. The metal rod carrying the small metal cup in which is placed the carbon to be burnt.

C. Glass tube bent at right angles and fastened by short rubber tube to

D. Three-way glass stop-cock by means of which communication can be made between A and the measuring tubes F and G, or between A and the reservoir H.

E. Capillary tube of fine bore bent at right angles connecting the three-way stop-cock with the measuring tubes.

F and G. Measuring tube and pressure adjusting tube by means of which any change in volume of the gaseous contents of A can be detected and measured. F is graduated to 1/10 cm.

H. Reservoir filled with potash solution for the absorption and measurement of the carbon dioxide found in A.

J. Screw or pinch clip for closing the india-rubber tube, which when quite filled with potash is slipped on to D.

Fig. 64.

its formation. Obviously if we are prepared to accept an approximation to this relationship as indication of identity, it is not necessary that the reaction should have been completed, *i.e.* that we should have used all the oxygen available, or that we should know the volume of carbon dioxide formed. If however we wish to measure the actual ratio $c : 1$, we must measure m, the small volume change resulting on combustion, and n, the volume of carbon dioxide formed, when we get the relationship: volume of carbon dioxide : volume of oxygen used in its production

$$= n : n \pm m = \frac{n}{n \pm m} : 1 = c : 1.$$

The apparatus depicted in fig. 64 can be easily set up with the resources of an ordinary laboratory; a Hempel gas burette is generally used for measuring any volume change which accompanies the formation of the carbon dioxide; a Lunge's nitrometer may be substituted for the Hempel gas burette, or an equally effective arrangement may be made from two burettes, as shown in the figure.

Procedure.

Some of the air in A is replaced by oxygen; though this is not absolutely necessary it is desirable in order to help the combustion by increasing the concentration of the oxygen; but it is better not to use pure oxygen lest the combustion should be too rapid and the heat evolution so great as to produce a very considerable increase of pressure in the closed system which might lead to leakage of the gas through the cork or at the tap D.

A small piece of charcoal is placed in B.

E is detached from F, and F and G are nearly filled with water.

F is connected to E, D placed so as to make communication between A and F, and a value obtained for the effect on F of pushing the cork into A; for this purpose the level of the water in F is read, the cork pushed into A, the pressure in F adjusted to that of the atmosphere and an accurate reading taken; this process is repeated a number of times so as to obtain a fairly reliable mean value.

We are now ready to burn the carbon in A and note the effect on the volume of the gas; with D still placed so as to make communication between F and A, and the cork not inserted in A, the level of the water in F is read—let this be a—the charcoal in B is ignited by a Bunsen burner, the cork is rapidly and securely pushed into A and D turned to an angle of $45°$ so as to close both C and F; when the combustion has ceased allow sufficient time to elapse to ensure that A has cooled to the temperature of the room; turn D so as to make communication with F and by raising or lowering G bring the pressure inside the apparatus to that of the atmosphere. Close the tap at the top of F and read the level of F—let this be b. It is desirable to close the tap of F as soon as the levels of F and G have been equalised so as to give as little chance as possible for the diffusion of carbon dioxide and its solution in F—m will represent any volume change that has occurred in A in the course of the experiment; the more successful we have been in eliminating experimental error the more nearly will $(b-a)-m$ approach to zero. Now measure the volume of carbon dioxide found; for this purpose fill H with moderately strong potash solution (one part of potash to two of water), get all the air in the long rubber tube replaced by the solution, let some of the solution flow out by opening the clip J for an instant; slip the rubber tubing over the nozzle end of D, twist A through 180° by twisting C in the rubber tube which connects A with D, turn D so as to make connection between A and H and let the potash run in—if necessary start it by cooling A by the spontaneous evaporation of a few drops of ether poured on the outside, and then assist the absorption by shaking. When absorption has been completed adjust the pressure to that of the atmosphere, close D, detach the tubes connecting D with F and H respectively and measure the volume of potash solution that has entered A and C by means of a graduated cylinder; let this volume be n, then as shown above

$$n : n \pm m = \text{carbon dioxide} : \text{oxygen}.$$
$$\text{(Standard value} = 1 : 1.)$$

(iii) Ratio hydrogen : oxygen uniting to form water. This has been determined in Experiment XXXII, p. 245, when its value was found to be in close approximation to the standard value 2·00 : 1.

Interpretation of the results. According to (i) and (ii) the ratio of the volumes A of hydrogen and B of oxygen combined with the same unknown amount of carbon C is given by the relations

$$A : BC = 2 : 1,$$
$$BC : B = 1 : 1,$$

and (iii) gives for the ratio of the volumes A of hydrogen and B of oxygen which unite to form the compound $A'B'$,

$$A' = B' = 2 : 1 = A : B,$$
$$\therefore \ A : B \text{ is the same in } AC \text{ as in } AB.$$

CHAPTER IX

COMBINING OR EQUIVALENT WEIGHTS

I. The conception of the combining weight.

This is the outcome of the recognition of the laws of chemical combination; it is a generalisation including all these laws, though it is sometimes quite justly connected specially with the law of permanent ratios, which for that reason is also designated as the law of combining weights. In dealing with the latter in the preceding chapter, it has been pointed out that it supplies a basis on which the composition of any compound whatever can be represented by the general formula

$$mA + nB + pC + qD + ...,$$

where if A denotes an arbitrarily chosen weight of any one element, then B, C, D,... are constants which denote values characteristic of and peculiar to the other elements in the compound, and m, n, p, q, ... are simple whole numbers. To these specific constants A, B, C, D, ... which represent the ratio of the quantities of the different elements which unite with one another, has been given the name *combining weight*. The term is used as if synonymous with *equivalent weight*, which latter name, as a matter of fact, is more frequently met with as the designation of the constants A, B, C, D, ... Fundamentally the two are not identical, combining weight referring to composition as expressed by the ratio B/A, and equivalent weight to the chemical equivalency in the combining power of two substances—elementary or compound—towards a third; but their indiscriminate use cannot really be objected to, because, according to the law of permanent ratios, the quantities of two elements equivalent in their combining power with a third are also the quantities which will combine with one another, and *vice versa*.

II. Consideration of the general formula

$$mA + nB + pC + qD + \ldots$$

which represents the quantitative composition of any compound M.

1. Choice of a standard.

Obviously, since there is nothing fixed about the amount of M, the constancy found holds, not for the absolute weights of the different component elements, but only for that of their relative amounts, *i.e.* the combining *ratios*; hence in order to secure uniformity for practical purposes, it becomes necessary to select once for all a common standard A, when the values B, C, D, etc. denote the constant quantities of the various elements which will combine with or which are equivalent to that definite fixed amount of A. The choice of A, like the choice of a standard to be used in the measurement of length or of mass, is a process which, though in principle absolutely arbitrary, is in reality dominated by considerations of convenience. Two things have to be chosen, (a) the element, and (b) the amount of it.

(a) Choice of the element.

The considerations set forth below under (2) show that if we determine the required ratio B/A indirectly by measuring B/C and assuming C/A as an antecedent datum, the error attaching to each of these ratios will tell in the value of the error of B/A, an error which will increase with the number of antecedent data involved. This means that, *in general*, a method will be the less trustworthy, the less direct it is. Hence it is obvious that the element chosen for standard should be one which gives opportunity for a large number of *direct* determinations, as will be the case if it forms suitable compounds with a large number of other elements. All elements except bromine and fluorine form compounds with oxygen; a very large number of these compounds can be obtained in a state of great purity, and consequently their composition can be ascertained with great accuracy. Hence oxygen, which at different times in the course of the last century was used as a standard element, is now definitely accepted as such by international agreement. On the other hand hydrogen, since it has the smallest combining weight, seemed to be marked out by nature as the standard, and has in fact held that place alternately with oxygen. When judged by the above criteria, however, its unsuitability is obvious: the number of elements which form simple compounds with hydrogen is small, and these compounds, either because they are too unstable to be properly purified or because the relative amount of hydrogen contained in them is so very small, do not lend themselves well to the purpose of accurate analysis.

(*b*) Choice of the standard amount of the standard element.

It has on practical grounds been found desirable to make the numbers expressing the combining ratios as small as possible subject to the limitation of none being less than unity. When, as was done at one time by Berzelius, oxygen was taken as 100, the values for many elements became clumsily high (*e.g.* $Pb = 1294.5$, $Au = 1243.0$); when Thomson used $O = 1$, the combining weights of some elements were represented by numbers smaller than unity. Early experiments had led to the belief that the ratio of Hydrogen to Oxygen in water was exactly $1 : 8$; hence $H = 1$ was accepted as the nominal standard and equivalent weights obtained from the composition of oxides were referred indirectly to it. But later more accurate experiments showed that the ratio was more nearly $1.0078 : 8$ or $1 : 7.938$. It has been decided by the International Committee on Atomic[1] Weights that the more conventional standard $H = 1$ should be abandoned in favour of $O = 16$ both for the reasons given above and also because the latter presents the practical advantage that for many elements the values derived from it for the atomic weights approximate so closely to whole numbers that in any but the most accurate work the whole numbers may be used.

2. Direct and indirect reference to the standard.

The reference to the standard may be direct as in the determination of the combining ratio between carbon or

[1] See below (p. 326) for methods of selection of atomic weights when the combining weights have been determined.

hydrogen and oxygen in the gravimetric synthesis of carbon dioxide and water respectively, or indirect, as in the determination of the combining weight of nitrogen by analysis of silver cyanide, the combining weights of silver and carbon being assumed as antecedent data. Thus, the combining weight of nitrogen has recently been ascertained with great accuracy by the analysis of nitrous oxide and nitric oxide (*ante*, p. 272). In these determinations the reference to the standard (oxygen) was direct, and hence the error in the final result solely that involved in the measurement of the ratio weight of oxide : weight of oxygen; the *accurate* analysis of these two oxides however only became possible after the introduction of methods of work at low temperature which made these substances in the liquefied and solidified state amenable to the ordinary processes of purification. Hence all the earlier methods used for the determination of the combining weight of nitrogen had of necessity to be indirect ones. Thus when in 1900 Dean found this value by determining the ratio silver cyanide : silver, two experimental measurements were necessary, viz. (i) the volume of a solution of potassium bromide required to precipitate all the silver in the silver solution obtained by the decomposition of a known weight of the cyanide, and (ii) the silver precipitating value of the potassium bromide solution, from which the weight of silver equivalent to the bromide solution used and hence to the weight of cyanide taken was found. Then the weight of cyanide which contains the combining weight of silver was calculated, and by subtracting from this the combining weights of silver and carbon the value for the combining weight of nitrogen was obtained. This value will be affected not only by the errors of the two sets of measurements, but also by those attaching to the values Ag and C, *i.e.* the combining weights of silver and carbon used as antecedent data. Thus *a priori* in theory the great superiority of direct methods of determination is obvious. But in the case of a student's experiment based on the foregoing, it will probably not matter in the least, as far as the final result is concerned, whether the method used is a direct or an indirect one, because it may be taken as practically certain that the experimental errors affecting the value obtained for $p : q$ will greatly exceed those attaching to the standard values used as antecedent data.

3. **Operation of the law of multiple ratios.**

The introduction of the factors $m, n, p, q, \ldots$ representative of simple whole numbers takes account of the law of multiple ratios, and shows that whilst the analysis of different compounds of the same element may yield different values for the combining weight, these are all simply related, being simple whole multiples or submultiples of one another. Thus the combining weight of carbon referred to oxygen $= 8\cdot00$ is 6 or 3 according as to whether it is determined from the composition of carbon monoxide or carbon dioxide; the combining weight of sulphur deduced from the equivalency ratio between oxygen and sulphur in the oxide and sulphide of lead is 16, whilst it is $\frac{16}{2}$ and $\frac{16}{3}$ if determined from the composition of sulphurous anhydride and sulphuric anhydride respectively. The difficulty thereby introduced into a system of symbolic notation will be discussed below.

III. Symbolic notation.

The recognition that to each element there belongs a characteristic constant value which represents the relative quantity of this element entering into any chemical combination whatever, makes it possible to devise a symbolic notation for the representation of chemical composition and chemical change which is both qualitative and quantitative.

Before entering into the discussion of this matter, it may be of interest to give in the merest outline the history of the development of chemical notation[1].

---

[1] With chemical symbols it has been the same as with chemical names; they have been but the expression of the chemical knowledge of the time. In the Dark Ages a scanty number of jealously guarded facts were disguised by mystical signs; later on the facts, confused and merely empirical as they were, were represented in a manner also confused and fortuitous; and finally, when a fuller understanding of chemical phenomena had been reached, the symbols became not only more rational, more simple and more convenient, but at the same time also more comprehensive and expressive of more complex views (Kopp, *Geschichte der Chemie*).

Kopp, in his history of chemistry, tells us that nothing certain is known about the inner meaning of that connection which was supposed to exist between the seven metals known to antiquity and the planets, and that it is equally uncertain what the individual signs for these were meant to symbolise; nor can any approximate date be fixed for the time when in dim antiquity they were first conceived. What is known is that the signs for

Gold	Silver	Mercury	Copper	Iron	Tin	Lead
Sol	Luna	Mercurius	Venus	Mars	Jupiter	Saturnus
○	☽	☿	♀	♂	♃	♄

were used freely by the alchemists of the thirteenth century along with others representative of the four Aristotelian elements, viz. :

Fire	Air	Water	Earth
△	△	▽	▽

As time went on, various commonly occurring substances were represented by symbols, but these were chosen arbitrarily and not subject to any general consensus ; such signs were :

Spirit of Wine (Alcohol)	Fixed Alkali (Soda or Potash)	Volatile Alkali (Ammonia)

Clay (Alumina)	Vitriolic Acid (Sulphuric Acid)	Acid of Fluor (Hydrofluoric Acid)

It was a step in advance when it was proposed to choose for analogous substances one characteristic sign—such as a circle for all metals—and to differentiate the special substances belonging to the class by some special additional sign—such as the first letter of the name of the metal placed within the circle. Real progress was made towards the end of the eighteenth century in the attempts to represent the qualitative composition of compounds by the juxtaposition of the symbols representative of their components, thus:

— ꜄ ꜀ © ꜅ ꜆ ⊙

Oxygen Hydrogen Carbon Copper Water Carbonic Acid Copper Oxide

Lavoisier's application of the law of conservation of mass to the representation by means of an equation of the quantitative aspect of a chemical change may be taken as the beginning of a notation, quantitative in so far as it represented the weight of the complex substance as the sum of the weights of the constituents.

An example of Lavoisier's notation is shown below, the expression being the first of a series by which he represented the solution of iron in nitric acid, and giving the composition of the mixture before reaction[1]:

$$(a\male) + \left(2ab\triangledown + \frac{ab}{q}\triangledown\right) + \frac{ab}{s}\oplus + \frac{ab}{t}\triangle,$$

where

$\male$ = iron, $\oplus$ = oxygen,

$\triangledown$ = water, $\triangle$ = nitrous air,

a = quantity of iron taken,

b = ratio between the quantities of iron and nitric acid which react,

ab = quantity of nitric acid required to dissolve a of iron,

$\dfrac{ab}{q}$ =   „   „   water contained in ab of acid,

$2ab$ = two parts of water with which the acid must be diluted to moderate the reaction,

$\dfrac{ab}{s}$ = quantity of oxygen contained in ab of acid,

$\dfrac{ab}{t}$ =   „   „   nitrous air contained in ab of acid.

In this equation the symbols for the components represented weights, but not fixed weights characteristic of the specific elements. This further development only became

---

[1] See Pattison Muir, *History of Chemical Theories and Laws*, p. 66.

possible after the discovery and the establishment of the laws of chemical combination, and its introduction is associated with the name of Dalton, the founder of the Atomic Theory. Whatever may have been the order of the historical sequence, whether Dalton proceeded inductively, grasping 'in his comprehensive survey as significant to him of deeper meaning than to his predecessors, their empirical laws of constant and reciprocal proportions, no less than his own law of multiple proportions...,' or whether, as seems nearer the truth, his atomic theory of chemistry was but an extension, an application of the idea of the atomic structure of matter which had risen in his mind as a purely physical conception, what we are concerned with here is that the atomic hypothesis promulgated in the first decade of the nineteenth century correlated all the laws of chemical combination and led as a consequence to a symbolic notation identical in principle with the one still used by us. Those speculations concerning the ultimate constitution of matter which are associated with the names of the Greek philosophers Leucippus, Democritus and Epicurus, and which form the subject-matter of the work *De Rerum Natura*, of the Roman poet Lucretius, see p. 104, were revived in a modified and extended form by Dalton who thus introduced into the science the conception of the atom as the unit of chemical action. According to Dalton matter is considered as of finite divisibility, and the atoms represent the limit reached in mechanical division, viz. the smallest portions of substances which still retain the chemical properties of the whole mass; if derived from an element the atom is truly indivisible (ἄτομος); if derived from a compound it has parts, but these are different from the whole; chemical change is due to certain attractions—termed affinities—inherent in these atoms, and consists in the combination between simple numbers of elementary atoms, each of which is endowed with a constant weight characteristic of and specific for the different elements; from such atoms are formed 'compound atoms' the weights of which are the sum of the weights of the constituent elementary atoms. A direct outcome of this manner of viewing chemical composition and chemical change was the system of notation devised by Dalton, in which the composition of the reacting unit of a compound —*i.e.* Dalton's compound atom—was represented by the juxta-position of a certain number of geometrical symbols each of which represented one atom, *i.e.* a definite fixed weight of the constituent elements. The determination of the *absolute* weights of the atoms—which are quite outside the realm of sensual perception—being an impossibility, the quantities represented by the symbols were *relative* atomic weights[1] expressed in terms of the hydrogen atom as unit, thus:

⊙	stands for an atom of	hydrogen	its relative weight	1
◉	,, ,,	carbon or charcoal	,, ,,	5
○	,, ,,	oxygen	,, ,,	7
⊙○	,, ,,	water or steam composed of 1 of oxygen and 1 of hydrogen	,, ,,	8
○◉	,, ,,	carbon monoxide composed of 1 of carbon and 1 of oxygen	,, ,,	12
○◉○	,, ,,	carbon dioxide 1 carbon and 2 oxygen	,, ,,	19
◉◉ ○○	,, ,,	acetic acid 2 carbon and 2 water	,, ,,	26

To Berzelius we owe the simplification resulting from the substitution of letters for Dalton's geometrical symbols; a certain characteristic quantity of each element, *i.e.* its re-lative atomic weight, is represented by the first or the first two letters of its Latin name, and a small number placed below indicates the number of times the quantity repre-sented by this symbol weight is present. In this symbolic notation devised by Dalton and simplified by Berzelius the qualitative and quantitative composition of compounds is represented in terms of units which they considered were the relative atomic weights. If the standard of reference is the same, *i.e.* if the weight of the hydrogen atom is arbitrarily called 1 or that of the oxygen atom 8 or 16, the numerical

---

[1] The analyses from which Dalton deduced his atomic weights were very far from correct; but this did not trouble him. He was great as a "thinker," inferior to many a predecessor and contemporary as an experimenter.

values of the symbols will either be the same for the atomic
and the combining weights, or at least related in a very
simple manner. The choice of one particular value for the
atomic weight from amongst two or more possible ones all
related in this manner is a problem which can be stated as
follows :—What are the criteria for the assigning of values to
the factors $m, n, p, q, \ldots$ in the general equation given at the
beginning of this chapter? Taking carbon for example, if
carbon dioxide were the only existing combination between
these two elements, it might be represented by CO with
$O = 16$ and $C = 6$ or with $O = 8$, $C = 3$; but the existence of
another compound called carbon monoxide in which the quantity of carbon combined with the same weight of oxygen is
twice as great as in the former compound confronts us with
facts which can be represented equally well either by

C_2O (monoxide) and CO (dioxide) when $O = 16$ and $C = 6$,
or $\qquad\qquad\qquad O = 8$ and $C = 3$,
or by

CO (monoxide) and CO_2 (dioxide) when $O = 16$ and $C = 12$,
or $\qquad\qquad\qquad O = 8$ and $C = 6$.

Further data are required before a decision can be made;
we must inquire what kind of data will serve the purpose,
and how they are to be used. The way in which Dalton met
the difficulty by laying down a set of arbitrary rules for fixing
the values of $m, n, p, q, \ldots$ in atomic combination is dealt
with shortly in the appendix to this chapter, where it is also
shown how the development of the atomic theory and its
deduction from the molecular theory led to the substitution
for arbitrary rules, of unequivocal criteria. What we are concerned with here is the consideration of the manner and the
extent to which the difficulty has been met in the case of a
system of notation in which the symbols stand for atomic
weights.

The great French chemist Laurent (1807—1853) in his
masterly treatment of the nature and the solution of this
problem has with the utmost clearness stated the principle
which should guide us in the selection of the most *suitable*

values for the various atomic weights. The decision as to which multiple or submultiple of the combining weight found by experiment shall be chosen for the atomic weight of the element depends on the production of a formula for its compounds which shall in the simplest possible manner convey the greatest possible amount of chemical information. This further information conveyed by the symbolic notation should be noted: it shows not only of what the substance is composed qualitatively and quantitatively, but also what it can do chemically, its history past and present, its relationships. Thus the formulae H_2O and HCl show that whilst the hydrogen in water can be replaced in two stages, that in hydrochloric acid can be replaced in one stage only; the formulae H_2O and H_2S express certain similarities between the properties of water and those of sulphuretted hydrogen, both of these substances giving derivatives in which either the whole or only half of the hydrogen is replaced by an alkali metal.

The average elementary student concentrating all his interest and attention on the modern atomic weight notation gives little thought to the historical development of that system. The empirical method used by Berzelius was not given up because of any error inherent in it but because it was recognised that its permanent object, viz. the establishment of chemically adequate formulae, could be achieved with greater certainty and greater directness in another way. The subject is one of great importance and interest; and although its adequate treatment is beyond the scope of this book, some help towards a clear realisation of how chemically adequate formulae are constructed should be supplied by the matter dealt with in the next section and in the appendix to this chapter.

IV.   The experimental determination of the *Combining* **or** *Equivalent* **Weight and the choice of the** *Atomic* **Weight.**

Considered from the purely methodological point of view, the determination of the atomic weight, a quantity derived from the combining weight, must come last; but in actual practice, in the investigations by which our knowledge of this special section of stoichiometry is being extended, no such strict sequence is observed ; in fact, in the preponderating number of cases the selection of a formula for the compound precedes the determination of its composition. In all the so-called atomic weight determinations done nowadays, the object is the attainment of greater accuracy in the measurement of a combining ratio a/b ; a difficult and exacting piece of work is undertaken, the result of which—even in anticipation—turns on the change by a very few units in the third or the fourth significant place of a number already known with a considerable degree of accuracy and derived from the study of a compound the formula of which is supposed known (*e.g.* in the accurate determination of the atomic weights of potassium, barium and iron from the analysis of the chlorides the values are calculated on the basis of the formulae KCl, $BaCl_2$ and $FeCl_3$). But even in those instances which nowadays are of such rare occurrence, viz. the first determination of the atomic weight of a newly discovered element, the investigation of its specific properties may proceed in an order such that marked analogies with other well-known elements are recognised by assigning the same general formulae X_mO_n and X_mCl_n (or X_mCl_{2n}) to the oxide and chloride, a classificatory process which may be carried out on the basis of qualitative reactions only, before the quantitative composition of these substances has been ascertained. The element radium supplies an example of this. At a quite early stage of the work connected with its separation from other substances and the study of its specific chemical nature, it became evident that its properties are very similar to those of barium.

From its chemical properties, radium is an element of the group of alkaline earths, being the member next above barium....The salts of radium, chloride, nitrate, carbonate and sulphate resemble those of barium.... In their chemical properties, the salts of radium are absolutely analogous to the corresponding salts of barium.   (Mme. Curie, 1903.)

Hence the new element was classed with the alkaline earth metals[1], and the formula $RaCl_2$ was assigned to the chloride

---

[1] A course justified by the fact that a vacant place in the group II, series 7 column of the periodic table was actually found to accommodate the new-comer.

before any combining weight measurements whatever had been made.

There is therefore no inherent objection to a text-book treatment in which the mode of selection of the factors m, n, p, q, ... is dealt with before the experimental work required in the measurement of the ratio $a : b$. In the following examples taken from standard work and in the students' illustrative experiments, the selection of the values m, n, p, q, ... will be supposed to have preceded the determination of the ratio a/b.

1. Selection of the atomic weight, *i.e.* the assigning of the values to be given to the factors m, n, p, q, ... in the general formulae given above.

From what has already been said in the preceding section, it follows that the selection of the atomic weight is the result of the extensive comparative study of a very large number of substances and of a very large number of reactions; it amounts to a choice of the formulae by which the various compounds of the element considered shall be represented. No specific rules can be laid down, and the work involved is too extensive to lend itself to being illustrated by any students' experiment. All that can be done to help to a proper understanding of the nature of this work is to give a historical example. We will consider the time-honoured one of how the great Swedish chemist Berzelius in 1826 halved most of the atomic weights which he had in 1818 assigned to the various metals.

Ten years of incessant work, begun in 1808, were devoted by Berzelius to the determination of the composition of a very large number of compounds; from these data he deduced values for the atomic weights of all the elements then known, a process which involved the assigning of formulae to the compounds investigated. This he did, subject to the general principle that the formula given to a compound should indicate its chemical behaviour, and above all that substances similar in their properties should be represented by similar formulae. Having ascertained experimentally that the quantities of oxygen combined with the same amount of iron in ferrous and ferric compounds respectively are as $2:3$, and applying the principle of 'greatest simplicity,' he assigned to these oxides the formulae FeO_2 and FeO_3 respectively. The chemical similarity of ferric compounds with those of chromium thus led to the formula CrO_3 for the basic oxide of chromium, whilst the oxides of zinc, manganese etc. were formulated on the type of ferrous oxide, *i.e.* ZnO_2, MnO_2, etc. This was in 1818, but in 1826 all these formulae were revised, in consequence of the following considerations : chromic anhydride for the same amount of chromium contains twice as much oxygen as chromic oxide, and if the formula of the latter were CrO_3 that of the anhydride would be CrO_6. But the chemical analogies between chromic anhydride and sulphuric anhydride are so strongly marked as to call imperatively for recognition by similar formulae ; the choice lay between CrO_6 and SO_6 on the one hand, and CrO_3 and SO_3 on the other, and the latter were accepted, because otherwise the formulae for all the sulphur compounds would have had to be changed from those which were then and still are current, and would have been made, not only more complex but also chemically less adequate (SO_6, SO_4, H_4S in place of SO_3, SO_2, H_2S, etc.). But if chromic anhydride is CrO_3, chromic oxide becomes Cr_2O_3, and the oxides of iron, manganese and aluminium whose properties are similar to those of chromic oxide must be represented by Fe_2O_3, Mn_2O_3 and Al_2O_3 ; accepting Fe_2O_3 as the formula for ferric oxide, that for ferrous oxide becomes FeO, and the whole series of oxides which give rise to salts with properties analogous to those of ferrous iron must be represented as composed of one atom of metal combined with one atom of oxygen, *e.g.* ZnO, MgO. Berzelius' own estimate of the course adopted by him in this matter was : '...to hit upon what is true is a matter of luck....Unfortunately in these matters the certainty of our knowledge is as yet so small that all we can do is to follow the lines of greatest probability.'

The reasoning followed in assigning the formulae $KClO_3$, $KClO_4$ and KCl to the chlorate, perchlorate and chloride of potassium (*ante*, Chapter II, p. 90 *et seq.*) will be given at a later stage of this chapter (*post*, p. 337).

2. The experimental determination of the combining
   weight.

Attention must be drawn at the outset to the fact that all
current periodicals and most of the text books in dealing with
this subject do so under the heading 'determination of the
atomic weight.' In the case of a reader who specially wants
to inform himself concerning the standard work done in this
department in the past and to keep up with the present-day
contribution towards it, this does not matter. Such a person
realises that the determination of an atomic weight is made
up of two absolutely distinct and fundamentally different processes, viz. : (i) the accurate measurement of the combining
weight and (ii) the determination of the numerical relation
between the combining weight and the atomic weight, which
involves the establishment, by special methods, of formulae
for the compounds investigated ; he knows that the information given under the name of 'determination of atomic
weight' is concerned *solely* with the first of these processes.
But from the point of view of the average elementary student
the custom is to be regretted, because, as the experience of
teachers shows, it is very apt to lead to or at least to foster
the erroneous belief that the combining weight is in some way
derived from the atomic weight instead of being the fundamental datum from which the latter is deduced.

The paramount importance which attaches to the knowledge of the accurate values of these fundamental units, and
the large amount of work that is being devoted to 'redeterminations,' makes it desirable to show by a number of examples
how the above general principle is applied. In the choice of
examples preference has been given to cases dealing with the
elements nitrogen, tellurium and radium, all of which have
of late for very different reasons attracted considerable attention. Nitrogen, being an element which enters into the composition of so many compounds of great importance, demands
the chemist's greatest attention. The value accepted until
1907 for its combining weight on the authority of Stas' work
was 14·04, the method employed—an indirect one—having

involved the complete change of a known weight of the chloride of sodium or potassium into the nitrate, the atomic weights of potassium, sodium and chlorine being used as antecedent data (*post*, p. 333). But the end of the nineteenth century brought an extraordinary perfection in the measurement of gaseous densities; values which were reliable enough to be used in the calculation of accurate molecular weights on the basis of Avogadro's hypothesis were obtained (*post*, p. 385) and accurate atomic weights were deduced from these. It was found that the value obtained for the atomic weight of nitrogen from the ratio, density nitrogen : density oxygen $= X : 16\cdot000$ was $14\cdot007$.

The discrepancy between this, the so-called 'physico-chemical' and the older 'chemical' values, has led to a number of redeterminations, the result of which has been the substitution of the value $14\cdot01$ for $14\cdot04$.

Mention has been made already of the anomaly presented by the atomic weight of tellurium in relation to the periodic classification; whilst according to its properties it should in the sequence of elements come in the sulphur group and so *before* iodine, yet according to its atomic weight it would come *after* iodine. The obvious way of accounting for this anomaly has been to assume that the material which has yielded the embarrassingly high atomic weight is not a pure substance but a mixture; hence the many attempts at resolution, the principle of which has been set forth in Chapter III. Tellurium compounds prepared from material obtained from different localities have been subjected to a change in physical conditions and thereby separated into parts; in the parts or fractions so obtained, if the original material had been a mixture, differences in the physical properties of constituents would have produced at least a partial separation; but these parts, when compared with one another and with the original substance, have so far never shown the difference so eagerly sought. The devices resorted to for producing a separation were most diverse (see *ante*, p. 138); the special test applied in each case to prove identity or difference in the various fractions was the accurate measurement of the combining weight.

Apart from the interest which centres round radium on account of its startling radio-active properties, the choice of this element may be justified on the grounds that (a) accurate knowledge of its atomic weight (derived from the combining weight value) is of importance in those theoretical considerations and speculations which deal with the production of elements of lower atomic weight through the disintegration of the radio-active atoms of higher atomic weight by the splitting off of helium atoms; (b) the demands made on the technique of the experimenter are of a special nature owing to the very small amount of material available.

(1) *General principle.*

The combining weight of the element is obtained by the determination of the composition of a compound, or of the ratio between the quantities participating in or resulting from the interaction of substances which, besides the element of unknown combining weight, contain only elements of accurately known combining weight. Thus

$$X \quad : \quad X \quad + \quad mA$$

Comb. wt. Nitrogen : Comb. wt. Nitrogen + Comb. wt. Oxygen

$$= \quad a \quad : \quad b$$
$$= \text{wt. Nitrogen} : \text{wt. Nitric Oxide}$$

or more generally

$$(X + mB + nC) : (pM + qN) = a : b.$$

(2) *Classification of the methods employed in the determination of combining weights.*

(a) Determination of the composition of a compound by means of analysis or of synthesis ; if the compound is the oxide, the method becomes one of direct reference with its attendant advantages.

In previous chapters accounts have been given of the methods followed and the results obtained in the most accurate determination of the composition of water (pp. 199 *et seq.*), hydrogen chloride (p. 201), silver chloride (p. 231), nitric oxide (p. 201), and nitrosyl chloride (p. 202), which among them give values for the combining weights (in terms of the oxygen unit) of hydrogen, chlorine (two independent values), silver and nitrogen (two independent values). To these may be added the following examples :

(i) The determination of the combining weight of tellurium from the composition of the acidic oxide, to which because of its chemical analogies with sulphurous anhydride the formula TeO_2 and the name dioxide is given. Both analysis and synthesis have been used, but the former has been the more common, the reduction of a known weight of the pure dioxide to the metal being effected by heating in a current of hydrogen or of carbon monoxide or by precipitation with hydrazine hydrate.

In one such experiment 2·99688 grams of the dioxide yielded 2·39585 grams of metal, which, with $O = 8·00$, gives 31·89 as the combining weight of tellurium in this compound. If the atomic weight of oxygen is taken as 16 and the formula TeO_2 assumed for telluric oxide[1], then

$$TeO_2 \quad : Te = \text{wt. of dioxide} : \text{wt. of metal} = a : b,$$
$$Te + 32 : Te = \quad 2·99688 \quad : \quad 2·39585,$$
$$\therefore \; Te = \mathbf{127·56}.$$

(ii) The determination of the combining weight of tellurium from the composition of the bromide, which was produced by the direct combination of a weighed amount of the metal with bromine at about 50°C. in an atmosphere free from oxygen, the excess of bromine being subsequently expelled by a current of oxygen.

0·80348 gram of the metal yielded 2·81715 grams of the tetra-bromide, which with $Br = 79·96$ as the antecedent datum for the combining weight of bromine gives

$$Te : TeBr_4 \qquad = \text{wt. of metal} : \text{wt. of bromide} = a : b,$$
$$Te : Te + (4 \times 79·96) = \quad 0·80348 \quad : \quad 2·81715,$$
$$Te = \mathbf{127·62}.$$

---

[1] In all the examples quoted the experimental data are taken from the original memoirs, in which the factors n, m, p and q (p. 314) are supposed known and the formulae for the compounds produced are assumed. Hence in each case the antecedent data and the results obtained are *atomic* weights.

(*b*) Determination of the equivalency ratio between two elements.

The simultaneous deposition by electrolysis of tellurium and of silver.

The weights of silver and of tellurium thrown down from solutions of the two substances in different receptacles by the same current were 1·041101 grams and 0·307117 gram respectively, which with Ag = 107·93 as the combining weight for silver[1] gave

4 Ag   : Te = Silver precipitated : tellurium precipitated = $a:b$,
4 × 107·93 : Te =   1·041101   :   0·307117,

$$Te = \mathbf{127 \cdot 35}.$$

(*c*) Determination of the composition of salts in terms of the component acidic oxide and basic oxide.

This has been effected by: either the decomposition of the salt by heat, when the acidic oxide is expelled and the non-volatile basic oxide remains behind, *e.g.* the heating of calcium carbonate or of gallium-ammonium sulphate, a substance isomorphous with iron-ammonium alum ; or a double decomposition by means of an acid which under the conditions is non-volatile and expels the more volatile acidic oxide, *e.g.* treatment of calcium carbonate at the ordinary temperature with hydrochloric acid when carbonic anhydride is driven off, or fusion of potassium nitrate with silicic anhydride (fine sand) when nitric anhydride is expelled. In either case the quantities measured are the weight of salt taken and the loss of weight of the system in which the reaction was made to occur. The knowledge of the antecedent data required includes that of the ratio between the amounts of oxygen contained in the acidic and the basic oxide.

(i) Determination of the combining weight of tellurium from the ratio between the weight of dioxide and that of the nitrate of formula $2TeO_2 . HNO_3$ from which it is obtained by ignition.

In one such experiment 2·28215 grams of the nitrate yielded 1·90578 grams of dioxide,

$$Te_2HNO_7 : 2TeO_2 = \text{wt. of nitrate} : \text{wt. of oxide} = a : b,$$

[1] With H = 1, Ag = 107·07, and Te = 126·34.

which on substituting the antecedent data $H=1\cdot008$, $N=14\cdot01$ in the equation

$$\frac{2Te+1\cdot008+14\cdot01+112}{2Te+48}=\frac{2\cdot28215}{1\cdot90578}$$

gives $Te=\mathbf{127\cdot47}.$

(ii) Determination of the combining weight of nitrogen from the ratio between the weight of potassium nitrate taken and the nitric anhydride contained in it, this latter being measured by the loss of weight produced by igniting a known weight of the nitrate with excess of pure silica. In three experiments a total of $7\cdot51196$ grams of potassium nitrate was found to lose $4\cdot01209$ grams of nitric anhydride when changed into silicate.

$2KNO_3 : N_2O_5 =$ wt. of nitrate : wt. of nitric anhydride $= a : b$, which on substituting the value of the antecedent datum $K = 39\cdot10$ in the equation

$$\frac{78\cdot20+2N+96}{2N+80}=\frac{7\cdot51196}{4\cdot01209}$$

gives $N=\mathbf{13\cdot99}.$

(*d*) Change of a compound into another by the substitution of one radicle for another, when besides elements of known combining weight one or both of the compounds weighed contains the element of unknown combining weight, thus :

$$X+B+C : X+B+D = a : b,$$
or $$X+B+C : E+B+D = a : b,$$

e.g. the change of barium nitrate to barium sulphate, when $X=$ barium, or the change of silver iodide to silver chloride, when $X =$ iodine.

We may take as a detailed example the determination of the combining weight of nitrogen from the ratio potassium (or sodium) chloride : potassium (or sodium) nitrate, the complete replacement of the one acid by the other being achieved by repeated evaporation to dryness with excess of nitric acid[1]. The equation is

$K[Na]Cl : K[Na]NO_3 =$ wt. of chloride : wt. of nitrate $= a : b$.

[1] Recall the experiment (*ante*, p. 238, Exp. XXX) when the reverse process was accomplished, potassium nitrate having been changed into the chloride.

The data obtained for the ratios measured, and the values calculated from them on the basis of the antecedent data $K = 39\cdot10$, $Na = 23\cdot01$ and $Cl = 35\cdot46$ are

		Wt. of Chloride a	Wt. of Nitrate b
(i)	Potassium	$50\cdot7165$ grams	$68\cdot7938$ grams
(ii)	Sodium	$47\cdot9226$,,	$69\cdot7075$,,

Atomic wt. of Nitrogen $= K[Na]Cl \cdot \dfrac{b}{a} - (K[Na]+3O)$

(i) $\quad = \dfrac{68\cdot7938}{50\cdot7165}(35\cdot46+39\cdot10)-(39\cdot10+48)=14\cdot037$

(ii) $\quad = \dfrac{69\cdot7075}{47\cdot9226}(35\cdot46+23\cdot01)-(23\cdot01+48)=14\cdot040$

Strictly speaking, the general method just stated and illustrated includes the determination of the ratio between the weight of a halide (chloride or bromide) of the element of unknown combining weight and the weight of silver halide obtained from it, but the great importance of this latter method justifies its being dealt with separately.

(e) Determination of the ratio between the weight of a halide of the element of unknown combining weight and either the weight of silver required to precipitate all the halogen or the weight of the corresponding silver halide obtained from it.

$X + Cl[Br] : Ag = $ wt. of halide taken : wt. of silver required for precipitation $= a : b$.

$X + Cl[Br] : Ag + Cl[Br] = $ wt. of halide taken : wt. of silver halide obtained $= a : b$.

Great as is the accuracy which can be attained in the measurement of the ratio $a : b$, this is equalled if not surpassed by that attaching to the standard values of the antecedent data involved, viz. the combining weights of silver and chlorine (bromine). Stas first, and recently T. W. Richards and his school, have elaborated the method by studying the conditions and devising ingenious apparatus suitable for the preparation of pure specimens of the various halides[1] and of

[1] The problem to be solved consisted in the devising of a method for getting the halides *dry*, and removing the water without the occurrence of hydrolysis. An account of how this has been accomplished may be found in the *Proceedings of the American Academy*, **32**, 59 (1896).

silver, as well as for the collection of all the precipitated silver halide and the estimation of the amount remaining in solution[1].

(i) Determination of the combining weight of nitrogen from the weight of silver which when in solution is required for the precipitation of all the chlorine (or bromine) present in a known weight of ammonium chloride (or bromide).

The data obtained for the ratios measured and the values calculated from them on the basis of the antecedent data $Ag = 107\cdot88$, $Cl = 35\cdot46$, $Br = 79\cdot92$ and $H = 1\cdot008$ are:

		$NH_4Cl[Br] : Ag =$	Wt. of ammonium halide taken	Wt. of silver required
			a	b
(i)	Chloride		4·78257 grams	9·64484 grams
(ii)	Bromide		4·84099 ,,	5·33177 ,,

$$\text{Atomic wt. of Nitrogen} = \frac{a}{b} Ag - (Cl[Br] + 4H)$$

(i)
$$= \left(\frac{4\cdot78257}{9\cdot64484} \times 107\cdot88\right) - (35\cdot46 + 4) = 14\cdot036$$

(ii)
$$= \left(\frac{4\cdot84099}{5\cdot33177} \times 107\cdot88\right) - (79\cdot92 + 4) = 14\cdot010$$

(ii) Determination of the combining weight of tellurium from the ratio between the weight of the double salt tellurium-potassium bromide, and the weight of silver bromide obtained from this. The data obtained and the calculation involved are given below; the antecedent data used are $Ag = 107\cdot88$, $Br = 79\cdot92$ and $K = 39\cdot10$.

	$K_2TeBr_6 : 6AgBr =$ Wt. of double bromide	:	Wt. of silver bromide	$= a : b$
	(total of 5 exps.)			
	16·299 grams		26·796 grams	

$$\text{Atomic wt. of Tellurium} = \frac{a}{b} \cdot 6AgBr - (2K + 6Br) = \left(\frac{16\cdot299}{26\cdot796} \times 1126\cdot8\right) - 557\cdot72$$

$$= \mathbf{127\cdot67}$$

(iii) Determination of the combining weight of radium from the ratio between the weight of the carefully purified chloride and that of the silver chloride obtained from it. The antecedent data used are $Ag = 107\cdot88$ and $Cl = 35\cdot46$.

---

[1] A description of the nephelometer (cloud measurer), the instrument used for estimating the amount of silver chloride (or bromide) in solution, is given by Richards in the *American Chemical Journal*, **31**, 235 (1904), and **35**, 510 (1906).

$$\text{RaCl}_2 \quad : \quad 2\text{AgCl} \quad = \quad a \quad : \quad b$$

Observer		
Mme. Curie (total of 3 exps.)	1·2005	1·1578
Thorpe (total of 3 exps.)	·2050	·1975
Hönigschmid (total of 6 exps.)	6·06756	5·85933

$$\text{Atomic wt. of Radium} = \frac{a}{b} \cdot 2\text{AgCl} - 2\text{Cl} =$$

226·35
226·58
225·95

3. The fundamental ratios.

In the determination of atomic[1] weights, a small number of values are to be regarded as fundamental. They are the standards of reference; and by comparison with them all the other atomic weights are established. Two of these values, the atomic weights of hydrogen and oxygen, are primary; that is, one or the other of them is the basis of the entire system; hydrogen as unity in the older arrangements; oxygen equal to sixteen in the more modern scheme....Comparatively few of the atomic weights, however, are fixed by direct comparison with...oxygen. In most cases other values intervene, and especially the atomic weights of silver, chlorine, bromine,...potassium. These constants are first to be determined, and their establishment may be compared to a primary triangulation, of which the hydrogen-oxygen ratio is the base line. (Clarke, *The Constants of Nature. A Recalculation of the Atomic Weights*, 1910.)

The atomic weights of silver, chlorine and bromine enter into the calculation of nearly all other atomic weights, and form so to speak the platform upon which the entire structure stands. (*Report of the International Committee on Atomic Weights*, 1905.)

Considering that these fundamental data have been used in a number of the examples given in the preceding section, it must seem as if a serious mistake had been made and the logical order of treatment reversed; but against the undeniable and obvious weakness of a course in which, whilst the mode of determination of certain combining weights is described, others—for no apparent reason—are assumed as known already, there is to be set a great advantage, viz. that a clearer insight into the nature of the work involved can be gained by the consideration of one separate determination than by that of a number of correlated ones, such as are often resorted to in the determination of the fundamental ratios. After this attempt to placate outraged logic, let us retrace our steps and examine the foundations of the building which we have entered from a higher floor.

---

[1] Here the term 'atomic weight' is used when really the actual experimental work (see above) is strictly confined to the determination of the combining weight.

(*a*) The classical researches of Stas, which were carried out about half a century ago, did not comprise the determination of the ratio hydrogen : oxygen in water, for which Dumas' value 1 : 8·00, then recently determined, was accepted. Since direct reference to the standard oxygen was a *sine qua non*, the substances employed had to include at least one compound containing oxygen and in the determination of the values for silver and the halogens Stas—as other investigators had done before him—used chlorates (or bromates or iodates). These are ternary compounds which can be easily deprived of the oxygen contained in them, the binary chloride being left; the composition of these halide compounds can be ascertained either by direct synthesis in the case of the silver compounds, or indirectly by change into the silver halide, the composition of which is determined as above, in the case of most other halides.

In pursuance of the general principle concerning the choice of chemically adequate formulae, it has been found that the following are the simplest formulae which can be chosen to represent the chemical relationships and the reactions of the chlorides and of the chlorates and perchlorates, salts which for the same amount of other elements contain quantities of oxygen in the ratio of 3 : 4 (*ante*, p. 94):

$KClO_3$	$KClO_4$	KCl	HCl	AgCl	$AgClO_3$
Potass. chlorate	Potass. perchlorate	Potass. chloride	Hydrogen chloride	Silver chloride	Silver chlorate

The same stoichiometrical relations might however be represented by other formulae such as

$$KClO \qquad K_3Cl_3O_4 \qquad KCl \qquad HCl_3 \qquad \text{etc.,}$$
$$\text{or } K_3Cl_3$$

in which the atomic weights taken for potassium and chlorine, viz.

$$K = \frac{39·1}{3} \; ; \; Cl = \frac{35·46}{3},$$

are sub-multiples of those usually employed, but these formulae would not only be less simple but also less adequate. For example, HCl_3 for hydrogen chloride would lead to the expectation of other halide compounds of hydrogen in which some of the chlorine had been replaced by bromine or iodine; K_3Cl_3 for potass. chloride is complicated and would make the mechanism of change from chlorate to chloride complicated also, whilst the alternative, the simpler KCl, would veil the relationship between that salt and the hydrogen chloride HCl_3 from which it can be derived by the simple process of hydrogen substitution.

The object aimed at here, which is not comprehensive treatment but merely elucidation of the principle involved by means of adequate illustration, will be attained by giving the data obtained by Stas in one series of experiments which led to the establishment of some of the fundamental ratios, viz. the determination of silver and chlorine directly referred to oxygen by (i) the analysis of silver chlorate and (ii) the synthesis of silver chloride. The results were:

	(i) Wt. of silver chlorate used	Wt. of residual chloride
	a	b
1st exp.	138·7890 grams	103·9795 grams
2nd exp.	259·5287 ,,	194·4455 ,,

Wt. of silver + chlorine combined
in the chlorate with 3 × 16 oxygen

$$= \frac{b}{a-b} \cdot 3 \times 16.$$

	(ii) Wt. of silver	Wt. of silver chloride
	c	d
1st exp.	91·462	121·4993
2nd exp.	101·519	134·861

Wt. of silver in the amount of chloride derived from
the chlorate by the loss of 3 × 16 of oxygen

$$= \frac{c}{d} \cdot \frac{b}{a-b} \cdot 3 \times 16.$$

Thus we have two equations representative of the relations of three elements of which one is the standard, which supplies us with sufficient data for calculating the atomic weights of the two elements silver and chlorine in terms of oxygen = 16:

Atomic weight of silver $= \dfrac{c}{d} \cdot \dfrac{b}{a-b} \cdot 3 \times 16 = 107\cdot96.$

Atomic weight of chlorine $= \dfrac{d-c}{c} \cdot \dfrac{c}{d} \cdot \dfrac{b}{a-b} \cdot 3 \times 16 = 35\cdot45.$

Another illustration is afforded by the series used for the determination of the three fundamental ratios Ag, Cl and K; compounds containing the four elements potassium, chlorine, oxygen and silver were employed, and the necessary three equations were supplied by the measurement of the ratios

potass. chlorate : potass. chloride $= a : b,$
potass. chloride : silver $= c : d,$
silver : silver chloride $= e : f.$

This series will be made the subject of a students' illustrative experiment in the next section.

(b) Recent redeterminations of the fundamental ratios. The bulk of the work has been done in America, and the contributions made from Harvard by T. W. Richards and his school supply a self-contained series. But whilst half a century ago one man, Stas, stood as the one authority on the subject, now the advance is made much more effective by co-operation; the same measurement is made by different observers using different methods. Thus the number of data available has grown apace, and the concordance of the results constitutes, if not absolute certainty, at least very great probability of the absence of constant errors in the diverse methods employed. The data available can be combined in a variety of ways. The following illustration in which the atomic weight of nitrogen is obtained indirectly relative to $O = 16$ is chosen because of its simplicity:

(i) Oxygen : hydrogen from the composition of water—**O = 16·000**

	a	b	c
H_2 : O : $H_2O =$			
Morley's complete synthesis	33·2435 grams	263·9387 grams	297·1822 grams
Noyes' complete synthesis	21·32179 ,,	169·24165 ,,	190·55240 ,,

$\therefore$ H = mean of the 3 values
H_2 : O, H_2 : H_2O and O : H_2O
$= 1\cdot0076.$

(ii) Hydrogen : chlorine from the composition of hydrogen chloride—

H = 1·0076

H : Cl : HCl=	*a*	:	*b*	:	*c*
Noyes' complete synthesis	6·41928 grams		225·86022 grams		232·27947 grams
Edgar's complete synthesis[1]	9·9913 ,,		351·6321 ,,		361·6174 ,,

∴ Cl = mean of the 3 values
H : Cl, H : HCl and Cl : HCl

= 35·456.

(iii) Silver : chlorine from the composition of silver chloride—**Cl = 35·456**

Ag : AgCl =	*a*	:	*b*	Ag
Richards' and Wells' synthesis	82·66877 grams		109·83951 grams	**107·89**

(iv) Silver : nitrogen from the composition of silver nitrate—**Ag = 107·89**

Ag : AgNO₃ =	*a*	:	*b*	N
Richards' and Forbes' synthesis, 1907	42·96172 grams		67·65615 grams	**14·008**

The agreement of this value for nitrogen determined indirectly with that deduced from the composition of nitric oxide is so close as to increase the confidence in the intermediate values.

NO : O : N =	*a*	:	*b*	:	*c*
Gray's complete analyses of nitric oxide	2·93057 grams		1·56229 grams		1·36819 grams

N = mean of the 3 values NO : O, N : O, NO : N

= **14·012.**

V. Students' illustrative experiments.

1. Utilisation for the purpose of combining weight measurements of data primarily obtained with a different object in view.

(1) *Some of the Fundamental Ratios.*

(*a*) The combining weight of hydrogen deduced from the composition of water.

---

[1] Hydrogen obtained from weighed palladium hydride was burnt in a glass globe containing chlorine weighed in the liquid form. The hydrogen chloride formed was condensed in a steel bomb or by absorption in water and weighed.

(i) Direct determination, measurement of the ratio $O:H_2O$ by the gravimetric synthesis of water (Dumas' method). Results obtained by students working in the author's laboratory showed that in ten experiments a total of 13·073 grams of oxygen—measured by the loss of weight of copper oxide heated in a suitable bulb in a stream of hydrogen—yielded 14·668 grams of water—absorbed in a U-tube filled either with granulated calcium chloride or with pumice stone soaked in concentrated sulphuric acid.

$$\text{Oxygen}:\text{water} = a:b = 13\text{·}073:14\text{·}668.$$

The amount of hydrogen combining with 13·073 grams of oxygen is $14\text{·}668 - 13\text{·}073 = 1\text{·}595$ grams.

Therefore the amount of hydrogen combining with 8 grams of oxygen $= \dfrac{8 \times 1\text{·}595}{13\text{·}073} = \text{·}976$ gram.

The value differs by more than $2°/_{o}$ from the standard one, but the experiment, though extremely simple in principle, is, in spite of the great attention it has received from teachers conducting laboratory courses, one which is likely to give erratic results in the hands of inexperienced students. These are due to :—(α) the difficulty of preventing the passage into the absorption bulb of water other than what has been formed by the combustion of the oxygen given up in the copper oxide bulb, which may however be overcome by thorough drying of the hydrogen used; (β) the difficulty of ensuring the complete retention of the synthesised water, due to inefficiency of the absorption bulb. If the two errors happen to be commensurate in their magnitude, they will of course counteract one another, but as a matter of fact in a students' experiment one or the other is apt to become very much the larger, producing a corresponding raising or lowering of the final result. Moreover, even if the error should not be very great, say a loss of 2 mgs. of water in a total of 1 gram formed, *i.e.* $0\text{·}2\ °/_{o}$ of water, the whole of this will throw itself on to the hydrogen, of which the relative amount present in the water collected is very small ; in this special case it will be 2 mgs. in

$\dfrac{1000}{9}$ mgs., or nearly $2\,°/_o$. Put quite generally, this means that the percentage error in the combining weight of hydrogen will be nine times as great as that made in the measurement of the water.

(ii) Indirect determination from the volume ratio in which oxygen and hydrogen combine (*ante*, p. 245) and the densities of the two constituent gases. From results obtained by students in the author's laboratory[1] we get:

Volume ratio Oxygen : Hydrogen $= a : b = 1 : 2{\cdot}00$
Density Hydrogen $\quad\quad = D_H = 0{\cdot}0882$ grams per litre
Density Oxygen $\quad\quad\quad = D_O = 1{\cdot}435$,, ,,

$\therefore$ Combining weight of hydrogen $= \dfrac{\cdot 0882 \times 2 \times 8}{1{\cdot}435} = {\cdot}984.$

(*b*) The combining weights of (i) silver, (ii) chlorine, (iii) potassium from a series of connected measurements comprising the three ratios :

Potass. chlorate : potass. chloride $= KClO_3$: KCl $= a : b$ (see p. 94).
Potass. chloride : silver $\quad\quad = KCl$: Ag $= c : d$ (see p. 241).
Silver $\quad\quad$: silver chloride $= Ag$: $AgCl = e : f$ (see pp. 97 *et seq.*).

The following are the calculations for obtaining the required atomic weights Ag, Cl and K, when for reasons of chemical adequacy the formulae assigned to the compounds investigated are $KClO_3$, KCl and AgCl (*ante*, p. 337) :

$$K + Cl = 3 \times 16 \cdot \frac{b}{a-b},$$

$$Ag = \frac{48b}{a-b} \cdot \frac{d}{c},$$

$$Cl = \frac{48b}{a-b} \cdot \frac{d}{c} \cdot \frac{f-e}{e},$$

$$K = KCl - Cl = \frac{48b}{a-b} \left(1 - \frac{d\,(f-e)}{c \cdot e} \right).$$

---

[1] The density of hydrogen was obtained by finding the loss in weight when a known weight of aluminium was treated with potash, the hydrogen being allowed to escape, precautions being taken to prevent escape of water, etc.

The density of oxygen was obtained by finding

(*a*) the loss in weight $\quad$ } when a known weight of potassium
(*b*) the volume of gas given off } $\quad$ chlorate was heated.

Specimen Results.

	Silver $=\frac{48b}{d}\cdot\frac{c}{a-b}$	Chlorine $=\frac{48b}{d}\cdot\frac{c}{f-e}\cdot\frac{e}{a-b}$	Potassium $=\frac{48b}{a-b}\cdot d\left(1-\frac{d(f-e)}{c\cdot e}\right)$	
	Combining weights			
Individual Students :	Ratio Potass. : chlorate $=a:b=100:x$	Ratio Potass. : Silver chloride $=c:d=x:100$	Ratio Silver : chloride $=e:f=100:x$	Silver	Chlorine	Potassium
A	$5.1961:3.1589$ $x=60.79$	$2.9692:4.3011$ $x=69.01$	$3.722:4.9425$ $x=132.79$	107.82	35.32	39.03
B	$4.5089:2.7379$ $x=60.72$	$2.9606:4.2876$ $x=69.07$		107.84	35.19	39.07
C	$1.1233:.6831$ $x=60.81$	$1.4733:2.1403$ $x=69.02$	$3.3944:5.5036$ $x=132.68$	108.2	35.48	39.28
Mean of Results of Class, 1912	$x=60.77$ 11 exps.	$x=69.25$ 25 exps.	$x=132.3$ 17 exps.	107.5	35.47	39.06
Standard values				107.88	35.46	39.10

(2)   *Ratios in which the reference is either (a) direct to the
standard or (b) indirect through one or other of the
fundamental ratios found above under* (1).

In the calculations involved, the values used for the antecedent data will be the *standard* ones and not those obtained
in the illustrative experiments and quoted in (1). Provided
that the illustrative experiment has led to a clear realisation
on the part of the students of the method used in the determination of these antecedent data, there would seem to be no
point in handicapping the following experiments, which in a
sense are quite distinct, by the use of values almost certain
to carry large experimental errors. After all, this is but what
is done in the larger number of researches in this domain.
Of the many chemists who are continually adding to our
knowledge of the fundamental ratios, which are the combining
ratios of the elements, the majority work by methods which
presume a knowledge of the fundamental ratios Ag, Cl, Br,
H, N, ... and they use the accepted standard values in the
establishment of which they may have had no part whatever,
provided they consider them to be accurate to a degree at
least as great as their own measurements.

(a) Determination of the combining weights of magnesium, zinc, tin and aluminium from : (i) the oxygen
combining power ; (ii) the hydrogen substituting power (*ante*,
Exp. XLV, p. 303).

(i)   The composition of the oxides, ascertained by synthesis, when the reference to the standard oxygen is direct.

The method of finding the combining weight of metals by
the synthesis of the oxide through the nitrate is one much
drawn upon for students' illustrative experiments.   Copper,
lead and iron may be used in addition to or instead of the
metals selected here ; the conditions to be considered in
deciding whether the combining weight of a metal can be
determined by this method are : (α) Whether the metal can
be procured in a state of purity ; (β) whether the nitrate
can be decomposed completely within the range of temperature it is possible to produce by the apparatus available in

an ordinary laboratory ; (γ) whether the oxide is stable and non-volatile at the temperature at which it is formed from the nitrate. The popularity which the method enjoyed at one time in research work has diminished since it has been shown by T. W. Richards that oxides prepared by the ignition of nitrates retain with great persistence small quantities of oxides of nitrogen. In the case of students' work, this error inherent in the method is not likely to exceed the experimental error due to impurities in the material and the mechanical loss through spurting incurred in the preparation of the oxide.

It deserves mention at this stage that the determination of the combining weight of carbon by the synthesis of carbon dioxide can easily be made into a good students' experiment; the combination with oxygen is accomplished by burning a

Apparatus for determination of combining weight of carbon

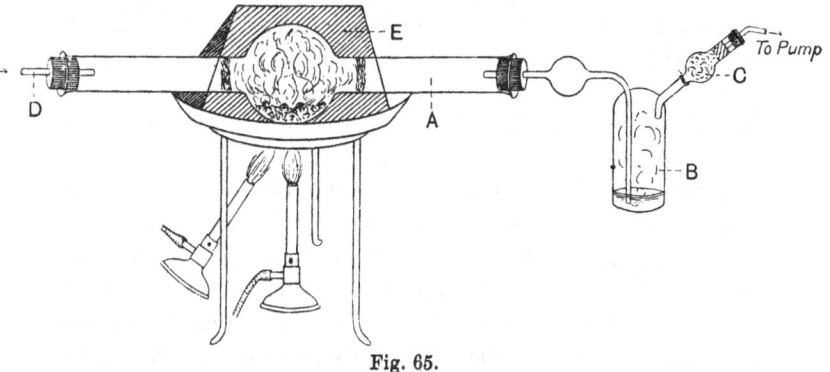

Fig. 65.

A. Bulbed tube containing carbon weighed before and after the experiment.
B. Vessel containing strong potash solution (1 : 1) for absorption of CO_2 formed.
C. Side tube containing solid caustic potash, and a plug of glass wool. It is connected with a filter pump by means of an indiarubber cork and glass tube.
D. Inlet for oxygen or dry air freed from CO_2.
E. Pent roof of asbestos.

known weight of carbon with oxygen and copper oxide and absorbing the volatile product carbon dioxide in potash. If a combustion furnace is not available, quite good results may be obtained by using a bulbed tube of hard glass and two good Bunsen burners, a kind of pent roof made from thin asbestos board helping towards the production of a sufficiently high temperature (see fig. 65). The experimental method is in principle identical with that employed in the estimation of the carbon contained in an organic compound (Clowes and Coleman, *Quantitative Analysis*, 10th ed., 1914, pp. 403 *et seq.*; Perkin and Kipping, *Organic Chemistry*, 1911, pp. 19 *et seq.*). There is of course a reversal of the known and the unknown quantities in the equation; in the organic analysis the unknown amount of carbon burnt is deduced from the weight of carbon dioxide formed, the composition of which latter is assumed as known; in the determination of the combining weight of carbon, the composition of the oxide termed carbon dioxide is ascertained by finding the weight of it produced by the burning of a known weight of carbon as pure as it is possible to procure.

(ii) The equivalency factor between metal and hydrogen, measured by the volume of hydrogen expelled by a known weight of metal acting on excess of acid (in the case of aluminium, soda or potash may be used instead), the density and the combining weight of hydrogen being required as antecedent data (see Exp. XLV, p. 303).

This method is applicable in the case of such metals as are obtainable in a sufficiently pure state, which evolve hydrogen from acids at a temperature not too high, and without the occurrence of secondary reactions : conditions which exclude copper, mercury, silver and lead. Now that calcium in a pure form has been put on the market, variety may be introduced and knowledge of descriptive chemistry furthered by using this metal for such a combining weight determination. The acids generally, used are hydrochloric and sulphuric; nitric acid, which yields instead of hydrogen red fumes of NO_2, colourless NO or N_2O or even nitrogen—reduction products of the acid—is of course excluded. On the other hand, water may in certain cases be substituted for acids, and it is said that water diluted by alcohol lends itself well to an interesting experiment for the determination of the combining weight of sodium, the metal being weighed with suitable precautions

Record of results.

Specimen note-book entry.

The determination of the combining (or equivalent) weights of certain metals.

(i) From the composition of the oxides :

	Magnesium		Zinc	Tin	Aluminium	
	I	II			I	II
Wt. of metal taken..............................=a	·263 gms.	·271 gms.	·718 gms.	·523		
,, oxide formed ....=b	·437 ,,	·453 ,,	·892 ,,	·664		
Combining wt. referred to 8·00 of oxygen $=\dfrac{a}{b-a}\cdot 8\cdot00$	12·0	12·0	33·0	29·7		
Formula assigned to oxide........................	Mg|O		ZnO	SnO$_2$		
∴ Atomic weight....................................	24·0	24·0	66·0	118·8		

(ii)   From the hydrogen evolved :

	Magnesium		Zinc	Tin	Aluminium	
Wt. of metal taken ....=a	·194 gms.	·189 gms.	·462 gms.	·4924	·160 gms.	·155 gms.
Vol. of expelled hydrogen collected over water...=V	197 c.c.	193 c.c.	173 c.c.	100 c.c.	212 c.c.	205 c.c.
Temperature ....=T	18° C.	18° C.	15° C.	13·0° C.	16° C.	16° C.
Barometer ....=B	756 mm.	756 mm.	747 mm.	744 mm.	756 mm.	756 mm.
Tension of aqueous vapour at temp. T ...=τ	16 ,,	16 ,,	13 ,,	11 ,,	14 ,,	14 ,,
Wt. of metal equiv. to 1·01 of hydrogen $=\dfrac{a\times1\cdot01}{V\cdot\dfrac{273}{273+T}\cdot\dfrac{B-\tau}{760}\cdot\cdot0000900}$	12·1	11·9	32·73	59·53	9·16	9·16
Formula assigned to oxide	Mg|O		ZnO	SnO	Al$_2$|O$_3$	
∴ Atomic weight	24·2	23·8	65·46	119·06	27·5	27·5

against oxidation by the air, and the hydrogen evolved by its action on the diluted water being measured in the usual way.

The results of (i) and (ii) are embodied in the table on p. 347.

(*b*) Determination of combining weight of iodine from the synthesis of silver iodide. With $Ag = 107.9$, on substituting the values found (*ante*, p. 217) in the equation we get:

$$Ag : I : AgI = a : b : c = .1726 : .2027 : .3754$$
$$\therefore \text{ I from } Ag : I \quad = 126.75$$
$$\text{,, } Ag : AgI = 126.60$$

2. **Provision of additional data for the purpose of combining weight measurements, by means of experiments done with this definite object.**

(1)   *The fundamental ratio chlorine : silver.*

(*a*)   From a combination of the two ratios

(i)   NaCl (or any other pure dry chloride) : Ag $= a : b$,

(ii)   NaCl   ,,   ,,   : AgCl $= a' : b'$,

the value for Ag *being supposed known.* What this amounts to is an indirect way of measuring the ratio Ag : AgCl, the determination of which by the quantitative synthesis of silver chloride has formed the subject of a previous experiment (*ante*, pp. 235 *et seq.*).

The principle and the technique of the measurement of these ratios have been dealt with fully on previous occasions (*ante*, pp. 241 *et seq.*), and practically nothing need be added. But in all work of this kind the trustworthiness of the result obtained depends, not only on accuracy in the measurement of the combining ratio, but also on the purity of the materials used, viz. the chloride investigated and the silver used in the precipitation. Stas devoted much labour to the production of what for a long time was considered silver as pure as it could be got; but in science—and that is what constitutes one of its greatest attractions—the achievement of one period constitutes but the starting-point for another, and T. W. Richards showed that Stas' silver was contaminated with occluded oxygen, and that the glass vessels in which the precipitation, evaporation and fusion were carried out and which were supposed to have remained unaltered in weight, had really been very appreciably attacked by the strong acids used, a source of error which can be avoided by the use of quartz or porcelain vessels. The metal that is sold under the name of 'foil for assay' is of a purity quite sufficient for the average student, who moreover is not very likely to be able to afford the time required for repeating in an illustrative experiment the various processes involved in the preparation of 'pure' silver. An account of Richards' method is given in Roscoe and Schorlemmer, *Treatise on Chemistry*, vol. II, Metals, p. 456.

As regards the production in the pure dry state of the chloride investigated, two things are involved, viz. (a) separation from other salts—*e.g.* chlorides of other metals, sulphates, nitrates—which is accomplished by repeated systematised recrystallisations, and (β) the complete removal of the solvent water. Mention has been made already (*ante*, footnote p. 334) of the difficulty of complete drying without the occurrence of hydrolysis, *i.e.* of a double decomposition in which volatile hydrogen chloride is formed by the combination of part of the hydrogen of the water with the chlorine, the place of the chlorine being taken by the oxygen and part of the hydrogen of the water, thus:

$$XCl_2 + 2H \cdot OH = X(OH)_2 + 2HCl.$$

This tendency to hydrolysis is possessed by the different metallic chlorides in varying degrees. Whilst the chlorides of the most strongly electropositive metals, sodium and potassium (and the other metals of the alkali group), yield on evaporation of the solutions of their chlorides substances neutral to litmus, calcium chloride is decomposed to a small degree, magnesium chloride by repeated evaporation yields all its chlorine as volatile hydrogen chloride, and bismuth (or antimony or stannous) chloride cannot even be dissolved in cold water without a considerable amount of decomposition occurring, which shows itself in the formation of the insoluble oxy-chloride (analytical test for bismuth or antimony, Clowes and Coleman, *Quantitative Analysis*, 8th ed., pp. 368, 369). The reaction is a reversible one subject to the law of·mass action, *i.e.* the greater the relative amount of water, the more decomposition of the chloride, and *vice versa*. The principle of the method used to produce a chloride quite free from water follows as a direct consequence of this, and consists in heating the chloride in a stream of pure dry hydrogen chloride to remove any water, the excess of the acid being swept out subsequently by heating in a stream of pure dry nitrogen, a gas quite inert under these conditions.

(i) The experimental procedure is identical with that described previously for the determination of the ratio potass. chloride : silver (*ante*, p. 241), and used again in the determination of the equivalency ratio between solutions prepared from a known weight of iodine and of silver respectively (*ante*, p. 217). Nothing requires to be added to what has already been said on the subject, and it is only the importance of the method that justifies the introduction of an experiment which, but for the interchange of what is assumed as known and what it is required to find, amounts to a repetition.

(ii) The technique of the work required in collecting for weighing in a Gooch crucible the silver chloride precipitated from a solution by the addition of excess of silver nitrate, is the same as in Exps. XXVI and XXXI.

The combined results give an independent value for the combining weight of chlorine in terms of silver thus:

If a grams of sodium chloride are required for the precipitation of b grams of silver, and if a' grams of sodium chloride yield on precipitation b' grams of silver chloride, then the combining weight of chlorine, with $Ag = 107\cdot9$ as antecedent datum,

$$= \left(\frac{b'}{a'} - \frac{b}{a}\right) \cdot \frac{107\cdot9}{\dfrac{b}{a}} = \left(\frac{ab'}{a'b} - 1\right) \cdot 107\cdot9.$$

Record of results.

Specimen note-book entry.

(i) The determination of the ratio $(Na + Cl) : Ag = a : b$,
with the antecedent datum $Ag = 107.9$.

	I	II
Wt. of dry sodium chloride taken .... $= a$	2·328 gms.	2·321 gms.
Wt. of silver foil dissolved in nitric acid ...... $= b'$	4·471 ,,	4·446 ,,
Determination of the excess of silver added ... $= c$		
Wt. of silver foil dissolved .... $= m$	1·083 gms.	
Vol. to which the solution of m grams of silver is made up .... $= V$	1000 c.c.	
Vol. of thiocyanate solution required for the precipitation of 25 c.c. of the above silver solution .... $= p$	$\begin{Bmatrix} 5\cdot15 \text{ c.c.} \\ 5\cdot10 \text{ ,,} \\ 5\cdot10 \text{ ,,} \end{Bmatrix}$ 5·10	
∴ Silver precipitated by 1 c.c. of thiocyanate solution .... $= \dfrac{25}{p} \cdot \dfrac{m}{V}$	0·0053 gm.	
Vol. of thiocyanate solution required to precipitate the excess of silver in the solution above the precipitate produced by the interaction between a grams of sodium chloride and b' grams of silver............ $= q$	32·1 c.c.	29·9 c.c.
∴ Excess of silver present ...... $= q \cdot \dfrac{25}{p} \cdot \dfrac{m}{V} = c$	0·170 gm.	0·158 gm.
∴ Silver used in the precipitation of a grams of chloride .... $= b' - c = b$	4·301 gms.	4·288 gms.
Chloride requiring for precipitation 107·9 grams of silver .... $= \dfrac{a}{b} \times 107.9$	58·39 ,,	58·42 ,,

(ii) Determination of the ratio $NaCl : AgCl = a' : b'$,
using the same antecedent data as under (i).

	I	II
Wt. of dry sodium chloride taken .... $= a'$	0·2143 gms.	0·5630 gms.
Gooch crucible + asbestos (i)	9·6100 ,,	10·0240 ,,
,, ,, (ii)	9·6098 ,,	10·0235 ,,
,, ,, (iii) .... $= m$	9·6098 ,,	10·0235 ,,
Gooch crucible + asbestos + silver chloride (i)	10·1355 ,,	11·4045 ,,
,, ,, ,, (ii)	10·1355 ,,	11·4045 ,,
∴ wt. of silver chloride obtained ...... $= m - n = b'$	0·5257 ,,	1·3810 ,,

Combining weight of chlorine deduced from the two ratios
$\dfrac{a}{b}$ and $\dfrac{a'}{b'}$, with the antecedent datum $Ag = 107.9$

$$= \left(\frac{b'a}{a'b} - 1\right) 107.9 = \text{(i) } 35.37,$$

$$\text{(ii) } 35.36.$$

From either (i) or (ii) the combining weight of sodium can be obtained thus:

From (i) combining weight of sodium

$$= \left(\frac{a}{b} \times 107 \cdot 9\right) - 35 \cdot 37 \qquad \textbf{23·02} \qquad \textbf{23·05}$$

From (ii) combining weight of sodium

$$= \left(a' \times \frac{107 \cdot 9 + 35 \cdot 37}{b'}\right) - 35 \cdot 37 \qquad \textbf{23·03} \qquad \textbf{23·05}$$

(*b*) From the ratio HCl : AgCl = $a : b$, the values for H and Ag being supposed known.

Fundamental ratios are ascertained **either** by a series of connected measurements in which n equations are used to represent in terms of the combining weights of the constituents the composition of substances containing $n + 1$ elements of which the standard oxygen is one (*ante*, p. 338), **or** by a series built up of successive steps, each of which, utilising the value or values found previously, adds knowledge of another one. Of course the order of sequence in the series is a more or less arbitrary matter, and one allowing of variety. Thus any of the following would be possible and legitimate:

A. (i) H from O : H; (ii) Cl from H : Cl;
 (iii) Ag from Ag : AgCl
  or X Cl : Ag and X Cl : AgCl.

B. (i) H from O : H; (ii) Ag from KClO₃ : KCl and KCl : Ag;
 (iii) Cl from HCl : AgCl.

C. (i) Ag from KClO₃ : KCl and KCl : Ag;
 (ii) Cl from Ag : AgCl; (iii) H from HCl : AgCl.

A priori there is nothing to indicate a superiority of any one of these schemes over any other, but on applying the usual criteria of efficiency it becomes obvious at once that the steps on which the others rest must be the most secure, and hence that the sequence should be as far as possible in order of decreasing accuracy: the most accurate value should come first, then the second best, and so on. This is the principle followed in the final combinations of the large number of data which many decades of continuous research work have brought together in this province of chemistry. Applying this principle to the deduction of a combining weight value from the ratio hydrogen chloride : silver chloride which forms the subject of the illustrative experiment about to be described, we are dealing with a case in which we have one equation connecting three elements, and hence must assume two combining weights as known in order to deduce a value for the third. Which shall be this third in our case? It has been shown (*ante*, pp. 340, 342) how the students' experiments dealt with in other connections incidentally supply values for H, Ag and Cl; hence the present experiment is an example of an alternative method, and we are at full liberty to use it as we like, *i.e.* to fit it into any series we think best of those given above. C is chosen in preference to B, because whatever experimental error attaches to the value of HCl throws itself completely either on chlorine or on hydrogen, according to whichever is supposed unknown, and H being only about $1/35$ of Cl, the proportionate effect will be 35 times as great.

Experiment XLIX.

The determination of the ratio hydrogen chloride : silver chloride.

The weight of the hydrogen chloride used is given by the increase in weight of an absorption bulb containing water into which the carefully dried gas is led (see fig. 66); great care being required to prevent any loss of weight due to evaporation of the water, the absorption bulb must be fitted with a chloride of calcium tube, the current of gas must be very slow and the absorption vessel must be kept cool[1]. For the same reason it is desirable not to connect the absorption bulb to the source of the gas until all the air has been driven out of the generating and drying vessels, since the bubbling through the water of a considerable volume of air which would enter C presumably quite dry and would emerge from it saturated with moisture would tax too severely—and quite unnecessarily—the powers of the drying tube attached to C. The insertion between B' and C of a three-way stop-cock leading to a separate large absorption vessel, an arrangement depicted and described on p. 42 and again on p. 237, would make this a very simple matter. When the amount of hydrogen chloride absorbed is considered sufficient (a current of gas passing at the rate of 20 to 30

---

[1] The heat produced by the solution of hydrogen chloride in water is very considerable.

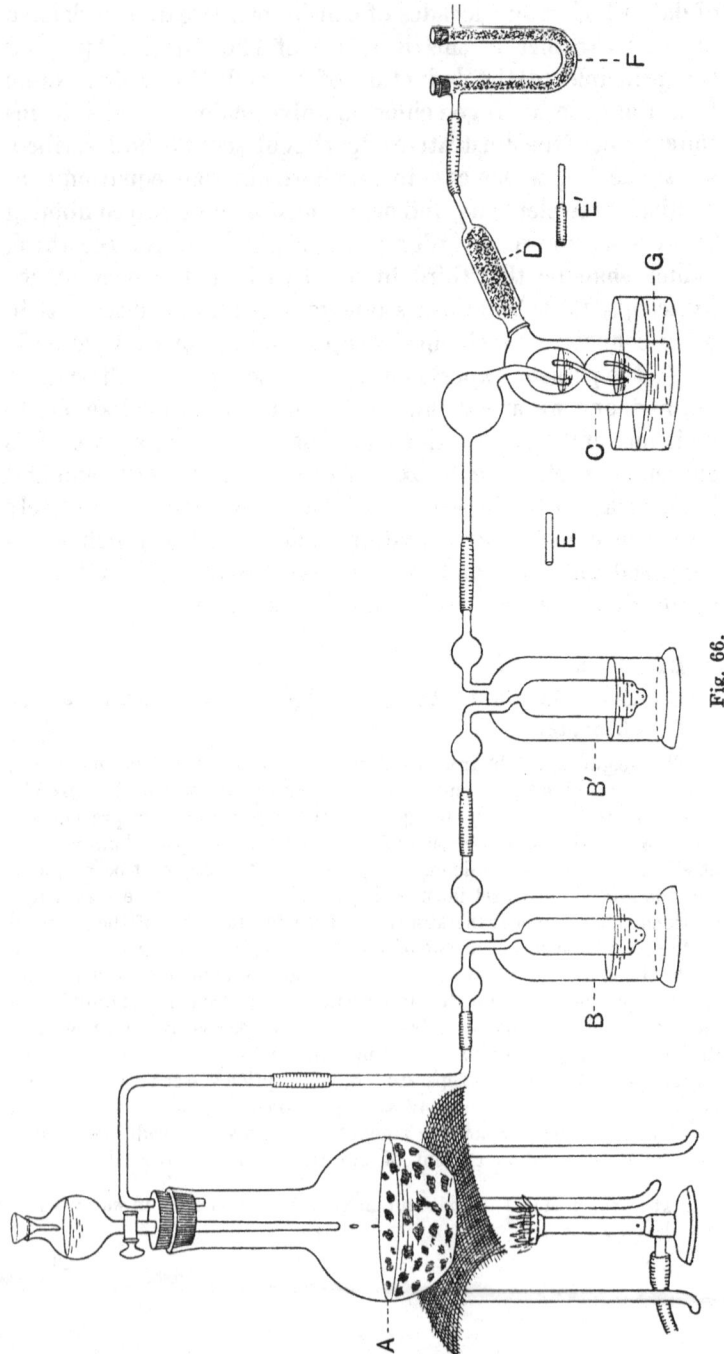

Fig. 66.

A. Round-bottomed flask of about 500 c.c. capacity, the lower half of which is filled with *lumps* of rock salt, on to which concentrated sulphuric acid is dropped from a funnel supplied with a tap.

B, B'. Gas wash-bottles containing concentrated sulphuric acid.

C. Weighed absorption bulb containing water, to which is attached the drying tube *D* filled with soda lime.

E, E'. Stoppers for closing *C*, *D* when detached for weighing.

F. Soda lime tube to protect *D* from outside moisture.

G. Bath of cold water.

bubbles per minute and maintained for 10 to 15 minutes will do, the quantities used in the specimen experiments given in the table being unnecessarily large), the solution is carefully transferred to a beaker, the bulb being washed out several times with successive small volumes of distilled water. The chlorine is then precipitated in the usual manner by the addition of excess of silver nitrate, and the silver chloride weighed in a Gooch crucible.

Record of results.

Specimen of note-book entry.

The determination of the ratio hydrogen chloride : silver chloride $= p : q$, and the utilisation of this value for the calculation of a value for the combining weight of chlorine, $Ag = 107{\cdot}9$ and $H = 1{\cdot}01$ being used as antecedent data.

The weight of the hydrogen chloride used	I	II	III
Wt. of absorption bulb + water + drying tube .... $=p'$	66·616 gms.	65·308 gms.	57·329 gms.
Wt. of above after absorption of the hydrogen chloride .... $=p''$	68·404 ,,	66·323 ,,	58·446 ,,
∴ Wt. of hydrogen chloride $=p''-p'=p$	1·788 ,,	1·015 ,,	1·117 ,,
The weight of the silver chloride produced			
Wt. of Gooch crucible + asbestos $=q'$	14·3195 ,,	14·3265 ,,	13·8548 ,,
Wt. of above + silver chloride : (i)	21·3687 ,,	18·3300 ,,	18·2386 ,,
(ii)	21·3677 ,,	18·3245 ,,	18·2385 ,,
(iii)		18·3245 ,,	
$=q''$			
∴ Wt. of silver chloride $=q''-q'=q$	·7·0482 ,,	3·9980 ,,	4·3837 ,,
The combining weight of chlorine calculated from the formula $HCl : AgCl = p : q$ $(1{\cdot}01 + Cl) : (107{\cdot}9 + Cl) = p : q$ $Cl = \dfrac{107{\cdot}9 \cdot p - 1{\cdot}01 \cdot q}{q - p}$	**35·35**	**35·35**	**35·49**

(2) *The analysis of salts containing a volatile acidic oxide which can be expelled either by the action of heat or by double decomposition with another acid, which under the conditions of the experiment is itself nonvolatile* (application of general method described *ante,* p. 332).

We may use this method **either** to find the combining weight of the metal contained in the basic oxide, in which case we require as an antecedent datum that of the non-metal contained in the acidic oxide, **or** *vice versa,* but in both cases we need to know the ratio of the number of combining weights of oxygen contained in the basic and the acidic oxide respectively, a value which for the same acid is a constant. Berzelius' work on the oxide, the sulphide and the sulphate of lead (*ante,* p. 293) serves as an example of the manner in which he was able to show that for sulphates the ratio is $1 : 3$; for carbonates it has been found to be $1 : 2$, for nitrates $1 : 5$.

Experiment L.

To find the combining weight of calcium from the determination of the ratio calcium carbonate : carbonic acid anhydride.

$$(CaO + CO_2) : CO_2 = a : b$$

with $C = 12 \cdot 00$ as antecedent datum.

Pure 'precipitated' calcium carbonate[1] spread in a thin layer on a watchglass is carefully dried by several hours' heating in an air bath kept at about 250° C. A known weight of the salt is treated in a suitable apparatus with dilute hydrochloric acid (see fig. 67); the carbon dioxide liberated is allowed to escape, after having been deprived of the moisture carried by it by passage through a calcium chloride tube; the loss in weight of the whole apparatus supplies the measure for the amount of carbonic anhydride contained in the amount of carbonate taken. This is the method commonly used in gravimetric analysis for the estimation of the carbon dioxide in a carbonate, and is described in detail in text books of quantitative analysis (Clowes and Coleman, *Quantitative Analysis,* 10th ed., 1914, pp. 108 *et seq.*).

---

[1] Obtained artificially by double decomposition between a soluble calcium salt (*e.g.* chloride) and a soluble carbonate (*e.g.* ammonium carbonate); marble used in the laboratory for the production of carbon dioxide is not a very pure form of calcium carbonate; Iceland spar is a very pure native form (*ante,* p. 250).

Procedure.

The procedure is the same as that described before in connection with the experiment on conservation of mass (*ante*, p. 210), with this difference, that the object now being the *complete* decomposition of the carbonate used

A. Erlenmeyer flask of between 150 and 200 c.c. capacity.

B. Pure calcium carbonate covered with water.

C. Small test-tube filled with concentrated hydrochloric acid, sufficient in quantity to decompose all the carbonate.

D. Chloride of calcium tube to retain the moisture carried by the escaping carbonic acid gas; any lime contained in the granulated solid must have been changed to carbonate.

E. Tube through which the air is drawn at the end of the experiment to expel all the carbon dioxide contained in the flask above the liquid and dissolved in the liquid.

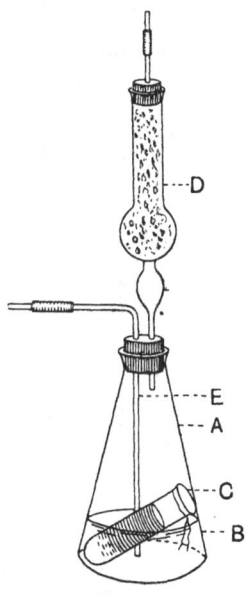

Fig. 67.

and the *complete* expulsion of all the carbonic anhydride liberated, care must be taken to have a sufficiency of acid in the small tube, and the removal of the carbon dioxide dissolved in the liquid may have to be helped by placing the whole apparatus for a few minutes on a previously heated sand bath ; the experimental difficulty consists in removing *all* the carbon dioxide without removing *any* water or hydrochloric acid.

Record of results.

Whilst the following figures, obtained by an average student working in the ordinary way, show that the experiment carried out with the simple home-made apparatus described can be made to yield concordant results agreeing with the standard value, it must be put on record that on the whole the values obtained have not been satisfactory, varying within the wide limits of 36 and 43 ; it would seem as if the experimental difficulty not satisfactorily overcome consisted in finding the exact conditions for completely expelling the carbon dioxide without losing any water or hydrochloric acid.

Specimen of note-book entry.

Determination of the combining weight of calcium from the ratio $(CaO + CO_2) : CO_2 = a : b$, the antecedent data being combining weight of carbon $= 12.00$; oxygen in acidic oxide : oxygen in basic oxide $= 2 : 1$.

		I	II
Wt. of pure dry carbonate .... $= a$		·522 gms.	·703 gms.
Wt. of whole apparatus :			
before the reaction .... $= b'$		77·007 ,,	78·272 ,,
after the reaction .... $= b''$ (i)		76·792 ,,	77·970 ,,
,, ,, (ii)		76·780 ,,	77·962 ,,
,, ,, (iii)		76·781 ,,	77·962 ,,
Loss of wt., representing carbon dioxide			
$= b' - b'' = b$		·229 ,,	·310 ,,
Combining weight of calcium ....$= \dfrac{44a}{b} - 60$		40·30 ,,	39·78 ,,
	Mean	**40·04**	

Experiment LI.

To find the combining weight of nitrogen from the determination of the ratio potassium nitrate : nitric anhydride (*ante*, p. 333).

$$(K_2O + N_2O_5) : N_2O_5 = a : b$$

with antecedent datum $K = 39.10$.

The principle of the experiment consists in the expulsion of volatile nitric anhydride by fusion of a known weight of the pure dry nitrate with excess of the non-volatile acidic oxide silica, the loss of weight resulting from the process of fusion being the measure of the nitric anhydride that had been present. A platinum crucible will be found most convenient, but a porcelain crucible can be made to do. The silica, which must be finely divided and free from any volatile impurity—including moisture—and which therefore should be previously ignited, has to be used in considerable excess. The mixing should be thorough, and is best done by adding the silica to the weighed nitrate contained in the crucible, stirring with a platinum wire or a glass rod drawn out to a fine point at one end.

Record of results.

Specimen of note-book entry.

Determination of the combining weight of nitrogen from the ratio $(K_2O + N_2O_5) : N_2O_5 = a : b$, with antecedent datum $K = 39{\cdot}10$.

Wt. of empty crucible ....$= a'$ ... 24·8418 gms.

  ,,   ,,   + potass. nitrate.........$= a''$ ... 25·5240 ,,

∴ wt. of potass. nitrate taken ......$= a'' - a' = a$ ... 0·6822 ,,

Wt. of crucible + contents after addition of silica :

    Before heating..................................$= b'$ ... 27·0590 ,,

    After heating (i)............................... ... 26·6962 ,,

    ,,   ,,   (ii) .... 26·6946 ,,

    ,,   ,,   (iii) ....$= b''$ ... 26·6946 ,,

∴ Nitric anhydride expelled............$= b' - b'' = b$ ... 0·3644 ,,

∴ Combining weight of nitrogen

$$= \frac{1}{2}\left(\frac{b}{a-b} \cdot K_2O - 50 \right)$$

$$= \frac{1}{2}\left(\frac{b}{a-b} \cdot 94{\cdot}2 - 80 \right)$$

    ... 14·005 ,,

APPENDIX TO CHAPTER IX

The Relation between the Combining Weight and the Atomic Weight.

In the chapter on Combustion it has been shown how a number of apparently diverse and unconnected phenomena came to be recognised as belonging together, and how when classification had done its work, in that case as in all similar ones, in response to a desire inherent in the human mind, the attempt was made to refer all the common effects observed to one common cause. It is at this stage that imagination steps in, shows its function and asserts its value in the development of a science. A guess is made, somebody assumes some property or properties as characteristic of the systems which exhibit the special phenomena considered, and the adequacy of the guess—called a working hypothesis—is tested inductively and deductively. Becher's and Stahl's guess, the conception of phlogiston (Chapter IV, p. 147) was of a nature such that, while it accounted for a number of the phenomena observed in combustion, it did not account for all, and in fact led to discrepancies between what according to the postulated properties and functions of the hypothetical phlogiston should happen, and what in reality does happen. Lavoisier, guided by a critical survey of all the facts previously known, together with those observed and established by himself, made the guess, or—what seems to be the proper scientific term—framed the working hypothesis, that air is made up of two different parts, one of which in combustion combines with the burning substance. This hypothesis—unlike that of phlogiston—was of a nature such that it could be put to an experimental test, which took the form of the experiment on the synthesis and the complete analysis of mercury calx, an experiment which has become classical; the result was the separation of air into two gases possessed of the postulated properties, and the re-formation of ordinary air when these two gases were mixed together. The hypothesis was *proved* to be true, and thereby it ceased to exist as a hypothesis and became an established fact. But it is not always thus : it is not a necessary attribute of hypotheses that the assumptions made should be capable of sensual realisation and therefore amenable to methods whereby their truth can be tested, established or disproved ; all that is required is, that the assumptions made should not be absurd, *i.e.* contrary to the known laws of nature and the established laws of thought ; hypotheses need not be true, their value depending merely on their utility, on their fitness for the task to accomplish which they have been created. What then, in a quite general way, is this task ?

...to reduce the number of laws, as far as possible, by showing that laws at first separated may be merged into one; to reduce the number of the chapters in the book of science by showing that some are truly mere sub-sections of chapters already written...

This is a purpose for which we may *imagine* a constitution of matter such that the isolated laws appear but as the necessary outcome of these fundamental properties of matter, and we may even go so far as to

...imagine properties which we should apprehend, if our senses were on quite another scale of magnitude and sensibility.

This is what Dalton did in the first years of the nineteenth century, when he supplied an *explanation,* or as some would prefer to have it called, a *condensed description* of the empirical laws of chemical combination, *i.e.* of regularities of so striking a nature as to produce in the human mind the desire to account for them or to represent them as due to some common cause, to some fundamental property of matter. Of this explanation it was said by Berzelius soon after it was offered that 'it represented the greatest advance that chemistry had ever yet made in the course of its development into a science.'

In an earlier part of this chapter (*ante*, p. 321) occasion has arisen for giving a short account of the nature of the Daltonian hypothesis, but our present purpose makes a certain amount of repetition necessary.

Dalton imagined matter to be not infinitely divisible, not continuous, but of grained structure; the constituent grains, which are so small as not to be capable of sensual realisation, are named atoms ; the atoms of one and the same substance are all identical and possessed of a characteristic constant weight, which is specific, *i.e.* different from the weight of the atoms of other substances ; chemical combination takes place between small numbers of the indivisible elementary atoms, resulting in the formation of 'compound atoms,' the weights of which are the sum of the weights of the constituent atoms, *e.g.* the combination of 1 atom of A weighing p with 1 atom of B weighing q, or with 2 or another small whole number of such atoms, forms 1 compound atom M which weighs $p + q$ or $p + 2q$, or quite generally, $mp + nq$, where m and n are simple whole numbers.

This is not the place to follow out in detail how this hypothesis supplies a simple, comprehensive and consistent explanation of the laws of chemical combination, viz., of conservation of mass and combination according to fixed, multiple and permanent ratios ; moreover the application is so simple and direct as to be quite obvious.   How Dalton applied his hypothesis to the construction of a system of notation at the same time qualitative and quantitative has been shown in an earlier part of this chapter.   What we are concerned with now is his conception of *atomic weight*, and the manner of measuring these quantities which are of such fundamental importance.   Recognising at the outset that the measurement of the *absolute* weight of the infinitely small atoms is not possible, and that what we are concerned with is the *relative* weight of the different kinds of atoms, Dalton expresses himself thus regarding the nature of the problem and the number of the facts he wishes to ascertain.

In all chemical investigations it has justly been considered an important object to ascertain the relative weights of the simples which constitute a compound, but unfortunately the enquiry has terminated here; whereas from the relative weights in the mass the relative weights of the ultimate particles or atoms of the bodies might have been inferred, from which their number and weight in various other compounds would appear in order to assist and to guide future investigations and to correct their results. Now it is one great object of this work to show the importance and advantage of ascertaining the relative weights of the ultimate particles both of simple and compound bodies, the number of simple elementary particles which constitute one compound particle...

The equation expressing the above relation is

$$mp : nq = a : b,$$

where p and q are the unknown absolute weights of the atoms of the elements A and B, p/q the required relative weight of A in terms of B as unit, m and n whole numbers which by hypothesis are always small, and $a : b$ is the experimentally found value for the ratio in which A and B are present in the compound M; thus if M represents water, p/q will be the atomic weight of oxygen referred to the weight of the hydrogen atom as unit. But besides this, the ratio of the weights of the elementary atoms, we wish to ascertain also m and n, the number of each kind of elementary atom present in one compound atom, and even if we content ourselves with a knowledge of the ratio m/n, the fact remains that the data available are insufficient, that we have only one equation containing the two unknown ratios p/q and m/n. To get over this difficulty Dalton framed a number of *a priori* arbitrary rules for assigning values to m and n, such as that if one compound only is known between two elements, it should be considered as binary, *i.e.* $m = n = 1$, that if two are known one should be considered as binary and the other ternary, etc. But these rules were not only arbitrary—really a number of hypotheses—but also insufficient, for there is always the possibility of further discovery, and why should the binary compound always be the first known? If of two combinations one is binary and the other ternary, which is to be which? in the case of the oxides of copper, the formulae CuO and CuO_2 for cuprous and cupric oxide conform to the rule just as well as Cu_2O and CuO. Without further enlarging on the theme, it should be clear that the fixing of values for m and n, on which depends that of p/q, in accordance with the rules laid down by Dalton was a process much influenced by the subjective element and the results of which were therefore open to doubt; hence in its original form the atomic hypothesis left the measurement of the relative atomic weights encumbered with the same uncertainty as that which we found attached to the selection of one of a number of combining weights as the atomic weight.

But in the first decades of the nineteenth century, as the direct result of the impetus given to the science by the Daltonian hypothesis, knowledge of the quantitative composition of inorganic compounds grew apace; and in 1818 Berzelius was able to publish a table giving the relative atomic weights of the elements then known in terms of the unit Oxygen$=100$ deduced, by more or less arbitrary methods, from a knowledge of the combining weights (see p. 326). This supplied material for the recognition of a connection between the values of the atomic weights thus derived and certain physical properties, such as the heat capacity of elements in the solid state or the crystalline form of their compounds. Thus Petit and Dulong (1819) showed that the specific heat of a solid element multiplied by its atomic weight gave an approximate constant value (about $6\cdot4$) known as the atomic heat; but in order that this constancy, found for a certain number of the elements, should hold generally, certain of Berzelius' atomic weight values had to be divided by 2 or $2/3$ or 4, *i.e.* the values arbitrarily chosen for m, n were replaced by others less arbitrary, in so far as their use led to the exhibition of a numerical regularity presumably expressive of some real common property belonging to the ultimate particles of the different elements. But when thus inductively established, the constancy of the value for atomic heat could then be applied deductively in assigning the values m and n, *i.e.* for elements whose specific heat in the solid state can be determined.

So much then for the origin, the nature, the limitations and the development of the Daltonian atomic hypothesis, in which the atom is the primary conception. But the atom holds a place and probably a more important one, by virtue of its being a derived conception, one following from that of the molecule as conceived and introduced into the science by Avogadro.

Whilst each solid and each liquid has its own characteristic coefficient of compressibility and of expansion the generalisations known respectively by the names of Boyle's Law ($pv =$ constant), Charles' Law ($v_0 : v_T = 273 : 273 + T$), and Gay-Lussac's Law (the combining volumes are simple whole numbers, *post*, Chapter X, p. 382) show : (i) that all the different gases behave alike in these respects ; (ii) that there exists a simplicity in the relations between *combining volumes* which has no parallel in those between *combining weights*. The inference is inevitable that in the constitution of all gases there must be some common feature, and that the gaseous state is marked by a simplicity of structure not met with in the solid or liquid state. The hypothesis put forward in 1811 by the Italian scientist Amadeo Avogadro (1776—1856) conceived this common simple structure of gases to consist in the presence in equal volumes—measured at the same temperature and pressure—of an equal number of independently existing particles ; put somewhat differently, this means that in the case of gases the volume occupied and the correlated properties of temperature and pressure depend, unlike other properties such as colour, heat capacity etc., not on the *kind* of constituent particles but only on their number. What is the nature of these particles and what their relation to the Daltonian atoms ? Using the present-day nomenclature for the designation of the different types of differently complex ultimate constituents conceived of by Avogadro, we have the *molecules*, the independently existing aggregations of matter, equal numbers of which occupy equal volumes ; the molecules of compounds are of necessity complex, being made up of two or more kinds of smaller particles or *atoms* which represent a more ultimate and simpler order of aggregation ; the molecules of elements may be simple or complex, each containing 1 or 2 or 3 or 4... atoms identical with one another. In the mechanism of chemical change the molecules are the acting units ; the change may consist in the splitting up of each of a number of complex molecules into two or more simpler ones, or *vice versa*, in the coalescence of two or more simpler into one more complex molecule, or what is the most common case in the formation of new kinds of molecules, by the interchange of the constituent atoms of the different kinds of molecules originally present.

The picture of the ultimate condition of matter thus created, not being capable of sensual realisation, constitutes a hypothesis the *truth* of which cannot be proved or disproved experimentally ; but this has not been a determinant factor in the history of its development : it owes the position of paramount importance which it holds and has held for more than half a century to the extensive use that has been made of it in the interpretation and correlation and discovery of facts. Originally devised for the purpose of correlating and explaining the laws characteristic of and peculiar to the gaseous state, it has furthered chemistry by its deductive application to the measurement of the comparative weights of the molecules, *i.e.* those quantities which constitute the units of chemical change, and has led—as will be shown presently—to an objective method of measuring relative atomic weights. As regards the principle underlying the molecular weight measurements, this is quite simple :

$$\frac{\text{Wt. of 1 molecule of substance } A}{\text{Wt. of 1 molecule of standard}}$$

$$= \frac{\text{Wt. of } X \text{ molecules of } A}{\text{Wt. of } X \text{ molecules of standard}},$$

but since by Avogadro's hypothesis X molecules of any gaseous substance whatever occupy the same volume

$$\frac{\text{Wt. of 1 molecule of } A}{\text{Wt. of 1 molecule of standard}}$$

$$= \frac{\text{Wt. of } m \text{ vols. of } A}{\text{Wt. of } m \text{ vols. of standard}}$$

$$= \frac{\text{gaseous density of } A}{\text{gaseous density of standard}}$$

If, as is the case now, the standard molecular weight is

oxygen = 32, we get by the substitution of this value and of 1·429 grams per litre for the gaseous density of oxygen :

$$\text{Molecular weight of } A_{O=32} = \frac{32}{1\cdot429} \times \text{gaseous density of } A$$

$$= 22\cdot39 \times \text{gaseous density of } A$$

$$= \text{weight of } 22\cdot39 \text{ litres of the gas } A \text{ measured at N.T.P.}$$

Verbally expanded, this means that the molecular weight of any substance is the weight of that volume of it which in the gaseous state occupies a volume equal to that filled at the same temperature and pressure by the molecular weight of the standard, *i.e.* by 32 grams[1] of oxygen, which is 22·39 litres[2]. It should be noted that this determination of relative molecular weight is a purely physical method based on a hypothesis which does not take into account specific differences between different kinds of matter but refers to a common property of all matter ; hence the composition of the substance is immaterial, need not be known and is not in any way revealed by the experimental work involved.

Though propounded early in the nineteenth century, it

---

[1] Stated thus, the molecular weight would appear as the weight in grams of 22·39 litres of the gaseous substance, and to do away with a possible source of confusion attention must be drawn to the fact that we are simply getting a set of comparable numbers; we do not attempt to ascertain the actual weight of the oxygen molecule, which whatever it may be belongs to a very low order of magnitude. What we say is that if we choose our unit of weight sufficiently small and such as to make the weight of the oxygen molecule 32 of these units, then the weight of any other molecule measured in these units will be related to 32 just as in the above equation and will be expressed by the same number.

'Imagine it could be proved that half a molecule of hydrogen weighed one millionth of a milligram'...—modern physical research has fixed this weight as about $\frac{3\cdot24}{10^{24}}$ gram—'then all the [values for the molecular weights obtained as above] become concrete quantities and express the absolute weight of the molecules in millionths of a milligram. The same would be the case if the common unit had any other concrete value. Thus these numbers are all comparable, whatever may be the concrete value of the common unit.' (Cannizzaro.)

[2] With hydrogen = 2 as standard, this value becomes $\frac{2}{\cdot08987} = 22\cdot25$ litres.

was only after 1858[1] that Avogadro's hypothesis was generally accepted and began to exert a dominating influence on chemistry. The chief reason for this lay probably in the fact that in the years immediately following upon the publication of Dalton's hypothesis attention was concentrated on the determination of atomic weights, which were mostly deduced from the composition of non-volatile metallic oxides, and that it was not until later that the phenomenally rapid development of organic chemistry brought within the scope of the science a very large number of substances which could be volatilised more or less easily. Then it was found that formulae such as C_2H_6O for alcohol, $C_4H_{10}O$ for ether, CH_4 and C_2H_4 for methane and ethylene etc. etc., which had been chosen on grounds of chemical adequacy, *i.e.* because they were best suited to represent the nature and properties of the corresponding substances, the manner in which they had been formed from other substances and the manner in which they could be decomposed, represented quantities which in the gaseous state occupy equal volumes, *i.e.* the formulae weights are in the ratio of the gaseous densities. Here then was a relation found empirically which inevitably led to the inference that of the smallest component particles of these different substances which are the units of chemical action, equal numbers must be present in equal volumes of the gaseous substances.

Moreover, the analyses of gaseous or volatile substances, when expressed not as parts per hundred but as parts of the molecular weights, *i.e.* of the quantities which in the gaseous state occupy equal volumes, lead to results such as the following[2]:

---

[1] It was in that year that the pamphlet by S. Cannizzaro, then Professor of Chemistry at Genoa, in which the nature and applicability of Avogadro's hypothesis were expounded in the most masterly fashion, reached a wider public.

[2] The table is a reproduction in a very slightly modified form of one given by Cannizzaro. (Ostwald, *Klassiker der exacten Wissenschaften*, No. 30.)

Name of the substance	Molecular weight referred to the weight of a half hydrogen molecule	Weight of the constituents contained in the molecular weight
Hydrogen	2	2 ($=2 \times 1$) hydrogen
Oxygen	32	32 ($=2 \times 16$) oxygen
Sulphur under 1000°	192	192 ($=6 \times 32$) sulphur
Sulphur above 1000°	64	64 ($=2 \times 32$) sulphur
Phosphorus	124	124 ($=4 \times 31$) phosphorus
Chlorine ...:	71	71 ($=2 \times 35\cdot5$) chlorine
Bromine	160	160 ($=2 \times 80$) bromine
Iodine	254	254 ($=2 \times 127$) iodine
Nitrogen	28	28 ($=2 \times 14$) nitrogen
Arsenic	300	300 ($=4 \times 75$) arsenic
Mercury	200	200 ($=1 \times 200$) mercury
Hydrochloric acid ...	36·5	35·5 chlorine + 1 hydrogen
Hydrobromic acid ...	81	80 bromine + 1 hydrogen
Hydriodic acid ...	128	127 iodine + 1 hydrogen
Water	18	16 oxygen + 2 ($=2 \times 1$) hydrogen
Ammonia	17	14 nitrogen + 3 ($=3 \times 1$) hydrogen
Arsine	78	75 arsenic + 3 ($=3 \times 1$) hydrogen
Phosphine	35	32 phosphorus + 3 ($=3 \times 1$) hydrogen
Mercurous chloride	235·5	35·5 chlorine + 200 mercury
Mercuric chloride ...	271	71 ($=2 \times 35\cdot5$) chlorine + 200 mercury
Arsenic chloride ...	181·5	106·5 ($=3 \times 35\cdot5$) chlorine + 75 arsenic
Phosphorous chloride	138·5	106·5 ($=3 \times 35\cdot5$) chlorine + 32 phosphorus
Ferric chloride ...	325	213 ($=6 \times 35\cdot5$) chlorine + 112 iron
Nitrous oxide ...	44	16 oxygen + 28 ($=2 \times 14$) nitrogen
Nitric oxide	30	16 oxygen + 14 nitrogen
Carbon monoxide ...	28	16 oxygen + 12 carbon
Carbon dioxide ...	44	32 ($= 2 \times 16$) oxygen + 12 carbon
Ethylene	28	4 ($= 4 \times 1$) hydrogen + 24 ($=2 \times 12$) carbon
Propylene	42	6 ($= 6 \times 1$) hydrogen + 36 ($=3 \times 12$) carbon
Acetic acid	60	4 ($= 4 \times 1$) hydrogen + 32 ($=2 \times 16$) oxygen + 24 ($=2 \times 12$) carbon
Acetic anhydride ...	102	6 ($= 6 \times 1$) hydrogen + 48 ($=3 \times 16$) oxygen + 48 ($=4 \times 12$) carbon
Alcohol	46	6 ($= 6 \times 1$) hydrogen + 16 oxygen + 24 ($=2 \times 12$) carbon
Ether	74	10 ($=10 \times 1$) hydrogen + 16 oxygen + 48 ($=4 \times 12$) carbon

...It is at once apparent that the several quantities of one and the same element contained in the various molecular weights are all whole multiples of one and the same quantity, which since it always enters into compounds undivided can justly be termed the atom....The different quantities of hydrogen contained in the molecules of the different substances are whole multiples of the quantity contained in the hydrochloric acid molecule, which justifies the choice of this amount as the common unit for molecules and atoms. The hydrogen atom is contained twice in the molecule of uncombined hydrogen. In like manner...the quantities of chlorine in the different molecules are all whole multiples of the quantity contained in the molecule of hydrochloric acid, *i.e.* of 35·5, and the various quantities of oxygen contained in the molecules are all multiples of that contained in water, *i.e.* of 16, a quantity which is half of that contained in the oxygen molecule....

In this manner we ascertain the smallest quantity of each element which always enters undivided into the molecules of the substance of which it forms a constituent, and which is justly named the atom. Hence in order to determine the atomic weight of each element, we must know the molecular weights and the composition of all or at least of most of its compounds.

(Cannizzaro, 1858.)

The above quotation shows in the clearest manner possible:

(i) How the conception of atomic weight is deduced from that of molecular weight, *i.e.* how it is arrived at as a conception derived from the data collected by the application of Avogadro's hypothesis; (ii) in what the experimental work involved in an atomic weight determination by this method consists. Further points of importance set forth in an equally lucid manner by Cannizzaro in the same pamphlet are:

1. The molecular weights of the elements themselves need not be known, only those of their compounds; *e.g.* carbon, which forms so large a number of volatile compounds is itself non-volatile, and all that we can say is that its molecular weight, if measurable, would be found $= 12 \times n$, where n is some simple whole number.

2. The value thus obtained, which for each element is the *least* amount contained in the molecular weight of any of its compounds, is a *maximum* possible value; if an element forms very few volatile compounds, there must be great uncertainty on this subject, as of course there is no *a priori* necessity that the special compound or compounds should contain in the molecule only *one* atom of the element considered. In such cases the results obtained by the method based on Avogadro's hypothesis must be compared and supplemented by those supplied by other methods, such as that of the constancy of the heat capacity (*ante*, p. 364) or that supplied by the periodic classification.

A further point of the utmost importance is that the data required for an atomic weight determination by this method, involving as they do a knowledge of the composition of a number of the compounds of the element, *include* a knowledge of the combining weight. Of course the compounds which happen to be volatile may not be such as to lend themselves to an *accurate* measurement of the ratio $a : b$, and hence the value of the combining weight, and with it that of the atomic weight, can only be found approximately by this method; but the approximation will be near enough to show which value to give to the factor m/n (p. 363), and hence to deduce an accurate atomic weight from the independently determined accurate combining weight[1].

The determination of atomic weights by the help of the periodic system also involves the finding of the value for m/n, *i.e.* the numerical relation to the empirical combining weight. The 'periodic system' is a method of classification of the elements based on the generalisation that if the elements are arranged in the order of their *atomic* weights, most of the properties of the elements and of their compounds show periodic variations.

I arranged the elements according to the magnitude of their atomic weights, when it became evident that there exists a kind of periodicity in the properties....At first the properties of the elements change as the atomic weights increase, but afterwards these repeat themselves in a new *series* or new period of elements.... (Mendeleeff, 1871.)

The members of the different *series* which are correlated by this recurrence of properties constitute the *groups*. In the usual tabular arrangement the *series* are the horizontal rows, each comprising a certain number of elements (8—11) arranged in arithmetical progression according to increasing atomic weight; the groups, which are the vertical rows, contain chemically similar elements. But at the time of the promulgation of the system, in order that such an arrangement should really express the relation between the atomic weights and the properties of the elements, it was found necessary : (i) To leave gaps which in the future might be filled in (*e.g.* radium found a place in group II, series 7, under barium, to which it is naturally related by common properties—*ante*, p. 325); (ii) To act on the supposition that the correct atomic weights of some elements were different multiples of the combining weights from those which had hitherto been

[1] For examples see *Study of Chemical Composition*, pp. 351 *et seq.*

assigned to them[1]. It is this last necessity that brings the periodic system within the scope of our present considerations. The regularities brought out by this system of classification occur only if the *atomic* weights are the one variable, not if the combining weights—whether the smallest or any other arbitrarily chosen ones—are used instead. The discovery of the relation between atomic weight and properties was made when for the great majority of elements the atomic weight values obtained by the application of the arbitrary methods of Dalton and Berzelius already described, had been accepted or modified by the chemists of the day after Cannizzaro's demonstration of the practical application of Avogadro's hypothesis to atomic weight determination. From the nature of the case a generalisation of such a comprehensive kind arrived at inductively could not have been made without a large number of available data ; but as soon as the relationship had been discerned and established, the value of its deductive application to the choice of atomic weight values became apparent. The study of the properties of an element shows its relations with others and indicates the group to which it belongs, and the legitimacy of this classification on the ground of properties will be shown by the occurrence of a vacant place in the group which will accommodate an element with the atomic weight equal to the combining weight $\times \frac{m}{n} (O=8)$. The element beryllium affords a striking example of the manner of using the periodic system for atomic weight determinations. The measurement of the ratio sulphate : oxide (by the calcination of the sulphate, a special case of the general method described *ante*, p. 332) gives the value 9·1 for the combining weight, and $9·1 \times \frac{m}{n}$ for the atomic weight, which accordingly might be $9·1 \times \frac{1}{2} = 4·55$, or $9·1 \times \frac{3}{2} = 13·65$, or $9·1 \times 2 = 18·2$, and

---

[1] Similarly Petit and Dulong, in order to obtain a constant for the atomic heat, had in the case of a number of elements to multiply or divide by some simple whole number the current values for the atomic weights.

so on, the formula for the oxide being correspondingly Be_2O, or BeO, or Be_2O_3 or BeO_2 etc. The measurement of the heat capacity does not in this case help towards the selection of the correct value for $\dfrac{m}{n}$, because beryllium, like carbon and boron—which also are solid elements of low atomic weight—is exceptional in that its heat capacity varies very much with the temperature (from 0·376 at 0° to 0·470 at 100° and 0·542 at 200°), and there is nothing to indicate an *a priori* superiority of any one of the various available values for use in the deduction of the atomic weight. Moreover, the study of the properties of the compounds of beryllium does not by itself lead to a definite conclusion :

...the magnesia formula [MgO] has been given to its oxide, but on the other hand beryllia has properties similar to those of alumina [Al_2O_3]....The periodic law affords...proofs in support of the formula BeO,

which requires that the atomic weight should be 9·1 and its position in the table between lithium = 7 and boron = 11 in a group with magnesium ; there is no place in the table for an element of atomic weight, 13·65, which would give an oxide of the formula Be_2O_3.

The detailed study of the various methods available for the determination of the atomic weight shows that in each there is a residual number of facts termed 'exceptions,' *i.e.* cases in which the results either are not concordant with those obtained by the other methods or lead to formulae which, when judged by the supreme test of chemical adequacy, are found unsuitable. Thus : (i) The evidence for 12 as the atomic weight of carbon, obtained by the vapour density method, is so extensive as to be conclusive, especially when taken together with the fact that with C = 12 the formulae for the various carbon compounds agree with those assigned on purely chemical grounds ; but the value of the atomic weight of carbon derived from the generalisation that the atomic weight is equal to the atomic heat divided by the heat capacity of the solid element between 0° and 100° would give $\dfrac{6\cdot4}{\cdot1893}$ (approx.), a much higher value. (ii) When ammonium chloride is volatilised, the density of the resulting gas ascertained, the molecular weight calculated from it in the usual manner (*ante*, p. 366) and the composition stated as parts per molecular weight, it is found that, whilst the amount of hydrogen present is twice that found in one molecular weight of hydrogen chloride, the amount of nitrogen and of chlorine is only about half that of the least quantity of these elements found in one molecular weight of any other nitrogen and chlorine compounds. Acceptance of these results would produce the formulae H_3N_2, HCl_2, H_2NCl for ammonia, hydrogen chloride and ammonium chloride respectively, and would necessitate a doubling of the numbers of nitrogen and of chlorine atoms in all the many compounds into the composition of which these elements enter ; the consequent obliteration of chemical analogies and the increase in complexity of the formulae are such that there has never been any hesitation in rejecting these numbers for the atomic weights of nitrogen and of chlorine. (iii) And finally, there is the case already referred to (*ante*, p. 329) of the anomaly presented by the position of tellurium in the periodic system ; from its close analogies with sulphur and selenium it belongs to group VI, and should therefore precede iodine and have a lower atomic weight than 126·92, that of iodine ; but its atomic weight (*ante*, pp. 331 *et seq.*) is 127·5, nearly a whole unit higher than that of iodine. Here, as in all similar cases, the contradictory and exceptional nature of the occurrence leads to special work, to a search for the cause of the apparent abnormality ; and the starting-point of many a research that has become classical may be traced back to some such purpose. As regards the immediate results in the instances at present under consideration : (i) An elaborate research on the specific heat of carbon showed that the temperature exerts a very great influence, and that, after increasing rapidly with the temperature, the specific heat of this element reaches at about 600°C. a constant value, which with $C = 12$ gives for the atomic heat 5·4, a number sufficiently near to the constant 6·4 to justify the selection of 12 as the atomic weight. (ii) In the case of ammonium chloride, the reconciliation between the atomic weight values for nitrogen and chlorine arrived at by the application of Avogadro's rule and by the requirements of chemical adequacy, has been achieved by the recognition that the heating of solid ammonium chloride does not yield this substance in the gaseous state but produces a mixture of hydrogen chloride and ammonia[1], which, since each molecule of the sal ammoniac has broken up into two, occupies twice the volume and yields a vapour density and molecular weight half of that required on theoretical grounds for NH_4Cl when $N = 14$ and $Cl = 35\cdot5$. (iii) All attempts to bring the atomic weight of tellurium to a value in accordance with the position in the periodic system to which its properties assign it, have so far failed (*ante*, pp. 137 *et seq.*).

If space permitted it would no doubt be advantageous to illustrate by as many examples as possible that weighing of pros and cons, and that making of compromises, which is often a feature of the process resulting in the fixing of an atomic weight; it is not a case for revision of experimental work faulty in principle or not sufficiently accurate, but of a mode of interpreting experimental results on the basis of certain hypotheses ; in assigning values to the relative weights of particles of matter which we believe to be all at one and the same stage of subdivision, we try to find those values which will in the most consistent manner fit in with the greatest number of empirical relations between these hypothetical quantities.

The reason for including in the scope of this book a consideration of the nature and the measurement of the hypothetical quantities termed atomic weights, and for introducing this subject here, is the desire to combat the confusion unfortunately very prevalent which exists in the minds of many elementary students on the subject of the relationship between the empirical combining weight and the atomic weight. It is hoped that what has been said will—either by itself or if necessary supplemented from books in which the various points are driven home by more examples—have brought out the fact that the determination of the combining

[1] On cooling, re-combination to solid ammonium chloride occurs.

weight, of a ratio $a : b$ representative of the composition by weight of a suitably chosen compound, is a purely empirical process, one in which every new determination has for its sole object increase in accuracy, and in which, as regards the attainment of that object, it is difficult to see finality ; in each generation there seems to be a limit set by the skill of the foremost workers, but each successive high-water mark is in its turn left below the crest of the wave which marks the best achievement of a new era. Thus, take the results obtained in the determination of two of the fundamentally most important ratios :

(1) Ratio Silver : Silver chloride $= a : b$.

The experimental work consisted in a synthesis, weighing the chloride b obtained from a known weight of pure silver a (*ante*, p. 231). Expressed as the weight of silver chloride obtained from 100 of silver, the results were :

			Probable Error
Berzelius	1820	132·757	± ·019
Turner	1829	132·832	± ·004
Penny	1839	132·836	± ·001
Marignac	1842	132·854	± ·002
Stas		132·8445	± ·0008
Richards and Wells	1905		
First Series		132·861	± ·0007
Second Series		132·867	± ·0005

(2) Ratio Carbon : Carbon dioxide $= a : b$.

This is found by the measurement—absorption in potash— of the weight of carbon dioxide b produced by the burning of a known weight a of pure carbon, generally diamond.

		$\dfrac{b}{a}$	Probable Error	Combining Weight of carbon calculated on basis of formula CO_2
Dumas and Stas	1840	2·6680	± ·0009	11·994
Erdmann and Marchand	1841	2·6655	± ·0008	12·005
Roscoe	1882	2·6673	± ·0006	11·997
Friedel	1884	2·6654	± ·0007	12·006
Van der Plaats	1885	2·6663	± ·0001	12·002

Inspection of the table shows that the differences between the results of the various successive observers are but small, and that they tend to get less as time advances. Moreover, if by the application of mathematics to the extreme deviations between the various results obtained in each separate research, we calculate the quantity called the *probable error*, *i.e.* the amount to which the given result is known to be uncertain, this is found to decrease correspondingly. And it may be taken as practically certain that the last entries do not represent the end ; either by the same direct method or by one more indirect, these ratios will be measured again, and whatever the result, it will differ only very slightly from the last of the values entered. No speculations, no hypotheses are involved in this work ; it is merely a case of measurement made more and more accurate: but with regard to the ratio $\frac{m}{n}$ by means of which the atomic weight is deduced from the combining weight, it is not a case of *gradually increasing approximation* to the correct value, it is a case of choice between whole numbers, the choice being determined by a combination of considerations into which the element of hypothesis enters. Take the case of carbon. In terms of $O = 16$ as standard, the value for the atomic weight will depend on whether carbon dioxide is CO, CO_2, C_2O, or quite generally, on the values of the simple whole numbers m and n in the formula C_mO_n. And having once found values for m and n these are practically final, for as long at any rate as we retain the molecular and atomic hypotheses in their present form. Repeating once again what has been said already (*ante*, p. 328), in the work generally described under the name of 'atomic weight determinations' almost invariably one only of the two factors involved in such measurements is dealt with ; the work reported upon is that of the measurement of the ratio $a : b$, a value not absolutely fixed but subject to slight variations affected by re-determinations ; no mention even is made of the manner of determining the other factor $m : n$, which is just assumed as an antecedent datum. But if a revolution in our present theoretical views of the nature of chemical composition should sweep away the atomic and molecular hypotheses, then, whatever the features of the new system might be, however much it might differ from the present one, however great an advance it might mark in increased comprehensiveness and increased simplicity, it would have to account for the facts of chemical combination as summarised in the law of constant combining weights, and the accurate knowledge of these constants would retain its importance ; the measurement of the combining ratios is a work done for all time.

In view of all this, what are we to say about the difficulty found by so many students—and not only those termed elementary—in realising the position? It has been the plaint of a great master of the science that '...in chemistry it happens only too often and almost as a matter of course, that hypotheses are looked upon as facts, or at any rate are dealt with as if they were facts....' It has to be admitted that in the teaching of chemistry there has within the last two decades been some improvement ; it is no longer the invariable rule for boys and girls to be introduced to the theoretical study of the subject by some text book which at or near its beginning provides pat definitions for the terms *atom* and *molecule, atomic* and *molecular weight* ; and though old customs die hard the practical training is no longer merely by test-tube work culminating in the analysis of a mixture cunningly made up to present complications not met with in *real* problems. But when the pupils have 'discovered the mechanism of combustion' and have 'done a research on water and on chalk,' then, in the case of those who later make acquaintance with the use of chemical notation, many more strange things happen ; both examiners and teachers are to blame, and it is difficult to say where the greater guilt lies. In papers set to elementary students, questions as to the meaning to be attached to a formula such as H_2O are frequently met with. Now it is the firm belief of the present writer that the formulation of a satisfactory answer to a question of this kind requires some maturity of mind, and should not be attempted until after the slow spontaneous clarification of ideas following on a period of longer or shorter but always inevitable confusion. To provoke the earner to commit himself to statements at an earlier stage can have but one result, viz. the perpetuation of error and the protraction of confusion. There are the extreme cases, such as when a first year student required to describe how the volumetric composition of water could be demonstrated experimentally wrote : 'Take one atom of oxygen and two atoms of hydrogen....' But worse, because affecting average students who are plastic material for the impress of right and wrong views, is the following case ; the instance was found in the laboratory note-book of a girl who during the end of her school course in preparation for college specialised in chemistry. The exercise set was to find a formula for carbon dioxide. The first impression produced on seeing this should be one of dismay and of pity ; a rapid summary in the reader's mind of the amount of work involved would comprise a determination of the gravimetric composition and the molecular weight, not only of carbon dioxide itself, but also of a number of other volatile carbon compounds, since who would suppose that in dealing with the problem the atomic weight of carbon was assumed as an antecedent datum ? But this was evidently the case. The experimental work consisted merely in what, if called by the right name, would have been a very valuable exercise, viz. the determination of the composition by weight ; this was followed by a *calculation* in which, in the usual manner, the percentages were divided by the atomic weights of oxygen and carbon, supposed known, and the formula then given as CO_2. Why not C_2O_4 or C_3O_6 ? and what can have been the pupil's—or for the matter of that the teacher's—idea as to the manner of determining the atomic weight of carbon, when, going round in a complete circle, they used that value as an antecedent datum in the representation of the composition of a substance which from its simplicity would be the first to be used in the collection of the data for finding the atomic weight of carbon ? But unfortunately that teacher and that pupil are not exceptional in their view of atomic weights as something known *a priori*, divulged to us by some special revelation.

CHAPTER X

THE LAW OF COMBINING VOLUMES

I. Historical.

The experiment (Chapter VI, p. 245) done to show that the composition of a compound is independent of the relative amounts of the constituents present at its formation consisted in the volumetric synthesis of water; measurement of the ratio of the *volume* of hydrogen to oxygen showed that in 11 experiments the mean value was 2·00, the greatest deviations having been 1·97 and 2·05. The principle of this experiment is the same as that of the classical research carried out in 1805 by Gay-Lussac and Humboldt, who when engaged in determining the composition of different samples of air by the method of explosion with excess of hydrogen, wished to find for themselves the correct value of the antecedent datum vol. hydrogen : vol. oxygen, a ratio which at that time was not known to any great degree of accuracy[1]. In two series of measurements in which they used excess of oxygen and of hydrogen respectively, they obtained for the ratio of the combining volumes numbers so near to 200 : 100 that the whole number could legitimately be taken as representative of their final result. Struck by the simplicity of this relationship, Gay-Lussac was led to the examination of pertinent data, collecting such as were available already and adding new ones, with the result that in 1809 he could generalise as follows :

...Gases[2],...in whatever proportion they may combine, always give rise to compounds whose elements by volume are multiples of each other,...[and] the apparent contraction of volume which they experience on combination has also a simple relation to the volume of the gases...so that representing one of the terms by unity, the other is 1 or 2 or...3.

Whilst Berzelius recognised directly the importance of Gay-Lussac's discovery and its bearing on the atomic hypo-

[1] The volume of hydrogen required by 100 of oxygen was, according to Monge 196, according to Davy 192 and according to Dalton 185.

[2] 'Combinaison des substances gazeuses,' *Mémoires de la Société d'Arcueil*, 1809, II, pp. 207—234.

thesis which had just then been presented to a wider public through the agency of Thomson, Dalton himself refused to accept it and threw doubt on the accuracy of the measurements on which the generalisation was based. This attitude of Dalton was due to the fact that an inevitable deduction from Gay-Lussac's law was incompatible with one of the postulates of his atomic hypothesis. The simplicity of the combining volume ratios finds a simple and natural explanation in the assumption that equal volumes of all the different gases contain equal numbers of those constituent particles which act as the units in chemical change ; to Dalton these units were atoms, which in the case of elements he conceived as indivisible (as indicated by the name ἄτομος = uncuttable). But whenever the volume of a compound gas is greater than the volume of a constituent (*e.g.* 1 volume of oxygen in 2 volumes of gaseous water ; 1 vol. of nitrogen and 1 vol. of oxygen in 2 vols. of nitric oxide ; 1 vol. of hydrogen and 1 vol. of chlorine in 2 vols. of hydrogen chloride), elementary atoms must have been divided in order to enter into the composition of a greater number of compound atoms : a fact necessitating a modification of part of his original hypothesis which Dalton was not prepared to concede. The difficulty was overcome by Avogadro in his assumption of two orders of particles, viz. the molecule and the atom, the first of which, even if belonging to an element, may be divisible in the process of chemical change (*ante*, Chapter IX, Appendix, p. 365). Berzelius, who supported the law of simple combining volumes, pointed out in a letter to Dalton that this was 'a matter in which it is only a case of measuring well or badly' ; Gay-Lussac's measurements have stood the test, and his generalisation ranks as one of the fundamental laws of the science.

II. Nature and scope of the law.

Whilst, except as far as the operation of the law of multiple ratios extends, there is no simple relation between the *weights* of A and B, the constituents of a compound C (*e.g.* 2 parts by weight of hydrogen to 15·88 of oxygen, 1 of hydrogen to 35·45 of chlorine), the very simple numerical relation formulated by Gay-Lussac holds for the *volumes* when two at least of the substances *A, B, C* are gaseous. This marks off sharply the gaseous from the liquid and solid state of aggregation, pointing to a greater simplicity of the structure of gases :

> These [simple] ratios by volume are not observed with solid or liquid substances, nor when we consider weights, and they form a new proof that it is only in the gaseous state that substances are in the same circumstances and obey regular laws. (Gay-Lussac, *ibid.*)

III.   Experimental basis and verification of the law.

The evidence on which Gay-Lussac had based his generalisation was extensive.   The data were derived partly from his own work and partly from that of others, either by the direct measurement of the combining volumes or by the calculation of these from the composition by weight of *C* (if not gaseous) and the gaseous density of at least two of the substances *A, B* and *C*.   The gaseous compounds thus dealt with were :

Compound	Constituents	
C	*A*	*B*
Ammonium carbonate	100 vols. carbonic acid	200 vols. ammonia
Ammonium chloride	100 vols. hydrochloric acid	100 vols. ammonia
200 vols. ammonia	100 vols. nitrogen	300 vols. hydrogen
Sulphuric anhydride	100 vols. sulphurous anhydride	50 vols. oxygen
100 vols. carbonic acid	100 vols. carbonic oxide	50 vols. oxygen
Nitrous oxide	100 nitrogen	49·5 vols. oxygen
100 vols. nitric oxide	100 nitrogen	100 vols. oxygen
Nitric peroxide	100 nitrogen	204·7 vols. oxygen

Obviously in most cases the numbers given are not those actually found, but the nearest whole numbers; there is nothing in the 1809 paper of Gay-Lussac to show how close the approximation was, and it is clear that great accuracy was neither attained nor even aimed at.

These values are all purely empirical (like combining weight ratios, specific gravities or specific heats), and thus there is a fundamental difference between them and the hypothetical quantities such as the relative molecular weights. The methods employed in the measurement of combining volume ratios make use of a great variety of experimental devices for attaining the end in view, viz. *either* the decomposition of C with the production of A or B or a mixture of both in the gaseous form, *or* conversely, the synthesis of C from A and B; the common feature of all such analyses and syntheses is that two at least of the three substances involved must be gaseous under ordinary conditions. Since the days of Hofmann, who devised ingenious methods and suitable apparatus for the quick and effective demonstration of gaseous analyses and syntheses illustrative of Gay-Lussac's law[1], the experiments known by his name are an essential feature of every course of lectures on general or theoretical chemistry, and the special apparatus is shown in every chemical apparatus maker's catalogue.

Students are strongly recommended to make a thorough and careful study of the experimental work by which the composition by volume of the most important simple gaseous compounds has been established. In doing so they should in each case consider the following points : the principle involved, the apparatus used, the experimental procedure, the results as obtained directly and their interpretation. It is of the utmost importance to be quite clear in each case, not only as to what the results obtained *do* show, but also as to what they *do not* show, and to guard against errors of which the following may serve as an example: on decomposing a known volume of hydrogen chloride by sodium amalgam, and finding that the volume of hydrogen produced is half that of the original gas, it is inferred that the volume of chlorine withdrawn is also equal to half that of the original gas, the two between them making up the volume of hydrogen chloride used.

The point at issue in the experimental determination of the volumetric composition of gases may be stated thus : If A and B are the constituents and C the compound, and if the corresponding volumes are m and n and p, then the data required are $m : n$ and $m + n : p$; and since volume, unlike mass, is not an additive property, 2 experiments involving at least 3 measurements are needed.

[1] Shown at a lecture given to the Chemical Society at the Royal College of Chemistry, February 2nd, 1865.

IV. Is the law exact?

Until comparatively recently, since most of the measurements were made with a view to supplying a lecture demonstration of the relationship (like those of Hofmann already referred to), great accuracy was not aimed at, and the question of the degree of exactness of the law was not considered.

A priori, knowing as we do that Boyle's and Charles' laws are not exact, but only true to a first approximation, it follows that, even if exact for one particular set of physical conditions, the law of combining volumes cannot be exact for others. Actual measurements recently made have shown that at 0° and a pressure of about 1 atmosphere, the deviations from simple whole numbers, though small, are quite definitely outside the limit of experimental error, thus :

Ratio by volume in which hydrogen and oxygen combine together to form water.

Scott	2·00285 : 1
Leduc	2·0037 : 1
	2·0024 : 1 (corrected by Morley for the deviations from Boyle's law not taken into account by Leduc).
Morley	2·00268 : 1

The table summarises our knowledge on this subject ; the measurements present within small compass a quite extraordinary variety of experimental methods and instances showing that with increased accuracy allowance has to be made for the influence of factors so small that the average student, when performing an illustrative experiment on the same lines, would never suspect their existence.   The knowledge thus gained of the existence and the recognition of 'complexity where at a first approximation simplicity is assumed' should widen the outlook of the students and should supply them with a standard for the classification of the experimental work which they read or hear about or do themselves.

V. Application of the law.

If m vols. of $A + n$ vols. of $B = p$ vols. of C then

$$\frac{n \cdot \text{density } B}{m \cdot \text{density } A} = \frac{\text{combining wt. } B}{\text{combining wt. } A}$$

or

$$\frac{p \cdot \text{density } C - m \text{ density } A}{m \cdot \text{density } A} = \frac{\text{combining wt. } B}{\text{combining wt. } A},$$

and since by Gay-Lussac's generalisation m and n and p are simple whole numbers, if the law were exact, the ratio of the *accurately* measured densities of gases would supply direct accurate knowledge of the combining weights. This method differs from those dealt with in the preceding chapter, in that when once the gases involved have been prepared in a state of sufficient purity, there is no further source of error due to loss of material or contamination with external impurities, such as is inevitably associated with the processes of analysis and synthesis required for the methods in which the weight ratio is determined directly. It is only comparatively recently, owing to improved methods of purifying gases and to the elaboration of the experimental processes for the accurate estimation of the quantities involved, that our knowledge of gaseous densities has become sufficiently exact to take a place among the standard methods for combining weight determinations[1].

But from the equation as given above it is apparent that the density ratios can only be used directly, if m, n and p are *exactly* whole numbers, which is not the case. Various ways of surmounting this difficulty have been resorted to:

(i) We must try to ascertain the values for the densities of the gases under consideration when these behave as perfect gases, when Boyle's and Charles' law and—the real point at issue here—Gay-Lussac's law of combining volumes, would be exact. Now from the connection between these laws we are justified in assuming that the conditions which would produce this effect on one of the gases would do so for all of them. But the result of many measurements has shown that the deviations from Boyle's and Charles' law diminish with decrease of pressure and rise of temperature, and it is therefore justifiable to believe that under these same conditions m, n and p would be sufficiently near to whole numbers to be considered as such, and that the values of the corresponding

---

[1] One result of this, termed by Lord Rayleigh, to whom the achievement is due, 'the triumph of the last place of decimals,' has been the discovery of argon, which in turn brought about the discovery of the other inert gases neon, krypton, xenon and helium.

densities would give directly the combining weight ratios sought. These so-called 'limiting densities' are found:

(a) By actual measurement of the densities at very low pressures or high temperatures. Obviously working under these conditions increases the experimental difficulty and involves a correspondingly large experimental error.

(b) By the application of corrections to the density values measured under ordinary conditions. The amount of the correction is in each case specific, and found through an equation which correlates the observed deviations from the coefficients of expansion and of compressibility with the deviation of the density from the so-called 'limiting value.' The following examples give an idea of the order of magnitude involved:

Density nitrogen$_{O=16}$

At ordinary temp. and pressure (Mean of various observers)	At 1067° (Jaquerod and Perrot)	At very small pressure (Rayleigh)
14·0036	14·0077	14·0090
(corrected) 14·0073		

Weight of 1 litre of hydrogen chloride found at 0° and ordinary pressure $= 1·63915$
   ,,   ,,   ,,   corrected for limiting density $= 1·63698$

(ii) Actual accurate measurement of the combining volume ratio at the ordinary temperature and pressure at which the densities are measured. The results of such accurate measurements have been given already (p. 384). The students' experiment referred to on p. 342, in which the combining weight of hydrogen was deduced from the density of hydrogen and oxygen and the combining volume ratio, is an illustration of the application of this method ; it should not be necessary to point out that in such an experiment the deviations from Gay-Lussac's law do not enter into consideration.

VI. Students' illustrative experiments.

1. Water.

(1) Ratio hydrogen : oxygen.

(a) By eudiometric synthesis. This experiment has been

dealt with in an earlier chapter, Exp. XXXII, p. 245; students are recommended to read the description of this experiment in different text books with the object of comparing the various types of apparatus used and of trying to form some opinion as to their specific advantages and disadvantages, with special reference to the following points: (α) initial cost of apparatus and material, (β) liability of apparatus to breakage, (γ) quickness of manipulation, (δ) calculations involved, (ϵ) accuracy of result.

(*b*) By analysis. Electrolysis of acidified water. This is an experiment invariably shown in lectures, and there is not much point in repeating it as a students' experiment except that it supplies practice in the setting up of a simple apparatus. Moreover the experiment is open to various objections. On the theoretical side, the fact that pure water does not conduct, and that the decomposition only occurs if an electrolyte (either acid, alkali or salt) is present in solution, obscures the issue; there is no justification for the *a priori* assumption that only the water is decomposed, and the experiment by which it can be proved that as far as the actual results go this is what the reaction amounts to, though fairly simple, diverts the students' attention from the point at issue to the various physical phenomena involved. On the practical side there is the drawback that the accuracy is not great, there being two sources of error, each tending to make the volume of oxygen too small:

(i) Oxygen is more soluble in water than is hydrogen:

	Oxygen c.c.	Hydrogen c.c.
At 0° and 760 mm. 1 c.c. water dissolves	·049	·0215
15° ,, ,, 1 ,, ,,	·034	·0190

(ii) Secondary reactions occur, such as the production of some ozone—a denser form of oxygen—and hydrogen peroxide, which remain in solution. If however it is decided to perform the experiment, a very simple form of apparatus which can be put together quickly and at small cost is shown below.

Simple form of apparatus to show electrolytic decomposition of water.

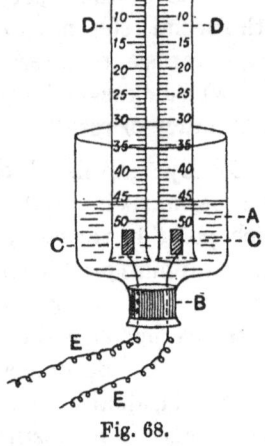

A. Inverted bottomless bottle containing acidulated water.
B. Paraffined cork through which platinum wires pass.
CC. Electrodes made of platinum foil.
DD. Graduated tubes for collection of products of electrolysis.
EE. Wires connected with battery.

Fig. 68.

(*c*) Indirectly from the composition by weight and the densities of the component gases. In Chapter IX, p. 342, a value for the gravimetric composition of water, and hence for the combining weight of oxygen, was deduced from the volume ratio hydrogen : oxygen, together with the densities of these gases. Conversely, by using the gravimetric composition as the antecedent datum, we can deduce the volumetric composition :

$$\frac{\text{Wt. hydrogen}}{\text{Density hydrogen}} : \frac{\text{Wt. oxygen}}{\text{Density oxygen}} = x : 1.$$

(2) Ratio vol. oxygen + vol. hydrogen : vol. steam. The measurement of this ratio is also almost invariably made the subject of a lecture experiment, and all text books give pictures and descriptions of the apparatus used ; the principle consists in the direct synthesis of steam from volumes of the two constituents in the ratio in which they combine at a temperature well over 100°, and the measurement of the volume formed ; knowledge of the combining volume ratio hydrogen : oxygen is required as an antecedent datum.

Data which can be used for the indirect measurement of this ratio have been found in previous experiments thus :

2 × density hydrogen + density oxygen = density steam

2 × ·09   +   1·43   = x × ·82

$$x = 2,$$

i.e. in the combination of 2 volumes of hydrogen with 1 volume of oxygen, 2 volumes of steam are formed, that is, the contraction is one-third the volume of the mixed gases.

2. Carbon dioxide.

Ratio oxygen : carbon dioxide determined by direct synthesis. Carbon is burned in a known volume of oxygen (pure or mixed with an inert gas), and we note the change in the total volume of the residual gas, or we actually measure the volume of carbon dioxide formed by the contraction produced on absorbing it with potash. The experiment has been dealt with in Chapter VIII, p. 311.

3. Methane.

(i) Ratio methane : hydrogen.

(ii) Ratio methane : carbon dioxide.

(i) Decomposition by sparking of a known volume of methane, and measurement of the hydrogen liberated.

(ii) Explosion of a known volume of methane with excess of oxygen, and measurement of the carbon dioxide produced by absorption with potash (*ante*, p. 310).

4. Hydrogen chloride.

(i) Ratio hydrogen chloride : hydrogen.

Found by analysis of a known volume of the pure *dry* gas from which the chlorine is withdrawn by sodium amalgam, the residual hydrogen being measured.

The principle of the experiment is the same as in Gray and Burt's accurate measurement in which heated aluminium was used instead of the liquid sodium amalgam.

Experiment LII. To find the volume of hydrogen liberated from a known volume of dry hydrogen chloride.

The sodium amalgam—a solution of metallic sodium in mercury—is easily made by picking up with a pestle very small shavings of sodium made by cutting up on a porcelain slab a lump from which the outside oxidised portions have been removed. About 50 c.c. of dry mercury are put into a small mortar, and 5 grams of the cut-up sodium are introduced gradually, by dipping the pestle with the adhering sodium under the mercury and rubbing gently against the bottom of the mortar until a slight crackling noise indicates that solution has occurred. If the pieces of sodium are made small enough, the reaction takes place without violence, otherwise there may be an occasional little flash ; after the first few pieces have been brought into solution, a process accompanied by the evolution of heat, the rise of temperature causes the reaction to occur much more quickly. The amalgam should as soon as made be transferred to a small, perfectly dry and well-stoppered bottle.

The hydrogen chloride is made in the usual manner by the action of concentrated sulphuric acid on lumps of rock salt. The need of very thorough drying should be borne in mind ; at least two gas wash-bottles with concentrated sulphuric acid should be employed.

The kind of tube used for the decomposition of the gas, and the measurement of the residual hydrogen depends entirely on the degree of accuracy aimed at. Considering the many sources of error, the experiment might be treated, not as an opportunity for practice in accurate measurement, but just as an illustration barely laying claim to be called a quantitative experiment, of the principles of a method at best carrying so many errors that it would not be employed in research work.

In this case an ordinary graduated tube, or even a large test-tube, may be used, the volumes it is required to determine being marked by means of a thin indiarubber band and subsequently measured with water and a graduated cylinder. The tubes, whose total volume is ascertained at the very beginning or the end of the experiment (in the case of the graduated tube this consists in finding the volume of the piece between the last of the graduations and the open end), are thoroughly dried[1] and filled by downward displacement with dry hydrogen chloride delivered at a slow rate, the delivery tube being slowly withdrawn (so as to get the volume vacated filled by the gas and not by inrushing air) when the appearance of fumes at the mouth of the tube indicates that it is full[2]. About 5 to 10 c.c. of the amalgam are poured in rapidly, and a cork pushed in straight and quickly, so as to

[1] Rinsing several times with small quantities of a mixture of alcohol and ether, and then sucking or blowing out the air whilst gently warming, is a quick and effective method ; of course the sucking or blowing must be continued until the tube is quite cold again.

[2] Students are recommended to try and get some definite idea as to the efficiency of this method of filling by means of a blank experiment, in which after the withdrawal of the delivery tube the vessel supposed to be full of hydrogen chloride is closed with the thumb (or with a well-fitting plate or a cork) and opened under water, and the volume of gas which remains unabsorbed ascertained as a fraction of the total volume of the vessel.

justify as much as possible the assumption that we are keeping in the tube the amount of gas which filled it right up to the open end. The volume V_S occupied by the amalgam is next read or marked, after which the tube is shaken vigorously for a couple of minutes or so, when it becomes very warm. The cork is next taken out under water which by being allowed to stand for some time has been brought to the temperature of the room, the amalgam drops out, the levels are equalised, and the volume of the hydrogen, V_H, is read or marked. All these operations should be performed very quickly, to reduce as much as possible the error introduced by the inclusion in V_H of some hydrogen produced by the action on water of any sodium amalgam clinging to the sides of the tube.

Record of results.

Total volume of the tube	Volume of amalgam	Volume of hydrogen chloride used	Volume of residual hydrogen	Volume of hydrogen obtained by the decomposition of 2 vols. of hydrogen chloride
V	V_S	$V - V_S$ $= A$	V_H $= B$	$\dfrac{2B}{A}$
$100 + 14 \cdot 5$	8 c.c.	$106 \cdot 5$ c.c.	$54 \cdot 5$ c.c.	$1 \cdot 02$

Summary of results obtained by a class of students.
I. With simple apparatus as above described.

	A	B	$\dfrac{2B}{A}$
Student A	102	$59 \cdot 5$	$(1 \cdot 17)$ *
B	$106 \cdot 5$	$54 \cdot 5$	$1 \cdot 02$
C	94	$48 \cdot 5$	$1 \cdot 03$
D	89	$45 \cdot 5$	$1 \cdot 02$
E	109	57	$1 \cdot 04$
F	91	48	$(1 \cdot 05)$
G	84	$42 \cdot 5$	$1 \cdot 01$†
		Mean	**$1 \cdot 025$**

II. With more elaborate apparatus as described further on.

	A	B	$\dfrac{2B}{A}$
Student A	200	109	$(1 \cdot 09)$‡
B	200	100	$1 \cdot 00$
C	200	$99 \cdot 5$	$\cdot 995$
D	200	$101 \cdot 0$	$1 \cdot 01$
		Mean	**$1 \cdot 005$**

* First tube filled after starting hydrogen chloride current ; evidently all the air had not yet been expelled from the apparatus.

† Last tube filled.

‡ From the fact that another tube prepared in the same way but not used was found to have deposited a coating of moisture, it is justifiable to assume that the tube used was not dry. The other tubes were then dried by the more thorough process of rinsing with alcohol and ether, etc.

Critical examination of these results shows that in the experiment with the simple apparatus the deviation of the values from each other and from the standard value are greater than in the series obtained when a more elaborate apparatus was used ; also that throughout the values are too high, indicating the presence of some constant error or errors. Students are recommended to consider this question, and to try to form an opinion on the basis of the following data :

Source of error.	Effect on final result.
(i) Gas used contaminated with air.	Volume of residual gas too great.
(ii) Tube not quite dry—an additional amount of hydrogen chloride present in solution.	Volume of residual gas too great.
(iii) Original gas measured is dry—residual hydrogen measured is moist, therefore pressure in first gas is that of barometer, B, and pressure of hydrogen $B - \tau$.	Apparent volume of hydrogen too great, in the ratio $\dfrac{B}{B-\tau}$, which at a pressure of 760 mm. and a temp. 18° is $\dfrac{760}{760-15} = \dfrac{760}{745} =$ about 1·02, an error of $+2\,^\circ/_\circ$.
(iv) The decomposition has been incomplete, owing either to insufficient amount of sodium or insufficient shaking, and some hydrogen chloride soluble in water remains behind.	Volume of residual gas too small.
(v) Amalgam clinging to sides of tube gives rise to evolution of additional hydrogen.	Volume of residual gas too great.
(vi) In pushing home the cork (simple apparatus) instead of compressing all the gas originally present into the tube, some escapes-into the air.	Volume of original gas is supposed greater than it is in reality, and hence volume of residual gas is by comparison made too small.

It becomes evident at once that when, as in the case of the experiment with the simple apparatus, *all* these errors are likely to be operative, the preponderating effect is to make $\dfrac{2B}{A}$ too large.

Several of the above errors can be completely eliminated, and all the others except (iii) reduced by the use of a more complicated apparatus, the construction and use of which should be fairly evident from fig. 69, very few points requiring special mention.

The graduated tube A specially made[1] for the purpose

---

[1] A large burette could easily be so adapted as to secure many of the special advantages of this arrangement.

of this and similar volumetric experiments holds between the two taps 200 c.c. Great care must be taken to have both taps air-tight; lubrication with some grease—lard or one or other of the mixtures specially sold for the purpose—is essential. The process of filling is identical with that described *ante*, p. 43, see also fig. 5, p. 42. The hydrogen chloride contained in the cup above the tap T_2 is removed by suction (water pump or aspirator) or by blowing (hand

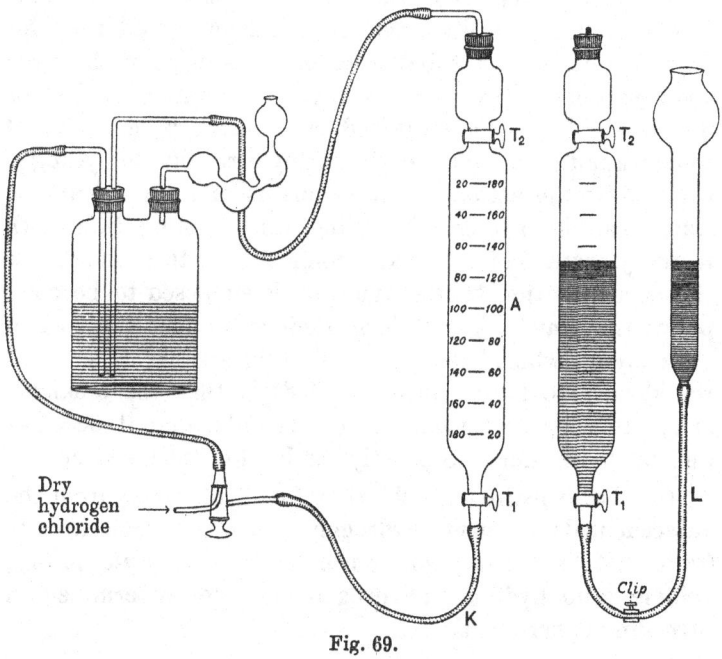

Fig. 69.

or foot bellows); the passage of the amalgam from the cup into the graduated tube is brought about by cooling the tube by means of a little ether dabbed on the outside. Of course care must be taken in manipulating the tap not to let *all* the amalgam drop into the tube, whereby an opening would be left through which air could pass in. The opening of the tap T_1 under water is done in a largish vessel just as with the simple apparatus, so as to allow the amalgam to fall out, and by rapid shaking of the water that has entered the tube as much as possible of the amalgam is washed off; the equalising of the pressure is effected by shutting T'_1, detaching the tube K from A and connecting with the levelling tube by slipping on the rubber tube L, which of course must be completely filled with water, after which T_2 and the clip are opened.

The advantages of this apparatus as compared with the very simple arrangement first described are the certainty of complete filling, reduction of the number of measurements required (no deduction for amalgam, no addition for ungraduated part of tube), ease in adjustment of the level and consequent great accuracy of volume reading. The error due to the hydrogen measured being moist is of course of some magnitude. This, as well as any error due to hydrogen evolved by the action of the remnants of the amalgam on water, can be avoided by filling the levelling tube with mercury instead of water and connecting with it as soon as the decomposition by the amalgam is supposed to be complete; the drawback to this procedure (besides the cost of the mercury, which before being used for another experiment would have to be carefully purified) is that the amalgam is apt to cling so tenaciously to the tube that it becomes difficult to set and read exactly the level of the mercury.

(2) Ratio hydrogen : chlorine, found indirectly from the antecedent data Density hydrogen D_H; density chlorine D_{Cl} (*ante*, p. 41); density hydrogen chloride D_{HCl} (*ante*, p. 41); volume ratio hydrogen : hydrogen chloride (determined in preceding experiment).

$$D_H + xD_{Cl} = 2 \cdot D_{HCl},$$

which by substituting the numbers actually obtained in the students' illustrative experiments (see pp. 342, 47, 13), gives for x

$$0\cdot88 + x \cdot 3\cdot228 = 2 \times 1\cdot688,$$

$$x = \frac{3\cdot376 - 0\cdot088}{3\cdot228} = 1\cdot02.$$

As we know the value for D_{HCl} to be unreliable (recall criticism in Introductory chapter), $1\cdot02$ is a very fair approximation to the whole number. The combination of (1) and (2), therefore, gives that complete information about the composition of the gas hydrogen chloride which (1) alone does not supply, and we can state that 2 vols. hydrogen chloride are formed by the combination of 1 vol. hydrogen and 1 vol. chlorine.

5. Ammonia.

(1) The ratio hydrogen : nitrogen can be found by a method which is known by the name of Hofmann, its originator. It consists in the action of excess of ammonia on a known volume of chlorine, which, abstracting the amount of hydrogen required by it (*i.e.* a volume equal to itself—see above), liberates a volume of nitrogen which in the ammonia was combined with that volume of hydrogen; hence

$$V_{\text{hydrogen}} : V_{\text{nitrogen}} = V_{\text{chlorine taken}} : V_{\text{residual nitrogen}}.$$

Experiment LIII. To show the general action of the halogen elements on hydrides, an action which is utilised in the experiments on the volumetric composition of ammonia.

(1) The action of chlorine on

(i) Water. The experiment of decomposing water by chlorine in sunlight and collecting the oxygen evolved is a very popular one and intended to explain the mechanism of the bleaching action of chlorine ; it is described in a number of text books (Newth, *Inorganic Chemistry* ; A. Smith, *Introduction to Inorganic Chemistry* ; Reynolds, *Experimental Chemistry*, vol. II, etc.) and generally takes the form of standing chlorine water contained in a suitable vessel in a window in bright sunlight, and testing the gas evolved after a sufficient amount of it has collected. There is no corresponding action produced if bromine or iodine are used instead of chlorine.

(ii) Sulphuretted hydrogen, which by analysis as well as by synthesis can easily be shown to be hydrogen sulphide.

Lead sulphuretted hydrogen gas into chlorine water, and note the separation of sulphur ; show that in the presence of excess of chlorine the action goes further, resulting in the production of sulphuric acid : using a small volume of sulphuretted hydrogen water (about 2 c.c. of a strong solution diluted to 20 c.c.) add strong chlorine water ; as soon as no more sulphur seems to separate out, boil the solution, adding more and more chlorine water until it has become quite clear ; then test for sulphuric acid in the usual manner. If an apparatus for the evolution of chlorine gas is available, use the gas instead of solution, whereby the reaction will be greatly accelerated.

(2) The action of bromine on

(i) Water—no effect, in accordance with the observation that bromine bleaches very little, if at all.

(ii) Sulphuretted hydrogen. Using bromine water, proceed as with chlorine. The action of bromine on the gas is similar to that of chlorine.

(iii) Ammonia itself or a derivative of ammonia, such as urea. Instead of bromine water, the bromine may be used in the form of sodium hypobromite, obtained by the gradual addition of bromine to a cold solution of soda (not too concentrated).

To a solution of ammonia add slowly a fairly strong solution of hypobromite; note the decoloration produced in the yellow hypobromite, also the evolution of a colourless gas which when collected (over boiled-out water) gives the characteristic negative tests for nitrogen. It will add to the value of the experiment if some of the students use a solution of urea instead of ammonia.

(3) The action of iodine on

(i) Water—no effect.

(ii) Sulphuretted hydrogen. Put a small amount (not more than 0·2 gram) of iodine into a flask, and pour on it about 200 c.c. of strong solution of sulphuretted hydrogen; note the separation of sulphur and the production of a deep coloured solution of the excess of iodine in the hydriodic acid formed. Pass in sulphuretted hydrogen until the solution is decolorised, and shake, when probably the colour will reappear, and repeat the process until all the iodine has been used up; remove excess of sulphuretted hydrogen by sucking air through the solution, filter from precipitated sulphur, and test the clear filtrate for (a) free acid, (b) iodide, (c) sulphate. Of this latter very little or none will be found, showing the action of chlorine and bromine to be different from that of iodine as regards the change of sulphur to sulphuric acid.

Experiment LIII. To find the volume of nitrogen liberated from excess of ammonia by a known volume of chlorine (Hofmann's experiment).

A graduated gas tube of 200 c.c. capacity (see fig. 70) is provided with a very well-fitting rubber cork through which passes a small funnel with a glass tap, made air-tight by careful lubrication (a dropping funnel with bulb of 25 c.c. capacity, part of the stem of which has been cut off, serves well). The volume of the tube between the end of the graduations and the place indicated in some manner certain not to be obliterated in the course of the experiment, is measured. The tube is filled with chlorine, all possible care being taken to ensure the absence of air; using the arrangement depicted on p. 42, it is possible to postpone the collection of the chlorine until there is definite evidence of all air having been expelled from the generating apparatus and the wash-bottles. As regards the actual filling of the tube, this may be done by downward displacement through a long glass

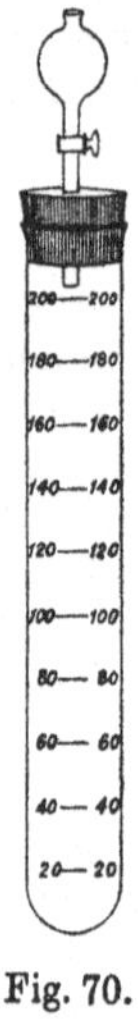

Fig. 70.

tube reaching down to the bottom, which is withdrawn when, judging by the colour, the graduated tube is supposed to be full. The solubility of chlorine in water is comparatively small, so that the absolute dryness of the tube is not so much a consideration as it was in the experiments with hydrogen chloride; moreover, the gas being practically insoluble in a saturated solution of sodium chloride, the tube can be filled with chlorine by collection over this liquid, in which case however some special precautions (*e.g.* closing under the liquid with a cork) will have to be taken in order to prevent loss of gas on inversion prior to the insertion of the cork and funnel. The tube having been filled and the cork and funnel (with the tap closed) pushed in to the mark, the tap is opened for a second to allow the excess of gas to escape. 2 to 3 c.c. of strong ammonia solution are then put into the funnel, the tube cooled by means of a little ether dabbed on with a piece of cotton wool, and the ammonia allowed to run in drop by drop ; this latter process requires very careful manipulation of the tap, and students will do well to practise in a blank experiment with a tube containing only air. As each drop falls in, there is a vigorous action marked by a flash of light, the formation of a dense cloud of white solid ammonium chloride, and the disappearance of the green colour of the chlorine. The careful introduction of ammonia must be continued until the green colour of the chlorine has completely disappeared ; when this has been accomplished, the gaseous contents of the tube are nitrogen, water vapour and ammonia. To remove the excess of ammonia, the pressure of which is considerable, the remainder of the ammonia is removed from the bulb and dilute sulphuric acid run in as long as it will do so ; owing to the initial considerable difference in pressure inside and outside the tube, the liquid tends to rush in, and again care is required in the manipulation of the tap to prevent the emptying of the funnel and consequent inrush of air. When the inflow of liquid has almost ceased, the tube (which has become hot owing to the heat evolved first in the burning of the ammonia and then in the neutralisation of the excess of ammonia) is clamped and left to take the temperature of the room, after which the tap is opened again to let the few drops of liquid flow in which are required to make up for the contraction of the gas on cooling. The volume of the residual gas is then read.

The majority of text books contain a description of this experiment as performed with the apparatus originally described by Hofmann. The present writer has not been able to find anywhere either actual numbers obtained or any information as to the accuracy attained or the special difficulties presented, and hence wonders if the experiment is really done as often in practice as it is described in print. In the writer's laboratory the experiment has for many years been included in the regular course, not as a demonstration but as a students' experiment, using the very simple apparatus depicted on p. 396. The results have not been uniformly satisfactory; the volume of the residual gas was found to be too large, probably in consequence of the difficulty attendant on : (i) filling the tube completely with chlorine ; (ii) getting a tap sufficiently well fitting to prevent the entrance of some air when, owing to the action that has occurred, the pressure inside the tube is only about 1/3 of that of the atmosphere. But provided these difficulties are realised from the outset and as far as possible guarded against, the experiment is well worth doing in the simple manner which does not require special apparatus.

Of the various errors affecting the value found, the following require consideration :

(i) The nitrogen measured is moist, and if the original chlorine was dry (as would be the case if collected by downward displacement) the volume of the nitrogen will appear too large.

(ii) The pressure under which the nitrogen is measured is that at the final closing of the tap, and is therefore made up of the atmospheric pressure *plus* that of a column of dilute sulphuric acid equal to the distance between the open end of the stem and the level of the liquid in the bulb of the funnel; this may amount to something between 50 and 100 mm., corresponding to 4 to 8 mm. of mercury, and whilst acting in a direction opposite to the error due to (i), will not completely counteract it.

Many of the difficulties encountered and the sources of error are removed by using, in place of the simple apparatus above depicted, the special gas tube described in connection with the experiment on the volumetric composition of hydrogen chloride (see fig., p. 393): the complete filling with chlorine, the neutralisation of the excess of ammonia, the adjustment to atmospheric pressure by the inflow of dilute sulphuric acid are all simplified, and in the writer's experience concordant results in good agreement with the standard value are obtained. There is the further great practical advantage that there is no trouble with the excess of chlorine, and no mechanical difficulties such as occur when the tubes must be filled in a draught cupboard, since the whole process can be carried out at the laboratory bench.

Specimen note-book entry and summary of results.

The Volumetric Composition of Ammonia; determination of the ratio hydrogen : nitrogen by Hofmann's chlorine experiment with the simple apparatus described on p. 396.

	Volume of *dry* chlorine used to decompose some of the ammonia present in excess, the action consisting in the formation of hydrogen chloride by the withdrawal from the ammonia of a volume of hydrogen equal to the volume of chlorine taken	Volume of *moist* nitrogen left behind as the result of the decomposition of an amount (unknown) of ammonia which yielded *a* volumes of hydrogen	Volume ratio Hydrogen : Nitrogen
	a	*b*	*a* : *b*
Student A	96·8	30·4	(3·19)
B	96·8	32·2	3·01
C	215	74	2·90
D	220	73	3·01
E	214·5	73·5	2·92
F	206	69	2·98
G	221	75	2·95
		Mean	**2·96**

(2) The determination of the ratio ammonia : nitrogen. The principle of the reaction used in the analysis is identical with that of the previous experiment, consisting as it does in the withdrawal of the hydrogen by means of a halogen and the measurement of the residual nitrogen ; but whilst in Hofmann's chlorine experiment a known volume of gaseous halogen acting on an unknown amount of ammonia present is made the measure of the hydrogen, in this case bromine in the form of a solution of hypobromite is made to act on a known volume of ammonia, from which it withdraws an unknown volume of hydrogen. The experiment is described by Sir William Ramsay (*Experimental Proofs of Chemical Theory*, p. 59), and also by Mr Vaughan Cornish (*Practical Proofs of Chemical Laws*, p. 72), who introduces some modifications. The main points are that a carefully dried graduated tube is filled by upward displacement with ammonia obtained by heating a strong solution of the gas (immersion of flask containing the solution in hot water) and subsequent drying by passage over soda lime (see fig. 71). The tube is then placed in a dish containing a solution of hypobromite of suitable strength

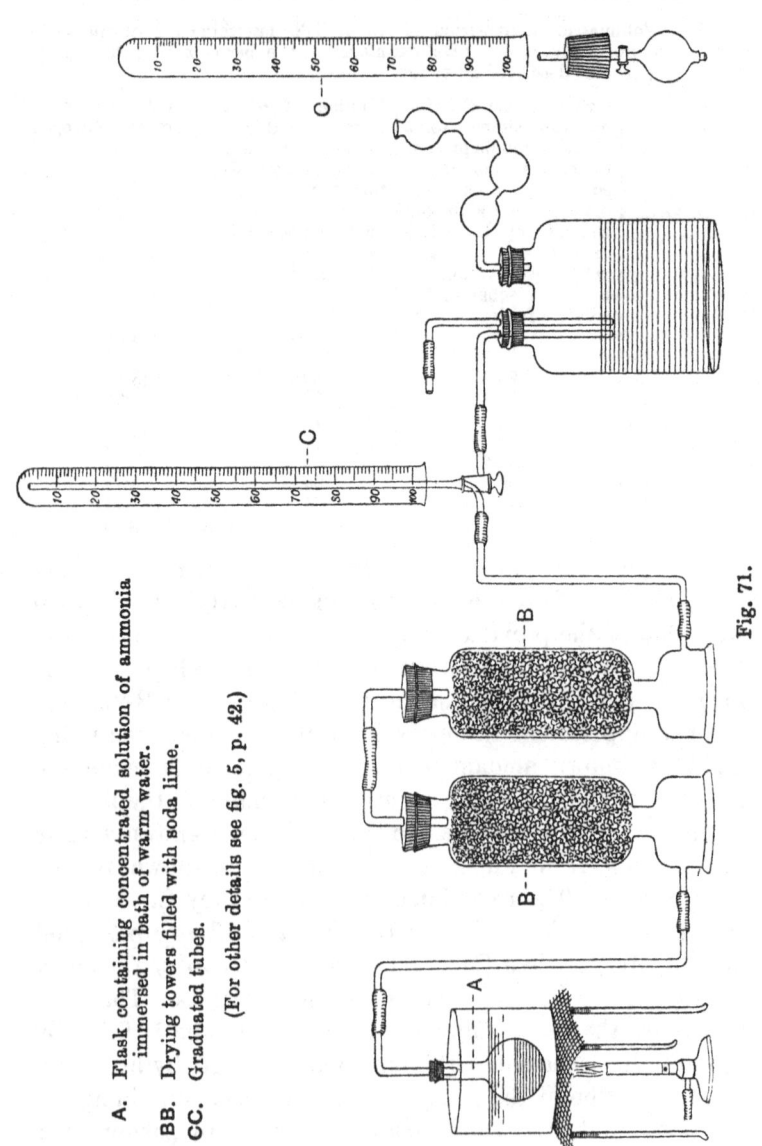

A. Flask containing concentrated solution of ammonia immersed in bath of warm water.

BB. Drying towers filled with soda lime.

CC. Graduated tubes.

(For other details see fig. 5, p. 42.)

Fig. 71.

(*ante*, p. 396), left until no further volume change seems to
occur, and transferred to a deep vessel containing water at
the temperature at which the ammonia was collected; the
tube is clamped and allowed to stand until its temperature is
that of the water, the levels are adjusted and the volume of
nitrogen is read. In the results quoted by Mr Vaughan
Cornish the ratios found were 1·983 : 1; and 2·016 : 1.
Discarding the last figure, which considering the probable
errors in the measurements has no meaning, the deviation from the whole number 2 is 1°/₀, which would be
quite satisfactory but for the fact that the volume comparison having been made between dry ammonia and moist
nitrogen, the volume of the latter is affected by an error
of about 2°/₀. No notice is taken of this in the discussion of
the result. Moreover, the fact of the nitrogen being in this
case evolved in the liquid, which unless allowed to stand for
a considerable time and agitated is apt to become super-
saturated, makes an appreciable error due to this cause more
than probable; the effect would be to make the volume as
measured too small, an error in the opposite direction to the
first one. As pointed out by Cornish, since the solubility of
the ammonia in the strongly alkaline hypobromite solution is
but small, the reaction occurs mainly near the surface of the
solution, which thereby is soon deprived of the active hypobromite (made evident by the loss of the yellow colour), after
which, if left to itself, the velocity of the reaction becomes
very small, and some mechanical help is necessary to mix the
solution and bring more hypobromite into contact with the
ammonia. This difficulty becomes more marked, the narrower
the gas measuring tube used ; which lands us on the horns of
the dilemma that whilst accuracy of measurement demands
a narrow tube, the desirability of completing the experiment
within reasonable time points to the use of a wide one. In
the writer's experience this difficulty has proved so serious
that the experimental procedure has been modified by using
a 100 c.c. tube fitted with a cork and funnel just as in the
previous experiment ; the tube is filled by downward displacement as described by Ramsay and by Cornish, and the

cork with the funnel-tap closed is pushed in from below, after which the pressure in the tube is brought to that of the atmosphere by opening the tap for a second. Some hypobromite solution is then run in, the tube being at first kept cool by means of ether. The hypobromite should be added about 5 c.c. at a time, and before each fresh addition the contents of the tube should be well shaken; the complete decomposition of the ammonia is shown by the liquid in the tube retaining the yellow colour of the last lot of hypobromite added. It remains to measure the nitrogen formed, for which purpose the pressure in the tube must first be brought to that of the atmosphere by opening the tap and allowing hypobromite solution, or preferably water, to run in as long as it will do so; but caution is necessary, and the running in of the liquid must be postponed until after several vigorous shakings separated by intervals of standing, the maximum possible amount of the nitrogen dissolved in the liquid in which it was formed has been expelled. Unless this precaution is taken, the inrush of liquid will be too great, and the nitrogen gradually escaping from the solution will either be retained in the tube at too high a pressure or some will escape when the tap is opened. The result as obtained by this experimental procedure is affected, just as the volume measurement of the nitrogen in Hofmann's chlorine experiment, by two constant errors acting in opposite directions, viz. the pressure of the gas inside the tube is made up of that due to nitrogen and to aqueous vapour, and is equal to that of the atmosphere as indicated by the barometer *plus* that of the column of liquid in the dropping funnel. This experiment also is improved by the substitution of the special gas tube above described for the simpler form of apparatus (see fig. 69).

INDEX

Printed in the United States
By Bookmasters